Graham N. George, Ingrid J. Pickering

X-ray Absorption Spectroscopy

Also of interest

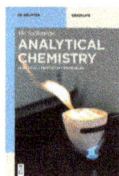

Analytical Chemistry.
Principles and Practice
Soffiantini, 2021
ISBN 978-3-11-072119-5, e-ISBN (PDF) 978-3-11-072120-1

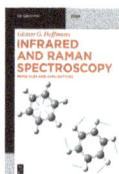

Infrared and Raman Spectroscopy.
Principles and Applications
Hoffmann, 2023
ISBN 978-3-11-071754-9, e-ISBN (PDF) 978-3-11-071755-6

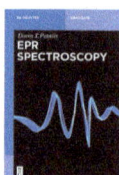

EPR Spectroscopy
Petasis, 2022
ISBN 978-3-11-041753-1, e-ISBN (PDF) 978-3-11-042357-0

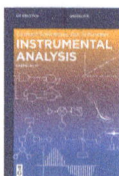

Instrumental Analysis.
Chemical IT
Schlemmer G, Schlemmer J, 2022
ISBN 978-3-11-068964-8, e-ISBN (PDF) 978-3-11-068966-2

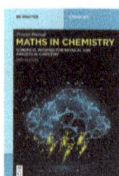

Maths in Chemistry.
Numerical Methods for Physical and Analytical Chemistry
Bansal, 2024
ISBN 978-3-11-133392-2, e-ISBN (PDF) 978-3-11-133444-8

Graham N. George, Ingrid J. Pickering

X-ray Absorption Spectroscopy

—

DE GRUYTER

Authors
Prof. Graham N. George
Department of Geological Sciences
University of Saskatchewan
Saskatoon SK S7N 5E2
Canada

Prof. Ingrid J. Pickering
Department of Geological Sciences
University of Saskatchewan
Saskatoon SK S7N 5E2
Canada

ISBN 978-3-11-057037-3
e-ISBN (PDF) 978-3-11-057044-1
e-ISBN (EPUB) 978-3-11-057057-1

Library of Congress Control Number: 2024940933

Bibliographic information published by the Deutsche Nationalbibliothek
The Deutsche Nationalbibliothek lists this publication in the Deutsche Nationalbibliografie;
detailed bibliographic data are available on the Internet at http://dnb.dnb.de.

© 2025 Walter de Gruyter GmbH, Berlin/Boston
Cover image: © Graham N. George, A 2-dimensional rendition of the probability for the final state wave
function of a photoelectron derived from 1s excitation in X-ray absorption spectroscopy. The X-ray electric
vector would be oriented vertically in the plane of the page.
Typesetting: Integra Software Services Pvt. Ltd.

www.degruyter.com

We dedicate this book to Farrel Lytle, one of the early pioneers of X-ray absorption spectroscopy (XAS) and a friend for many years. Farrel's numerous scientific and technical contributions, together with those of other pioneers, have impacted so many fields of research that it will be hard for anyone to follow in his footsteps.

The content of this book grew out of a course on XAS that we have taught jointly at the University of Saskatchewan for nearly 20 years. This in turn grew from a short course of lectures that we co-taught for the inorganic chemistry honours students at the University of Sydney in 1994. We also dedicate this book to the many students who have taken our course, and with whom we have greatly enjoyed interacting over the years.

Acknowledgements

We are grateful to our families for their support and to our mentors and teachers over the years, especially our XAS colleagues and friends Roger Prince and Steve Cramer, and XAS leaders Keith Hodgson and Britt Hedman at the Stanford Synchrotron Radiation Lightsource (SSRL). We express particular gratitude to all the staff of SSRL; the field of XAS would have significantly diminished impact without their tireless enthusiasm, support and "can-do" philosophy. We also appreciate past and present members of our joint research groups, and acknowledge the energy and enthusiasm of Fellows in the INSPIRE and THRUST synchrotron training programmes. It is a privilege for us to have been part of the growth of the Canadian Light Source (CLS) and appreciate the dedication of its staff and other champions.

Much of the data included in this book was collected at SSRL and the CLS. Additional data are from the National Synchrotron Light Source (NSLS) and from the Advanced Photon Source (APS). SSRL, SLAC National Accelerator Laboratory, is supported by the U.S. Department of Energy (DOE), Office of Science, Office of Basic Energy Sciences. The SSRL Structural Molecular Biology Program is supported by the DOE Office of Biological and Environmental Research, and by the National Institutes of Health, National Institute of General Medical Sciences. The APS and NSLS are U.S. DOE Office of Science user facilities respectively operated for the DOE by Argonne National Laboratory and Brookhaven National Laboratory. The CLS is a national research facility of the University of Saskatchewan and is supported by the Canada Foundation for Innovation, the Natural Sciences and Engineering Research Council of Canada, the National Research Council, the Canadian Institutes of Health Research, the Government of Saskatchewan, and the University of Saskatchewan.

https://doi.org/10.1515/9783110570441-202

Contents

About this book

Our book is primarily intended for students at the senior undergraduate and graduate levels. It should also be useful for established researchers wanting to use XAS in their research; there may even be something in this volume for practicing professionals such as beamline scientists. The chapters are designed to be read in order. The course that we teach at the University of Saskatchewan has enjoyed an incredibly wide intake, with students whose backgrounds and initial training were in physics, chemistry, geology, engineering, medicine, pharmacy, biology, toxicology, agriculture and environmental science. Because of this, we have included a fair amount of background material that may be elementary for those who have studied chemistry or physics.

https://doi.org/10.1515/9783110570441-204

About the authors

Graham George and Ingrid Pickering are both Professors and Tier 1 Canada Research Chairs at the University of Saskatchewan which is home to the Canadian Light Source (CLS). They are Professors in the Department of Geological Sciences, College of Arts and Science, Associates in the Department of Chemistry, and members of the Toxicology Center and of the Division of Biomedical Engineering. George and Pickering are both Fellows of the Royal Society of Canada, while George is also a Fellow of the Royal Society of Chemistry. Pickering serves as Chief Scientific Officer of the Canadian Light Source. Prior to moving to Canada, they were staff scientists at the Stanford Synchrotron Radiation Laboratory (now Lightsource, SSRL). George and Pickering are both from the UK but met in Eastern USA, where they were initially independent but soon collaborative users at the National Synchrotron Light Source (NSLS). They were married in 1992 and have three grown children, who indulge their dinner-table discussions of synchrotron light.

https://doi.org/10.1515/9783110570441-205

1 Introduction

This book is aimed to provide a comprehensive but accessible discussion of X-ray absorption spectroscopy (XAS), starting from the sources of synchrotron light, which have enabled such measurements to be so widely applied. Due to the importance of such sources, we begin with a brief and somewhat personal history of the development of these vital large-scale facilities.

1.1 First light

In the spring of 1972, more than 50 years ago, as we write, the electron-positron collider, known as SPEAR at the Stanford Linear Accelerator Center (SLAC), first began to operate. Although distinct from the motivation for its construction, SPEAR operations heralded the dawn of hard X-ray synchrotron radiation research. While the story of those early years begins well before both of us came on the scene, we have the privilege of knowing many of the individuals concerned. SPEAR – or **S**tanford **P**ositron **E**lectron **A**symmetric **R**ing[1] – was the brainchild of Burt Richter; the fact that it was built at all is a measure of Burt's tenacity and vision. In the 1960s the standard approach to high energy physics was essentially to slam a highly accelerated particle beam onto a fixed target and analyse the spray of sub-atomic debris that emerged. Burt and others realized that there could be a better way. If one could collide a particle beam with an anti-particle beam, while controlling the energies of those beams, then the energy of the matter anti-matter annihilation could be tuned to produce particular sub-atomic particles. The first colliding beam machine, an electron-electron collider that had been built not long before on Stanford University's main campus in a collaboration with Princeton University and funded by the Office of Naval Research (grant to Pief Panofsky), consisted of two rings in a figure-8-type configuration, with electrons in both. To develop SPEAR, funding from the US Atomic Energy Commission was repeatedly requested, but was always denied. But Richter was undeterred – it transpired that if SPEAR was labelled as an experiment, with no permanent buildings, then it could be built out of SLAC's normal operating budget. So, SPEAR was built in just two years, and completed four months ahead of schedule (Figure 1.1). Moreover, the approach was a huge success, resulting in November of 1974 in what became known as the **November revolution**, when separate experiments using SPEAR and at Broo-

---

1 Early designs for SPEAR – Stanford Positron Electron Asymmetric Ring - used individual *asymmetric* vacuum chambers for positrons and for electrons intersecting in two places for the beams to collide. While this design was never built, the acronym SPEAR was retained.

https://doi.org/10.1515/9783110570441-001

khaven Laboratories independently discovered the J/ψ meson,[2] for which Burt Richter and Sam Ting shared the 1976 Nobel Prize in Physics. Hard on the heels of the Stanford researchers were the German team at DESY (Deutsches Elektronen-Synchrotron) Hamburg, who were constructing their own electron-positron storage ring called DORIS (**Do**uble **Ri**ng **S**tore), somewhat larger at 300 m in circumference, compared to SPEAR's 234 m. The construction of DORIS actually began ahead of that of SPEAR, in 1969, but it did not begin operations until 1974.[3] Before turning from high energy physics to focus more upon the topic of this book, we need to mention that there was a second momentous early discovery at SPEAR, which was the 1976 discovery of the τ lepton, for which Martin Perl was awarded the 1982 Wolf Prize and shared the 1995 Nobel Prize. Two Nobel Prize-winning discoveries, hard on the heels of each other

Figure 1.1: The SPEAR storage ring (lower right) before construction of the Stanford Synchrotron radiation Project (SSRP). The large red-roofed building on the left is end-station A, containing equipment for experiments using the Stanford linear accelerator or Linac. The Linac was also used to fill SPEAR with electrons and positrons, injected, respectively, through the right and left tangential conduits at the top left of the picture of the SPEAR storage ring. Photograph courtesy of Helmut Wiedemann, SSRL.

---

2 The J/ψ discovery supported the idea of a previously unknown type of quark called charm, adding a fourth to the three others that were then known; up, down, and strange. The bottom and top quarks complete the total of six in what is called the standard model of Physics, and were discovered in 1977 and 1995, respectively.

3 DORIS also made important contributions to our understanding of the charm quark, in particular through the discovery of excited charmonium states.

from a project that was repeatedly denied funding, should be an encouragement to those who occasionally struggle to fund research ideas.

1.2 The origin of photon sciences

As early as June 1968, Stanford University Professors Bill Spicer and Stig Hagström had been discussing the possibility of using "cyclotron radiation" (later renamed synchrotron radiation) from the planned SPEAR storage ring for experiments in Physics, and on completion in 1972, SPEAR included a beam port, specifically for these experiments. First light in that first beamline was achieved in April 1973. The previous year, a proposal to the National Science Foundation (NSF) was submitted by Seb Doniach and others, resulting in the Stanford Synchrotron Radiation Project (SSRP) being funded by NSF in 1974, with Seb Doniach as director, Bill Spicer as deputy director and Herman Winick as associate director. In May of 1974, SSRP began operations 8 months ahead of schedule with five different experimental stations sharing the single beam port (Figure 1.2).

In those early days, all of the beamtime was parasitic, meaning that the machine was controlled entirely by the high energy physicists. The discovery of a sharp J/ψ resonance, with SPEAR operating at an electron beam energy of 1.55 GeV [1], was bad news for researchers wishing to use synchrotron X-rays because the X-ray flux at this energy was too low for most experiments. On November 11, 1974, SSRP was given a single prime shift (8 h) of beamtime with an electron beam energy of 3 GeV – perhaps the first dedicated experimental run – and the synchrotron experimenters were described as being very happy. Slightly better news came with the November 20th discovery of a second sharp resonance at 1.85 GeV, which gave some more, but still insufficient, X-ray flux. SSRP's response to this lack of X-rays was the invention of the wiggler by Herman Winick, with a wiggler beamline proposed to NSF in 1977 as part of an SSRP expansion. In 1978, the first wiggler was tested, heralding another new era – that of synchrotron radiation from insertion devices. Meanwhile, the competition was heating up, since in Hamburg, the DORIS storage ring commissioned an operational X-ray beamline in 1976, and in 1975, construction of the Synchrotron radiation source (SRS), the world's first dedicated source of synchrotron radiation, began at Daresbury laboratory in Cheshire, UK.

1.3 Early synchrotron XAS facilities

The roller coaster of early discoveries from those heady days included many that were directly relevant to XAS. In May of 1974, the world's first XAS beam line became operational, SSRP's beamline 1–5. It used a channel-cut silicon crystal monochromator, which was operated using a Siemens diffractometer, on loan from Boeing Corporation; the first, and nearly the last, XAS experiments on SPEAR are eloquently and humorously described by Lytle, and were of the Cr K-edge of a piece of stainless steel [2]. This experimental revolution was partnered by new understanding of XAS itself by Sayers, Stern and Lytle [3], published just a few years before the first synchrotron experiments, so that the springboard was ready for the scientific revolution that was to come.

In 1976, a second beamline (beamline 2, Figure 1.2) became operational, and in 1977, SSRP became the Stanford Synchrotron Radiation Laboratory (SSRL), with Artie Bienenstock succeeding Seb Doniach as director in 1978. We would be remiss if we did not mention a few more pioneers; early XAS researchers included Bob Shulman, Peter Eisenberger and Keith Hodgson, with then graduate students Steve Cramer, Pat Frank, Tom Eccles and Jim Phillips, who studied metalloproteins, while Grayson Via of Exxon Research and Engineering began to apply XAS to the study of catalyst systems. SSRL, now the Stanford Synchrotron Radiation Lightsource, celebrated its 50th Anniversary of First Light in 2023.

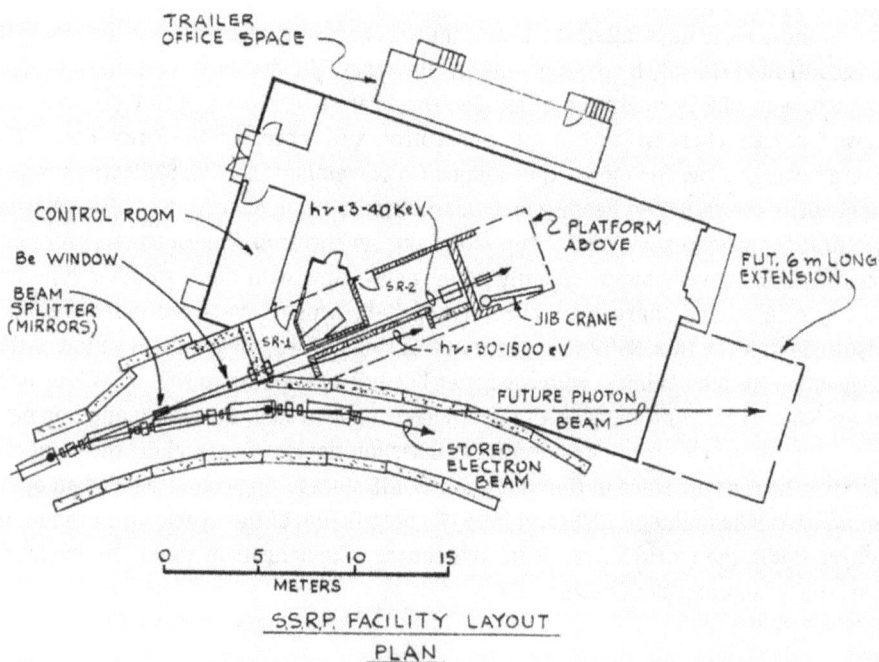

Figure 1.2: Plans for the first SSRP experimental hall, showing the position of beamline 1, and the planned location of beamline 2 ("future photon beamline"). Illustration reproduced here courtesy of Pierro Pianetta and Herman Winick of SSRL.

1.4 A first visit to SSRL

Here, we cannot resist inserting a personal note. One of us (George), although not part of any pioneering work, entered into the field of synchrotron radiation research reasonably early, in April 1982, with a visit to the EMBL outstation on DORIS while a graduate student at the University of Sussex, UK. Although the UK's Daresbury SRS had been operational since 1981, only bend-magnet sources were available, and as our interest was in Mo K-edge XAS, at an X-ray energy of ~20 keV, DORIS, then operating at 4.0 GeV, was a much better source for our experiments. Several visits to Hamburg yielded some interesting results and the good fortune to collaborate with some of the European XAS pioneers, notably Joan Bordas and Christoph Hermes. But in December 1983, a first visit to SSRL, to work with Steve Cramer on one of SSRL's wiggler stations, beamline 7-3, yielded results that were nothing short of spectacular for the time. In modern times, the beamline equipment typically is set up by skilled staff, and ready for the experimenter when the beamtime starts. But in 1983, this was not the case – all that was provided was a bare hutch with (if you were lucky) X-rays entering through a beam pipe. Nearly everything else needed to be set up by the experimenters, including X-ray detectors and electronics, a sample cooling system and computer interface hardware. Being a new user, starting an experiment was rather like learning to swim by being thrown in at the deep end of a pool. On his first day of SSRL beamtime, in December 1983, following an 11-hour flight from the UK to California the previous day, George spent hours, helping to wheel racks of electronics from storage, running cables between detectors, electronic racks and computer, and checking and calibrating components using an oscilloscope. By early evening, when the setup was finally ready, George was told that due to his 8 h of jetlag, he should take the night shift . . . but he was delighted because his samples were amongst the first to be run. And Steve brought bagels to the beamline for breakfast the next day, an exotic food not then available in the UK. While beamline setup may no longer be as intense, graduate students and other trainees still often take the night shift to ensure smooth running of experiments. Now that we are professors, we are especially grateful for their diligence.

1.5 Rapid development

The early days of synchrotron radiation research were distinguished by an increasingly rapid realization of how incredibly useful these fantastic sources of X-rays were. Figure 1.3 gives a flavour of the rate of rate of progress that synchrotron facilities experienced in the early days. The leftmost picture shows the experimental floor around the National Synchrotron Light Source vacuum ultraviolet (VUV) ring at Brookhaven National Laboratories in 1982. The ring can be seen towards the center of the picture, with a few experimental stations already operating. The center panel of Figure 1.3

shows the situation just three years later in 1985, and the right panel shows a very crowded experimental floor in 1989.

Figure 1.3: Views of the NSLS VUV storage ring dated (from left to right) 1982, 1985 and 1989, taken from approximately the same vantage point, showing the dramatic proliferation of hardware on the experimental floor. Photographs courtesy of Helmut Wiedemann, SSRL.

1.6 The generation game

Synchrotron radiation sources are frequently described in terms of generations. **First-generation** sources were facilities that were constructed to answer questions in high energy physics, where the synchrotron researchers depended upon "parasitic" beamtime. The **second-generation** facilities were designed specifically for synchrotron radiation experiments. The inaugural second-generation source world-wide was the 2 GeV SRS in Daresbury Laboratory. The first to be constructed in the USA was the National Synchrotron Light Source (NSLS) at the Brookhaven National Laboratory in New York, consisting of two storage rings, one for low X-ray energies (called the VUV or vacuum ultraviolet ring) became operational in 1982 (Figure 1.3) and the other hard X-ray ring, which became operational in 1984. NSLS was the first of the double-bend achromat lattices (see Chapter 2), a ground-breaking design by Renate Chasman and George Kenneth Green at the Brookhaven National Laboratory. Like the second generation, **third-generation** sources were also dedicated storage rings, but designed to produce smaller electron beams, yielding smaller X-ray beams, and incorporating long straight sections to contain insertion devices such as wigglers and undulators. Among these are many national synchrotron facilities around the world, each with their storied histories of champions who strove to build expertise abroad while advocating tirelessly for a national facility at home, including the Australian Synchrotron and the Canadian Light Source [4], both of which became operational in the 2000s. Now, a fourth generation of Synchrotron radiation sources, with substantially more compact electron beams and vastly superior performance, is being constructed.

1.7 X-ray absorption spectroscopy

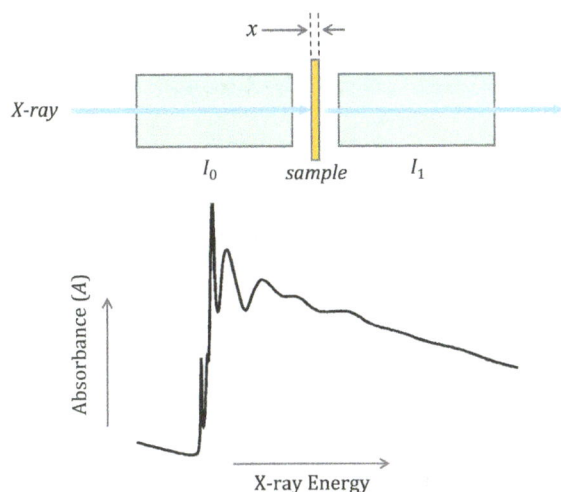

Figure 1.4: A simple X-ray absorption spectroscopy experiment.

XAS is, in principle, no different from many other types of spectroscopy; for example, UV-visible spectroscopy monitors the absorption of light as a function of wavelength, and likewise with XAS, the absorption of X-rays is monitored as a function of the incident X-ray energy. A simple setup, similar to that used in early experiments is shown in Figure 1.4. The monochromatic X-ray beam enters on the left, passes through a gas ionization chamber, a simple detector that monitors the incident X-ray intensity to give the incident X-ray intensity I_0, through the sample and downstream of the sample through a second gas ionization chamber to give the transmitted X-ray intensity I_1. The X-ray absorbance A of the sample is then monitored as a function of the incident X-ray energy by $A = \log I_0/I_1 = \mu x$, where x is the sample thickness and μ is the X-ray absorption coefficient. In some early experiments, the monochromator was scanned from high to low energy, but in modern experiments, the scanning direction typically increases the X-ray energy.

In this book, we aim to provide an overview of the many aspects of modern XAS, its strengths and its limitations, as well as experimental pitfalls and problems. As a preliminary to the main topic, we will also discuss synchrotron radiation, how it is produced and used, together with relevant aspects of beamlines.

References

[1]   Augustin, J.-E.; Boyarski, A. M.; Breidenbach, M.; Bulos, F.; Dakin, J. T.; Feldman, G. J.; Fisher, G. E.; Fryberger, D.; Hanson, G.; Jean-Marie, B.; Larsen, R. R.; Lüth, V.; Lynch, H. L.; Lyon, D.; Morehouse, C. C.; Paterson, J. M.; Perl, M. L.; Richther, B.; Rapidis, P.; Schwitters, R. F.; Tanenbaum, W. M.; Vannucci, F.; Abrams, G. S.; Briggs, D.; Chinowsky, W.; Friedberg, C. E.; Goldhaber, G.; Hollebeek, R. J.; Kadyk, J. A.; Lulu, B.; Pierre, F.; Trilling, G. H.; Whitaker, J. S.; Wiss, J.; Zipse, J. E. Discovery of a narrow resonance in e^+ e^- annihilation. *Phys. Rev. Lett.* **1974**, *33*, 1406–1409.

[2]   Lytle, F. W. First X-ray absorption spectroscopy at SSRL in 1974. *Synchrotron. Radiat. News* **2015**, *28*, 30–33.

[3]   Sayers, D. E.; Stern, E. A.; Lytle, F. W. New technique for investigating noncrystalline structures: Fourier analysis of the extended X-ray absorption fine structure. *Phys. Rev. Lett.* **1971**, *27*, 1204–1207.

[4]   Bancroft, G. M.; Johnson, D. D. *The Canadian Light Source. A Story of Scientific Collaboration*. University of Toronto Press: Toronto, Buffalo, London, **2020**.

2 Sources of synchrotron radiation

2.1 Introduction

The topic of sources is not strictly needed to apply XAS to various research questions, but we consider it to be an essential background material, as any investigator using XAS would be well advised to at least understand some of the fundamentals. The properties of the various sources are also important for the experimenter when making choices about which experimental facilities to select.

2.2 The Lorentz force

If we consider a particle of charge q moving with velocity $\mathbf{v}$ in a magnetic field $\mathbf{B}$ and electric field $\mathbf{E}$, then, when using S.I. units, the force $\mathbf{F}$ on the particle is given by

$$\mathbf{F} = q\mathbf{E} + q(\mathbf{v} \times \mathbf{B}) \qquad (2.1)$$

This equation tells us that what is known as the Lorentz force $\mathbf{F}$ is aligned **with** the direction of the electric field and **perpendicular** to the direction of the magnetic field. The Lorentz force will thus act to accelerate a charged particle when an electric field is aligned with the beam axis, and this acceleration will be independent of the velocity of the charged particle, acting on a charged particle that is at rest. As we will see, it is the electric fields that are used to accelerate charged particles – typically electrons – to relativistic velocities, while magnetic fields are used to deflect the trajectory of the electrons, accelerating them in the process, which generates synchrotron X-rays.

2.3 Electromagnetic radiation from accelerated charged particles

If we consider an electron or any charged particle at a stationary position in space, or moving with uniform velocity, then electric field lines will emanate from the charge at the speed of light to infinity. If we now subject our charged particle to acceleration, the emanating field lines are also accelerated. Close to the charged particle, the field lines will be directed towards where the particle now is, whereas further away, the field lines are directed towards where the particle was before it was accelerated. Between these two, the field lines are distorted, and this distortion travels out from the accelerated charged particle at the speed of light, with the magnitude of the distortion being proportional to the acceleration. Because of Maxwell's relationships, we know that the electric field must have an accompanying magnetic field so that the distortion is really electromagnetic in nature, and the resulting electromagnetic wave is what we call light. These ideas are summarized in Figure 2.1. Thus, charged particles do not

https://doi.org/10.1515/9783110570441-002

radiate, when stationary or moving with a uniform velocity, but do so when they are accelerated. The spatial distribution of the radiation will show no intensity in the direction in which the charged particle is accelerated, and show a maximum at 90° to this direction. In three dimensions, the spatial distribution of radiated power will form a donut shape (Figure 2.2).

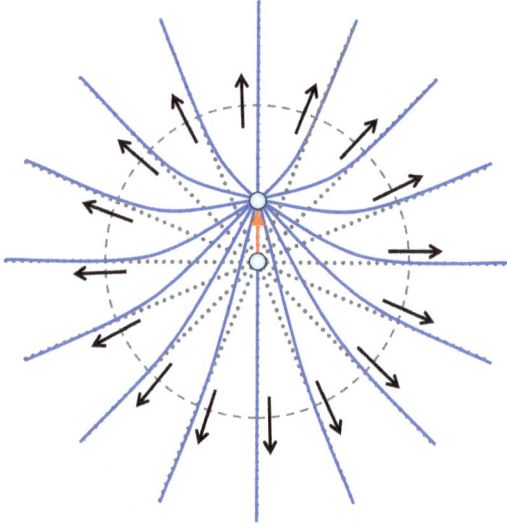

Figure 2.1: Electric field lines for an accelerated charged particle. The red arrow shows the direction of acceleration, with dotted lines indicating the particle position previous to acceleration. The radial electric field lines propagate from the charged particle at the speed of light, with the acceleration causing a wave propagating outwards.

We now consider a moving charged particle subjected to a centripetal acceleration so that it travels in a circular orbit. Here, the direction of acceleration is towards the centre of the circle and the donut spatial distribution of radiated light is arranged as shown in Figure 2.2. We now consider relativistic effects, which occur when velocities v approach the speed of light c, so that $(v/c) \rightarrow 1$. The ratio (v/c) is known as relativistic beta, which relates to the Lorentz factor, γ, an important quantity that specifies the changes in properties of an object that is moving. From this point onward, we will consider electrons as our charged particles, because almost all synchrotron radiation facilities use electrons:

$$\gamma = \frac{1}{\sqrt{1-\beta^2}} = \frac{E}{m_e c^2} = \frac{E[\text{GeV}]}{0.511 \times 10^{-3}} \qquad (2.2)$$

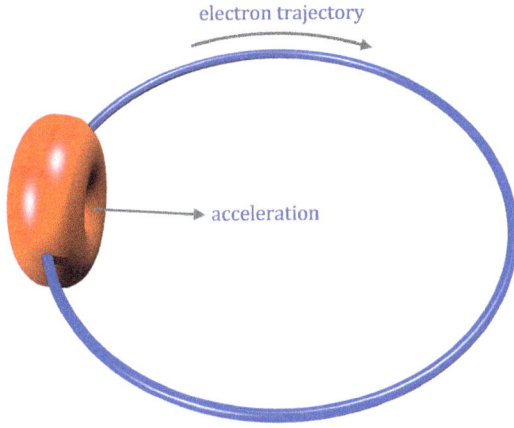

Figure 2.2: Radiation profile (red surface) from a charged particle moving in a circular trajectory (shown in blue).

Equation (2.2) gives γ both as a function of β and in terms of electron energy E, where m_e is the electron rest mass. For a 3.0 GeV synchrotron, we can compute $\gamma = 3.0/0.511 \times 10^{-3} = 5{,}871$.

If our accelerated electron is moving at relativistic velocities, then for an observer at rest, the angular distribution of the radiated power changes from the non-relativistic donut shape. For a circular trajectory of radius ρ, the radiated power P per unit solid angle Ω is given as follows:

$$\frac{dP}{d\Omega} = \left(\frac{1}{c^3 \mu_0} \frac{e^2}{(4\pi\epsilon_0)^2} \right) \frac{\beta^4}{\rho^2} \left(\frac{(\beta^2 - 1)\sin^2\vartheta\cos^2\varphi + (1 - \beta\cos\vartheta)^2}{(1 - \beta\cos\vartheta)^5} \right) \qquad (2.3)$$

where e is the electronic charge, μ_0 is the magnetic constant and ϵ_0 is the permittivity of a vacuum; ϑ and φ specify the direction of synchrotron light emission, relative to the direction of electron motion (ϑ) and the direction of acceleration (φ). From the perspective of the stationary observer, the angular distribution becomes sharply folded forward in the direction of electron motion, with the radiation mainly concentrated within a cone of opening angle $\pm 1/\gamma$. From the electron's frame of reference, the angular distribution would appear identical to the simple non-relativistic donut shape (e.g. Figure 2.3, $\beta \to 0$). The $\pm 1/\gamma$ cone will be important to us in developing an understanding of many aspects of generation of synchrotron radiation, so the reader is advised to remember this as we proceed.

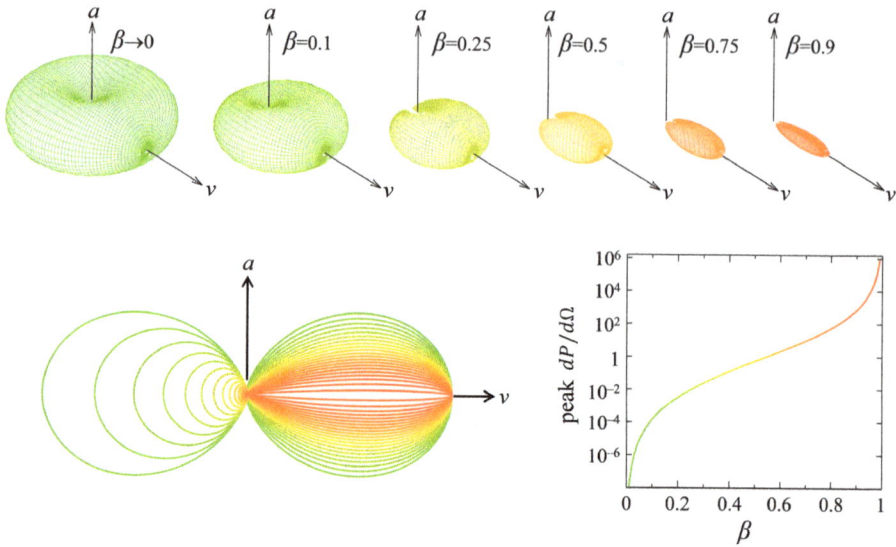

Figure 2.3: The upper panel shows normalized three-dimensional renditions of $dP/d\Omega$ for various values of β, with the direction of acceleration (a) and of motion (v) shown by the arrows. **lower left**: normalized two-dimensional profiles corresponding to cross-sections of the three dimensional case, showing how the radiated power folds sharply forward in the direction of motion as $\beta \to 1$, **lower-right**: relative value of $dP/d\Omega$ at $\vartheta = 0$, showing the dramatic increase in radiated power as $\beta \to 1$; the colours of profiles in the upper panel and lower-left panel correspond to the line colour for different β.

2.4 Synchrotron radiation spectrum and time structure

We consider what might be seen by an observer looking at synchrotron radiation emitted from electrons travelling in a circular path. The observer will first see the leading edge of the $1/\gamma$ cone of synchrotron light, and continue to observe it while $1/\gamma$ cone sweeps past the field of observation as the electron continues on its circular path, past the point at which the trailing edge of the $1/\gamma$ cone is last observed. The photons travel to the observer at the speed of light, whereas the electrons are moving slightly slower. The duration of the observed flash of synchrotron light will be the difference of the time taken by the electron and the photon moving between the points of first and last observation, which we can call δt, and it can be shown that $\delta t = 4\rho/3c\gamma^3$. If we assume $E = 3$ GeV, so that $\gamma = 5{,}871$ and bending radius $\rho = 12.7$ m, then $\delta t = 3.8 \times 10^{-19}$ s, or 0.38 attoseconds. This incredibly short pulse translates to a broad frequency spectrum, effectively a bell-shaped spectrum, peaked at a wavelength λ_c with a spectral width of approximately λ_c. This spectrum is frequently plotted as intensity or brightness vs photon energy, typically using a $\log_{10}$-$\log_{10}$ scale, which gives a characteristic output spectrum with a near-plateau at low X-ray energies, rising to a maximum and then falling off (see Sections 2.9 and 2.10), which experimenters using synchrotron radiation exploit to give a tunable source of X-rays. We will consider bending magnet sources in Section 2.11. The actual time structure of synchrotron radiation sources is a function of the ring and the radio frequency system, and will be discussed in Section 2.9.2.

2.5 Beam current

The beam current is the electrical current that is carried in the storage ring during normal operations. The spectral brightness is, under ideal circumstances, directly proportional to the beam current. Typically, beam currents are in the hundreds of mA, for example, Stanford's SPEAR-3 storage ring operates in top-up mode with a beam current of 500 mA, and the Canadian Light Source at 220 mA, also in top-up mode. The current flowing in an old-style hand-held flashlight with an incandescent bulb would be very similar. The heat load on the X-ray optics is also directly proportional to the beam current, and in many cases, increasing the beam current causes subtle distortions of these optics, so that the photons arriving at a sample do not increase linearly with the beam current. Individual electrons in a typical storage ring can travel millions of kilometres, and any collisions with residual atoms from the atmosphere will cause losses, so the vacuum insider the storage ring must be very good indeed, typically in the range of 10^{-10} torr[1] or lower. Nonetheless, electrons are gradually lost from the storage ring with time, and unless the storage ring is refilled with electrons, the beam current will decay exponentially with time. The beam has a characteristic **lifetime**, defined as the time for current to reduce by a factor of $1/e$, which is typically measured in hours. It is instructive to consider how many electrons circulate in a typical storage ring; 500 mA corresponds to 0.5 coulombs per second, which, given the electronic charge of 1.602×10^{-19} C, would correspond to 3.121×10^{18} electrons per second. If we again consider Stanford's SPEAR-3 storage ring with 234.137 m circumference, a round trip for an electron will take 781 ns, which means that under normal operating conditions, the storage ring contains about 2.44 trillion electrons (corresponding to 4 pico-moles of electrons) or 8.71 billion electrons in each of the 280 bunches.

2.6 Electron beam emittance

The brightness of the synchrotron radiation is determined by a quantity, known as the electron beam emittance. The horizontal emittance ε_x is typically much larger than the vertical and is defined as the product of the transverse electron beam size σ_x and the angular divergence of the beam in each direction transverse to the direction of electron

---

1 Torr is close to 1 mm of mercury, or 1/760 of atmospheric pressure at sea level (101,325/760 Pa).

motion σ'_x, with typical units of nm-rad, or pm-rad for the newest sources. A small emittance translates into a small, low-divergence beam of synchrotron radiation:

$$\varepsilon_x = \sigma_x \sigma'_x \tag{2.4}$$

The electron beam emittance in a storage ring has contributions from two competing processes; the first corresponds to a perturbation in the electron trajectory due to emission of a photon of synchrotron radiation, which causes a loss of energy so that the electron follows a different orbit, and this gives an increase in the electron beam emittance. The second process is a damping effect from loss of transverse momentum, upon emission of synchrotron radiation. The balance of the two, effectively yields an equilibrium emittance; the approach to equilibrium is relatively slow, of the order of milliseconds or tens of milliseconds, and is an important factor in the design of an injection system. Aggressive deflections from strong bending magnets tend to increase ε_x, and a well-known relationship is that the horizontal electron beam emittance scales as the cube of the deflection angle of the bending magnets, or, for a given storage ring lattice (see Section 2.9.3), the inverse cube of the total number of individual dipole bend magnets. The horizontal emittance also scales as the square of the electron beam energy so that low-emittance beams are easier to produce with low-energy storage rings:

$$\varepsilon_x = F \frac{E^2}{J_x N_b^3} \tag{2.5}$$

where F is the function of the magnet lattice, E is the electron beam energy, J_x is the horizontal damping function and N_b is the number of dipole or bending magnets in the ring.

2.7 Spectral brightness

For a number of years, the synchrotron radiation community disagreed about the correct name for describing the properties of a source, although there was widespread agreement that this should have units of photons s^{-1} mm^{-2} mrad^{-2} (0.1% bandwidth). Brightness, spectral brightness and brilliance had been used to describe source properties, but there is now consensus that **spectral brightness** is the best description. In the experimental hutch, important measures are photon flux, which is typically specified in units of photons s^{-1}, and flux density would be a measure of the photon flux hitting a sample per unit area, photons s^{-1} mm^{-2}.

2.8 Polarization and coherence

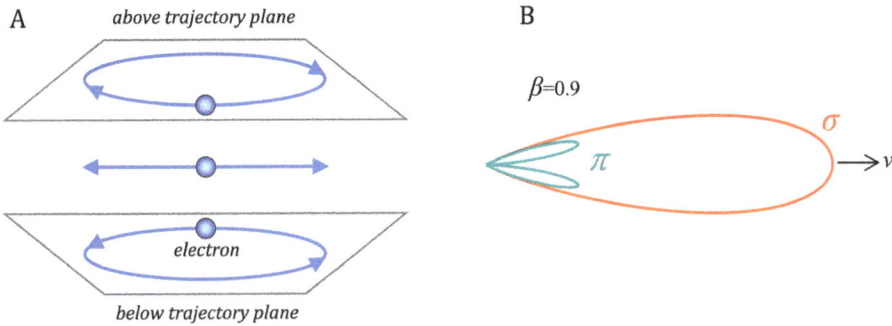

Figure 2.4: A. Polarization of synchrotron light from an electron in a circular orbit, when observed from different viewpoints. B. Polar plot of calculated polarization intensities for an electron in a circular orbit, with direction of motion shown by v and with $\beta = 0.9$, where σ and π, respectively, correspond to polarizations in the plane normal and parallel to the deflecting magnetic field.

We again consider an electron subjected to centripetal acceleration so that it moves in a circular orbit. We can compare views of the resulting synchrotron light from above, within and below the plane of the orbit, as shown in Figure 2.4A. When viewed within the plane of the ring, the circular trajectory appears like a straight line, and the resulting light will be plane polarized, but above and below the plane of the ring, it will be elliptically polarized with opposing polarizations. More rigorously calculated synchrotron radiation emission profiles with σ and π polarizations, which respectively correspond to polarizations in the plane normal and parallel to the deflecting magnetic field, confirm this simple view, as shown in Figure 2.4B. Similarly, the in-plane synchrotron radiation from bend magnet sources (Section 2.11), and plane wigglers and undulators (Section 2.12) is plane polarized, with the plane of polarization parallel to the synchrotron experimental floor. The importance of polarization will be discussed in later chapters of this book.

A second important property of synchrotron radiation is coherence, which occurs when wave sources have identical frequencies and waveforms. Two kinds of coherence relevant to synchrotron radiation can be distinguished. **Temporal coherence** occurs when the bunch of electrons is about the same size as the wavelength of the emitted light, and **spatial coherence** occurs with longer bunches but with a transverse beam emittance that is smaller than the wavelength of the emitted light λ, or more specifically $\varepsilon_x \leq \lambda/4\pi$, which is known as the diffraction limit. Both depend upon the wavelength of the emitted light, and it is therefore much easier to obtain synchrotron radiation with a degree of coherence at low X-ray energies, and much more difficult at high X-ray energies. We note that modern low-emittance synchrotron radiation sources have substantially improved coherence properties.

2.9 Components of synchrotron radiation facilities

Synchrotron radiation facilities consist of a really large number of components and sub-systems; here, we choose to dramatically simplify this and to discuss just a few. These are source of the electrons, the radio frequency system, the magnetic lattice and the sources of synchrotron radiation. A highly stylistic picture of a storage ring is shown in Figure 2.5.

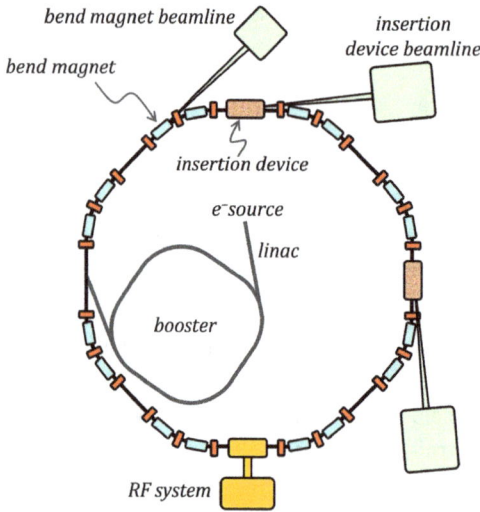

Figure 2.5: Idealized plan view of a synchrotron radiation light source, showing the various components discussed.

2.9.1 The electron source, linear accelerator and booster synchrotron

The source can be an electron gun, familiar to some in old-style cathode ray tubes. Details vary from facility to facility, and here we describe the Stanford's SPEAR booster system as an example. The electron gun is an RF gun with a thermionic cathode that operates at the same frequency as the linear accelerator (2.856 MHz); the energy of the electrons as they emerge is about 2.5 MeV, and this feeds the linear accelerator or **linac**, which uses RF technology (see Section 2.9.2) to accelerate the electrons to 150 MeV. At SSRL, the linac is comprised of three sections, each driven by a separate klystron RF source. Electrons from the linac are fed to the booster synchrotron, which accelerates them to 3 GeV through a simultaneous ramping of booster synchrotron magnet strength and RF cavity fields, before injection into the storage ring. Most modern synchrotron facilities have implemented what is known as top-up mode (Figure 2.6). Prior to top-up implementation at many synchrotron facilities around the world, the storage ring would be filled to capacity and then allowed to decay over perhaps 12–24 h, and much effort was focused on improving the lifetime of the stored beam. There are a number of disadvantages with such fill-and-decay operations, amongst which are the changing heat-loads on the beamline optics. With top-up mode, frequent fills of the storage ring are carried out (e.g. every 5 min) so that close to the maximum beam current is maintained, and the heat load on beamline X-ray optics is nearly constant.

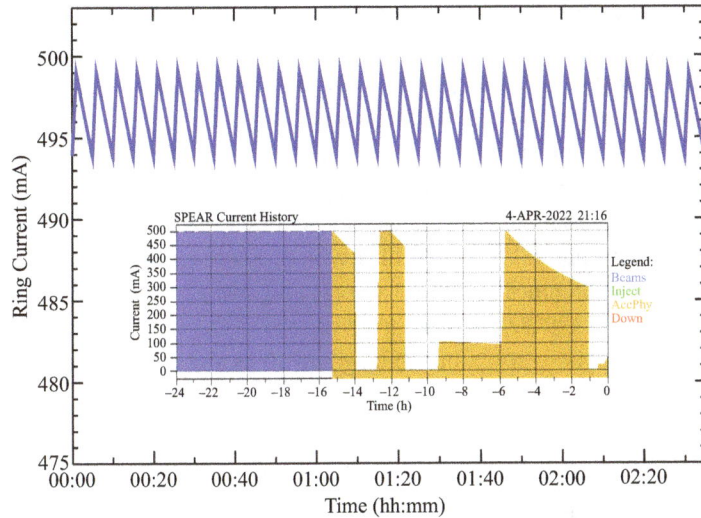

Figure 2.6: SPEAR storage ring history plot. The main graph shows the top-up fill pattern (blue line), with frequent fills every 5 min. to keep the ring current close to 500 mA. The inset shows the history, with solid blue (left) indicating user beam in top-up mode and orange showing accelerator physics, called machine studies at other synchrotron facilities.

2.9.2 The radio frequency (RF) system

This is responsible for accelerating the electrons in the ring and for replenishing the energy that is lost through synchrotron radiation. The oscillatory nature of radio waves precludes a continuous beam of electrons, and the electron beam is instead contained in what are called **bunches**, each of which is separated by the RF wavelength, with a bunch length that is much shorter than the RF wavelength. The path length travelled by the electrons around the storage ring must be a multiple of the RF wavelength. The RF system actually provides potential wells called RF **buckets**, each of which can be vacant or be occupied by an electron bunch, and these potential wells effectively rotate around the ring, driven by the RF system. It is the synchronization of the RF system with the passage of electron bunches that is responsible for the name synchrotron. With an appropriately sophisticated injector, it is possible to fill all the buckets in a storage ring to fill just a single bucket or perhaps to generate a specific pattern to give a particular time structure. The radio frequency radiation interacts with the electron bunches in RF cavities, the design of which can be quite sophisticated and may use superconducting technology to assist with stability. The cavities, of which there may be many positioned within the storage ring, are typically cylindrical in shape, with internal structures to direct the electric field in the direction that the electrons are desired to travel. As the electrons travel along the cavity, those with slightly lower velocities tend to experience slightly higher accelerations than those with higher velocities, and the result is that electron bunches are compressed. The bunch length in a typical ~3 GeV storage ring is 30 to 200 ps and the buckets are separated by 2 to 3 ns, depending upon the RF frequency. The source of the RF can be a klystron, which is a linear-beam vacuum tube amplifier, an inductive output tube or IOT (related to a klystron), or arrays of solid-state high-power amplifiers. The RF is carried to the cavities from the source via a waveguide, an often rectangular conduit, specifically designed to carry the RF, typically filled with pressure-controlled dry air or other gas. As an example, Stanford's SPEAR-3 storage ring has a circumference of 234.137 m, an RF frequency of 476.137 MHz and a fill pattern of 280 bunches, distributed in four groups of 70 bunches each, a bunch length of 20 ps and a bunch spacing of 2.1 ns.

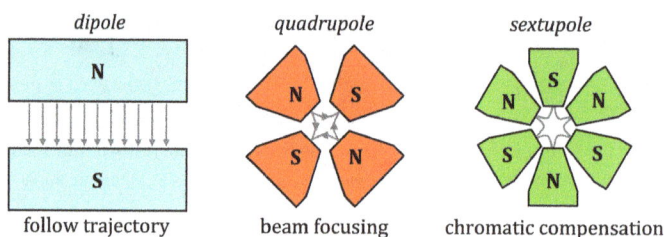

Figure 2.7: On-axis views of the different magnets typically found in a storage ring lattice.

2.9.3 Magnetic lattice

The arrangement of magnets inside the storage ring is called the lattice. Three basic types of magnets are typically used, as shown in Figure 2.7. Dipole magnets are the simplest in design, and are used to change the trajectory of the electron beam, as in a bend magnet, or in the injection system from the booster to the storage ring. Dipoles tend to increase the horizontal electron beam emittance – this will be considered below but we can describe it here as being a measure of the size and shape of the particle beam within the storage ring. Put simplistically, there will be a distribution of electron velocities within the ring; on deflection with a dipole, faster electrons will take the outside track; and slower electrons, the inside track, smearing out the beam in the horizontal direction. Other magnets systems are needed within the lattice to focus the electron beam. Quadrupole magnets consist of groups of four magnets arranged such that the dipole terms cancel, with the lowest significant terms in the field equations being quadrupole. The magnetic field that they create grows rapidly in magnitude as one goes away from the central axis, and so these magnets are used in beam focusing. Without going into details, Maxwell's equations stipulate that a quadrupole cannot focus in both horizontal and vertical planes at the same time, and hence two types of quadrupole are needed – called F and D quadrupoles, which focus in the horizontal and vertical, respectively, and de-focus in the other direction. If placed immediately next to one another and F and D quadrupole cancel, but if there is space between them (which must be carefully chosen), then focusing can be achieved in both horizontal and vertical planes. Quadrupole magnets have a focusing strength, which is dependent upon the electron energy, so the higher energy (faster) electrons have longer focal lengths than lower energy (slower) electrons. The effect is that the quadrupole magnets tend to expand the beam with distance, and this needs to be corrected, which can be done using sextupole magnets, which (with careful positioning) can cancel the tendency of the quadrupoles to spread out the electron beam.

Various types of lattice can be found in different synchrotron radiation sources. A common early lattice was the FODO lattice, named for the arrangement of spaced F and D quadrupoles – in which the Os stand for a non-quadrupole element in the cell, such as a space (nothing) or a dipole. Nearly all third-generation synchrotron sources used what is called a double-bend achromat (DBA) lattice,[2] or Chasman-Green lattice, named after its inventors Renate Chasman and George Kenneth Green of Brookhaven National Laboratory. Each cell within a DBA lattice contains focusing quadrupoles symmetrically located between pairs of dipoles. The most advanced and possibly the last DBA synchrotron facility to be constructed was the National Synchrotron Light Source (NSLS) II.[3] NSLS-II has a circumference of 792 m with 30 DBA cells, potentially accommodating more than 60 beamlines. The horizontal emittance of NSLS-II is remarkable for a third-generation source, at 1 nm-rad.

More recently, a fourth generation of storage rings have begun to be constructed, based upon a new type of lattice – the multi-bend achromat (MBA). MBA lattices use a larger number of dipoles per cell, and can produce beams with remarkably smaller horizontal electron beam emittance, which give a nearly round profile. The first MBA synchrotron radiation facility to become operational was the MAX-IV in Lund (Figure 2.8),

---

2 An exception to this is the third-generation Advanced Light Source (ALS) at Berkeley, California, a triple-bend-achromat. As we write, the ALS is currently preparing to upgrade to a fourth-generation source, called ALS-U.

3 Strictly speaking the middle eastern synchrotron light source, SESAME, may be the last DBA. NSLS-II achieved first light on October 23, 2014, with ground-breaking June 15, 2009, while SESAME saw first light on November 22, 2017, with ground-breaking January 6, 2003. The first MBA facility MAX-IV saw first light November 3, 2015, with ground-breaking November 22, 2010.

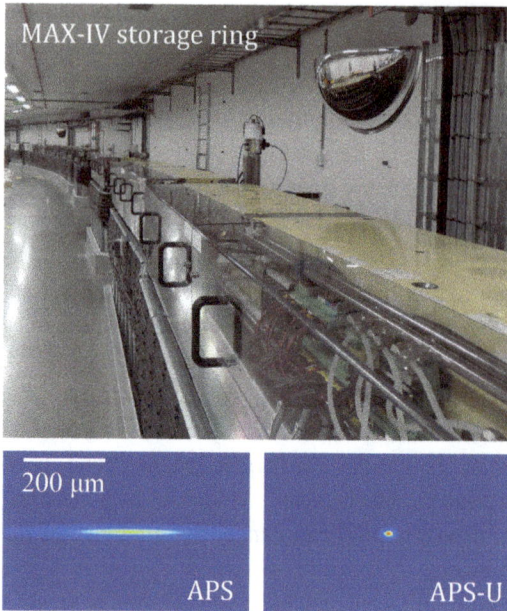

Figure 2.8: The upper panel shows the MAX-IV storage ring, which uses the world's first multi-bend achromat (MBA) lattice, a 3 GeV, 528 m circumference storage ring with 20 7-bend achromat cells, separated by 4.6 m straight sections. This facility gives a horizontal electron beam emittance ε_x of 330 pm rad or less. The lower panels show the calculated improvement in the electron beam profile expected for the MBA upgrade of the Advanced Photon Source (APS-U).

Sweden, with clusters of seven small bends, quadrupoles and sextupoles, and as we write, the promise of transformative scientific capabilities is already being realized.

2.10 Beam orbits and beam losses

The reader will probably realize that the picture that we have thus far presented is deliberately simplified. We will amplify slightly on some aspects of beam perturbations here. Among the most important beam perturbations from the ideal orbit are what are known as **betatron oscillations**, which are oscillations in a direction transverse to the equilibrium electron orbit, and can be either horizontal or vertical. The number of betatron oscillations around the storage ring is called the **tune**, or the **betatron tune**, which is a characteristic of the magnetic lattice that is primarily defined by the strength of the quadrupole magnets. Oscillations occur in the other direction too, along the trajectory of the beam, and these are called **synchrotron oscillations**, which can act to modulate the tunes. Both horizontal and vertical tunes cannot hold integer or half-integer values, because if they do, any small perturbations in the beam will build, causing unstable beam dynamics and near-immediate loss of stored electron beam.

2.11 Bend magnet sources

In a synchrotron facility, the simplest source of synchrotron radiation is the bend magnet. As we have discussed above, these magnets are responsible for the electrons forming a closed path around the ring, and before the invention of the insertion device, they were the only source of synchrotron radiation for experiments. We note that some fourth generation sources do not use light from bend magnets at all. An important quantity is the bending magnet critical energy E_c. This is defined as the energy that separates the bend magnet spectrum into two halves of equal radiated power. A good rule of thumb is that usable flux for XAS experiments can be obtained from a bending magnet source with photon energies up to $4 \times E_c$:

$$E_c = \frac{3\hbar c \gamma^3}{\rho} = \frac{2.218 E^3}{\rho} = 0.665 B_0 E^2 \tag{2.6}$$

$$E_c[\text{keV}] = 0.665 B_0[\text{T}] E^2[\text{GeV}]^2 \tag{2.7}$$

2.11.1 Superbend magnets

Superbends can be used in place of conventional bend magnets. They use higher magnetic fields B_0 to give a smaller ρ and can significantly shift the critical energy to higher energy. Superbends must be typically installed symmetrically around a ring, in pairs or threes. They have been used on low-energy storage rings to enable experiments in the hard X-ray regime, for example at the original Advanced Light Source 1.9 GeV storage ring, three 5T superbends magnet were used in place of three 1.3T (out of a total of 36) conventional bends, shifting the critical energy and resulting in an order of magnitude increase in X-ray flux at photon energies of 12 keV, with adjustments in the lattice, but with the cost of a **ca.** 20% increase in horizontal electron beam emittance. Hardbends are similar to superbends, but are slightly weaker field devices, and are based on strong field permanent magnet materials.

2.12 Wigglers and undulators

Wigglers and undulators both use periodic magnetic fields to accelerate the stored electron beam and both are **insertion devices**, in that they can be inserted into the storage ring in straight sections, where the electron beam would otherwise go straight. The pe-

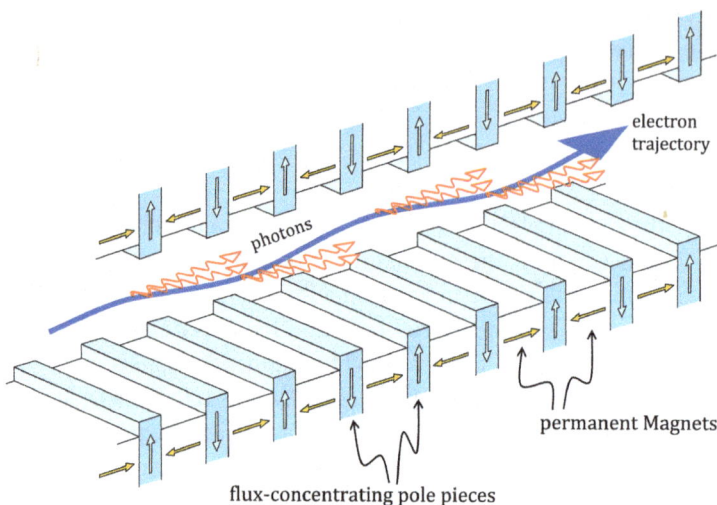

Figure 2.9: Schematic diagram of a plane wiggler magnet, showing the trajectory of the electron beam and the emission of synchrotron radiation from each excursion of the beam.

riodic magnets deflect the trajectory of the electrons, with the acceleration at each excursion giving rise to synchrotron radiation. A schematic diagram of an insertion device is shown in Figure 2.9. We will recall from Section 2.3 that synchrotron radiation is produced with a characteristic $1/\gamma$ opening angle, and both plane wigglers and undulators have a vertical beam divergence of $1/\gamma$, but the horizontal divergence differs. In a wiggler, the transverse excursions of the electron beam are made to be larger than $1/\gamma$, and a wider beam of synchrotron radiation results. In an undulator, the transverse excursions of the electrons are set to be on the order of $1/\gamma$, as shown in Figure 2.10. In both wigglers and undulators, the strength is defined by what is known as the deflection parameter K, which is related to the horizontal opening angle $\alpha = K/\gamma$:

$$K = \frac{e}{2\pi m_e c} B_0 \lambda_p = 0.934 \; B_0[\text{T}]\lambda_p[\text{cm}] \tag{2.8}$$

where B_0 is the peak magnetic field and λ_p is the period length.

The wiggler or undulator field can be controlled by moving the two halves of the device towards or away from each other to create stronger or weaker magnetic fields. The separation between the two halves is called the insertion device gap. The wiggler is the high-field case; the individual $1/\gamma$ fans of synchrotron radiation from each excursion of the electron beam do not line up, and the device produces a beam with large angular spread. Unlike bend magnets, where the need to form a closed electron trajectory around the storage ring is limiting, with insertion devices a wider range of applied fields are possible, and can be very high – permanent magnet wiggler fields of 2 T are now common. Radiation from each period of the wiggler adds to the inten-

sity to produce an extremely intense beam of radiation. A calculated comparison between synchrotron radiation from bend-magnet and wiggler sources is shown in
Figure 2.11.

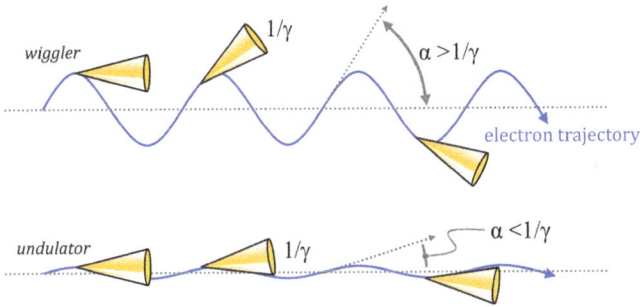

Figure 2.10: Schematic view of electron trajectories within wiggler and undulator magnets. For the
wiggler high-field case, the opening angle $\alpha > 1/\gamma$, with the consequence that the $1/\gamma$ cones of
synchrotron radiation from the individual-induced excursions of the electron beam do not align. For the
undulator low-field case, the $1/\gamma$ cones do align, causing interference that results in quasi-monochromatic
source properties.

Wigglers with low numbers of long periods produce broad fans of radiation, which allows several experimental stations to be simultaneously illuminated (e.g. Figure 2.12).
Wigglers with large numbers of short periods produce narrow fans of intense radiation,
which broadens at low X-ray energies, so that a single insertion device beamline might
accommodate a hard X-ray end-station and a soft X-ray side station. A special type of
wiggler is the wavelength shifter, or three-pole wiggler. Many early wavelength shifter
devices used superconducting electromagnets and were employed on low-energy storage rings to generate X-ray flux in the hard X-ray regime, but permanent magnet devices can give useful photons on mid-range energy (e.g. 3 GeV) storage rings. In essence,
the design consists of a single central pole with a high magnetic field, flanked by com-
pensating poles, and it is the central pole that is used to generate useful synchrotron
radiation. The third-generation storage ring NSLS-II has a number of permanent mag-
net-based three-pole wigglers, with a peak central field of 1.35 T, giving a critical energy
of 6 keV and an emitted horizontal fan of 2 mrad.

Undulators are the weak field case in which the individual $1/\gamma$ fans of synchrotron radiation from each excursion of the electron beam partially align. The radiation
from successive periods of the device interferes to produce a quasi-monochromatic
source. For wigglers, the deflection parameter $K \gg 1$ (e.g. $K = 14$) and for undulators,
$K \approx 1$. For an undulator, the wavelength of the radiation is given by

$$\lambda_n = \frac{\lambda_p}{2n\gamma^2} \left(1 + \frac{K^2}{2} + \gamma^2\theta_{obs}^2 \right) \tag{2.9}$$

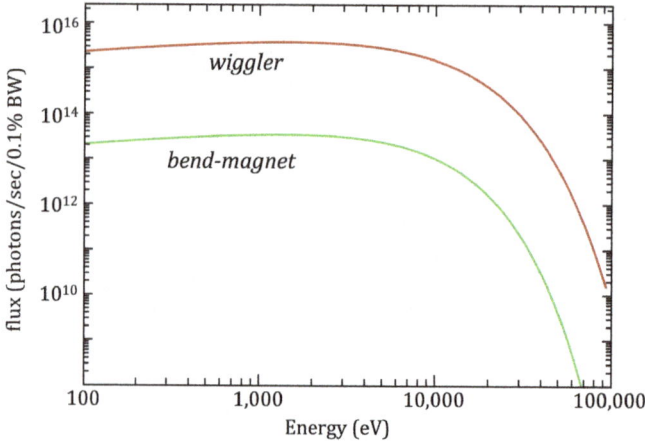

Figure 2.11: Comparison of the photon flux generated by a bend magnet and a wiggler, computed for an electron beam of 3 GeV and 500 mA, with bend magnet ρ of 12.7 m, and wiggler λ_p with 12 periods and $K = 14$. The more than two orders of magnitude increase in photon flux with the wiggler, relative to the bend magnet, is apparent from the plots.

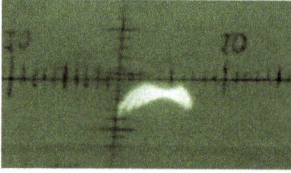

Figure 2.12: An in-hutch photograph of the footprint of the X-ray beam on a fluorescent screen from SSRL's beamline 9 wiggler. This picture was taken in the XAS side station beamline 9-3, which views the wiggler source at ~5-mrad off-axis. The bow-tie appearance of the beam is consistent with the focusing optics of the beamline (see Chapter 3) and the banding seen in the figure is consistent with the effect of the separated pole pairs in the wiggler, viewed from one side at the ~5-mrad observation angle. The banding was observed, following the SPEAR 3 upgrade, and was not visible with the previous higher emittance SPEAR 2 lattice (e.g. see Figure 3.22). Photograph courtesy of SSRL.

in which, λ_n is the wavelength of the nth harmonic ($n = 1,2,3,4, \ldots$), θ_{obs} is the observation angle in the horizontal plane. Alternatively, using practical units and neglecting the $\gamma^2 \theta_{\text{obs}}^2$ term, we can write an expression for the X-ray energy E_n n^{th} harmonic:

$$E_n[\text{keV}] = 0.950 \frac{nE^2[\text{GeV}]^2}{\lambda_p[\text{cm}]\left(1+K^2/2\right)} \tag{2.10}$$

In a typical undulator source, only the odd harmonics are used as X-ray sources for experiments, as these have peak intensities on-axis, whereas the even harmonics have near-zero intensity on-axis. Notice the E^2 dependence in eq. (2.11); for a synchrotron radiation source with high electron beam energy (with perhaps E of 6 GeV), the most intense harmonic, called the fundamental, with $n = 1$, can fall in the hard X-ray regime, but for a medium-energy synchrotron (e.g. with E of 3 GeV), the fundamental falls at lower energies, in what is called the tender X-ray regime, and higher orders are used for experiments needing hard X-rays. The energy of the various harmonics can be adjusted by changing the gap on the insertion device, changing B_0 and hence K (eq. (2.8)). Figure 2.13 shows a calculated emission spectrum from an undulator operating on a 3 GeV storage ring, with the inset in the figure showing the effect of scanning K on the undulator third harmonic.

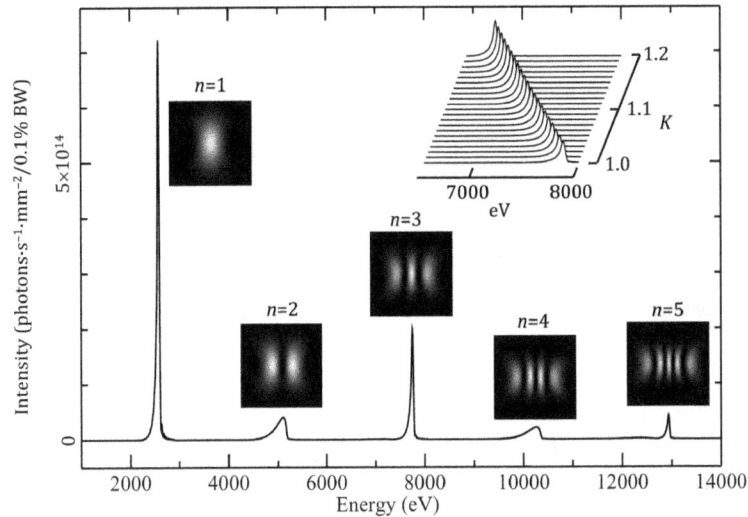

Figure 2.13: Calculated flux through a 1 mm pinhole at 30 m from the source for an undulator operating in a 3 GeV storage ring containing 500 mA. Parameters used were $\lambda_p = 2.2$ cm with 66 periods and $K = 1.0$. The calculated odd on-axis intensity profiles for harmonics $n = 1$ through 5 are shown and the inset shows the effect of K (varied by changing the undulator gap) upon the position of the third harmonic ($n = 3$).

Figure 2.14 shows what are known as undulator tuning curves, which were calculated for the same undulator parameters used for Figure 2.13. Tuning curves typically plot the intensities of the odd undulator harmonics versus the peak energies for a range of undulator gaps. Note that by a careful choice of undulator harmonic and gap, almost the complete range of energies can be covered, with a notable inaccessible region between the upper limit of the first harmonic and the lower limit of the third harmonic. This inaccessible energy gap is common on many undulator beamlines, although devices using some of the more modern high-remanence magnet materials may be able to eliminate this gap entirely.

We note here that a traditional undulator, with a motorized variable gap between upper and lower halves, may not be the best source for X-ray spectroscopy. This is because the undulator is a substantial device, typically requiring a hefty stepper

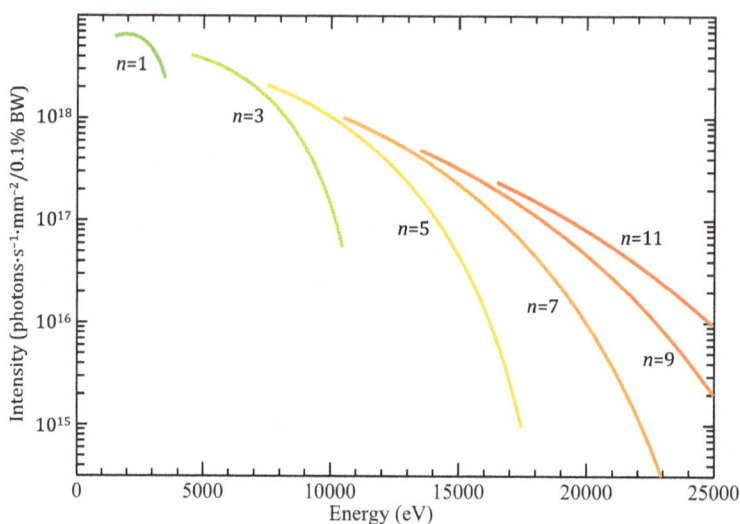

Figure 2.14: Calculated integrated intensity tuning curves for the undulator of Figure 2.13 showing odd harmonics $n = 1$ through 11. Note that the intensities differ from those of Figure 2.13 as a pinhole was not used for the tuning curve calculations.

motor to translate the two halves vertically, and this motor may lack the fine control that is required for the monochromator. As a consequence, the method of moving the undulator gap in harmony with the motions of the X-ray monochromator may cause problems because the undulator will only move every thousand or so monochromator motor steps. Moving the gap can cause a vertical movement of the beam, perhaps with a consequent loss of alignment for the most sensitive experiments. In particular, the so-called QEXAFS (quick-EXAFS) experiments, where the monochromator is continuously moved and data collected on the fly, have not met with much success when using variable gap undulator sources. Measures such as tapering the gap by moving different ends of the insertion device in different directions can help, as this destroys some of the properties of the device, broadening the harmonics considerably, which may allow the gap to remain fixed. However, there is an innovative solution that was invented in 1992 by Roger Carr and colleagues at SSRL, which is the adjustable phase undulator. Instead of moving the gap to adjust the field, the magnetic field is varied by altering the relative longitudinal position of the arrays of magnets, keeping the gap fixed. Figure 2.15 shows a schematic of how these devices work.

While, with a wiggler, the vertical divergence of the X-ray beam remains unchanged from the bend magnet case, $\sigma'_v = 1/\gamma$, with a horizontal divergence of $\sigma'_h = 2K/\gamma$, an undulator can give very small divergences in both the horizontal and vertical directions, which can be measured in μ-radians. The undulator divergences are approximately given by eq. (2.11), in which n is the harmonic number (eq. (2.9)) and N is the number of periods

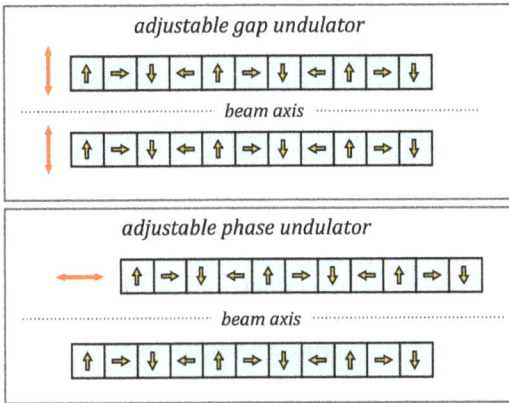

Figure 2.15: Side-view schematic diagram comparing conventional (variable gap) and adjustable phase undulator designs. The yellow arrows show the direction of magnetization in the undulator magnet blocks, and the red arrows the direction of motion of the halves of the undulator. For the adjustable gap design (upper), both sets of poles are moved vertically so as to change the field, but for the adjustable phase design, one (or both) sets of poles may be moved parallel to the electron beam axis.

that undulator possesses (the number of magnet poles will be twice the number of periods, or $2N$):

$$\sigma'_h = \sigma'_v \approx \frac{1}{\gamma\sqrt{nN}} \tag{2.11}$$

Thus far, we have considered only what are known as plane wigglers and undulators, which depend on electron undulations in the transverse direction. Early studies using circular polarized X-rays used bend-magnet sources, with separate optical paths above and below the plane of the ring for left and right circularly polarized light. These suffered from very low intensities, which limited the scientific applications. In the early 1990s, reports appeared of undulator designs specifically tailored to produce spiral electron motions, in order to produce an intense beam of elliptically polarized X-rays. One early model was tested on SSRL's beamline 5, and had four sets of movable arrays of permanent magnets; it could produce plane-polarized X-rays in either the horizontal or vertical direction and close to 100% left and right circularly polarized radiation. Such devices are extremely useful for polarization spectroscopy applications.

Finally, we discuss undulator structure from wiggler sources. The presence of this should be no surprise, as plane wigglers and undulators are simply high- and low-field cases of the same device type. As K increases substantially, above 1.0, the undulator structure becomes merged until the smooth output of a wiggler is finally obtained. Nonetheless, wigglers frequently show residual undulator structure and this can be a problem for spectroscopy applications. Figure 2.16 shows the output of the SSRL beamline 6-2 54 pole wiggler, which was the world's first permanent magnet wiggler, and has λ_p = 7.0 cm, operating with K = 6.538. In general, residual undulator problems are worse at lower X-ray energies and become washed out at very high energies due to the loss of structure with very high n. It is worth considering why residual undulator structure can cause problems, after all if incident intensity, called I_0, is measured and the data is normalized using I_0, then there should be no observable issues. The answer is that a beam showing residual undulator structure has structure, particularly in the horizontal plane (e.g. see Figure 2.13). If focusing optics are used, then these will help to obscure horizontal intensity variations from residual undulator structure, but even so, any heterogeneity across the sample can result in a signal that does not ratio properly with I_0, and consequently the undulator structure can appear in the data. In many cases, such structure might be very small but, as we will see, the EXAFS analyses changes in absorption cross-section that are also very small.

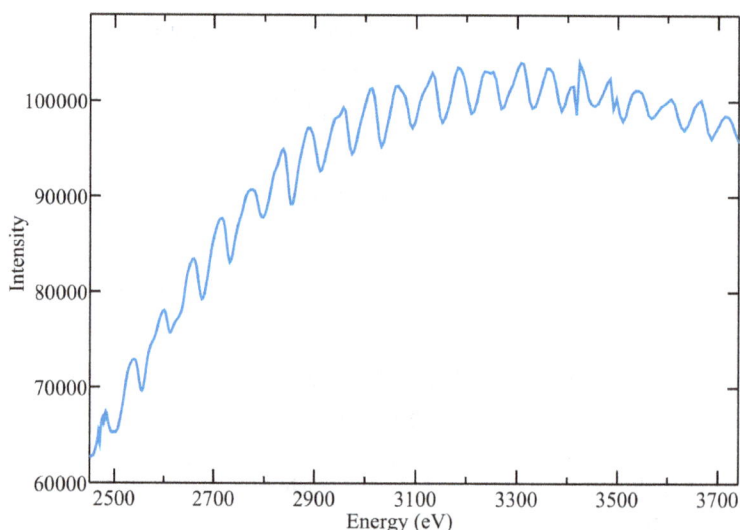

Figure 2.16: Residual undulator structure from the SSRL beamline 6 wiggler operating at a wiggler field of 1.0 T (K = 6.538); the observed periodicity of the oscillations in beam intensity correspond nicely to eq. (2.10), which predicts undulator features at 54.59 n eV, so that the energy range of the figure approximately corresponds to $n \approx$ 46 to 67.

The authors first observed undulator structure in SSRL's beamline 6-2 in the early 1990s (Figure 2.16), and various measures were investigated to mitigate this, including tapering the insertion device. Here, the motors controlling the insertion device gap are deliberately opposed so that the gap is narrower at one end of the device than the other. This can blur the undulator structure but will probably never remove it all together.

In closing this chapter, we note that the X-ray intensities available from insertion devices can be very impressive; Figure 2.17b shows the airglow caused by the passage of X-rays from a modern undulator source through the atmosphere.

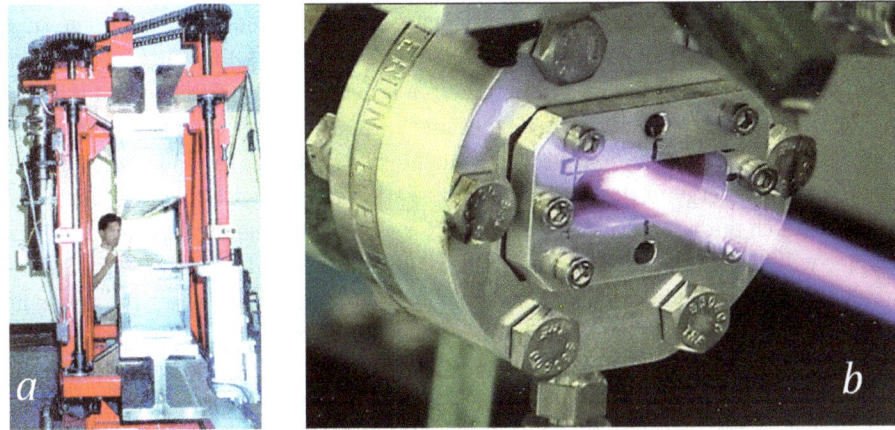

Figure 2.17: Panel a: The SSRL beamline 9 wiggler, the world's first 2T permanent magnet device, before its installation into the SPEAR storage ring (photo courtesy of Stanford Synchrotron Radiation Lightsource), looking along the length of the insertion device. The motors and gears used to move the upper and lower halves of the insertion device are clearly visible. Panel b: the airglow from the white beam of the in-vacuum hybrid undulator of the BioXAS-imaging beamline at the Canadian Light Source (photo courtesy of Canadian Light Source).

Further reading

[1]   Wiedemann, H. *Synchrotron Radiation*. Springer-Verlag: Berlin, Heidelberg, Germany, **2010**.
[2]   Winick, H. *Synchrotron Radiation Sources. A Primer*. World Scientific Press: Singapore, New Jersey, London, Hong Kong, 1994.

3 X-ray beamlines

3.1 Introduction

As we discussed in Chapter 2, the beam of synchrotron radiation from a storage ring can have a broad distribution in energy, and will diverge from the source. The direct beam of synchrotron radiation from the source is often called "white" in that it is not monochromatic. For a spectroscopy experiment we need to reject photons that have energies distant from that desired, with good energy resolution, and to have the ability to scan the energy of the incident photons. We may also wish to shape the X-ray beam to match the sample and optimize the X-ray flux available for experiments. This chapter will discuss the various optical elements that are used to achieve this. Subsequent chapters will address detectors, sample environments and other downstream aspects of the beamline. Figure 3.1 shows a schematic representation of a typical hard X-ray XAS beamline, with the major optical components indicated. If radiation damage is an issue, as it often is with biological samples (for example) the beamlines may deliberately lack focusing optics (X-ray mirrors) to reduce the flux density on the sample. We will therefore first discuss the X-ray monochromator, which is used to select a narrow range of X-ray energies or wavelengths.

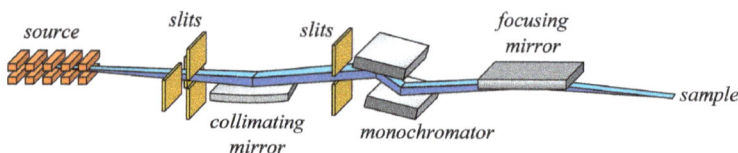

Figure 3.1: Schematic diagram of the major components of a beamline. Typical dimensions for such a beamline are worth noting. The source might be ~17 m from the collimating mirror, which in turn might be separated from the monochromator by 2–10 m, and the focusing mirror relatively close (separated by ca. 2 m) to the monochromator, and the sample 8–10 m downstream of this. The total length of a typical beamline from source to sample might be 25–35 m.

3.2 Hard X-ray monochromators

The monochromator takes a polychromatic or "white" beam of synchrotron radiation and rejects all but a narrow band of energies to form a monochromatic beam. For spectroscopy, there are two principal types of incident beam monochromators, which are *grating monochromators* used for low energy X-ray experiments and *crystal monochromators* used for hard X-ray experiments. Here we consider the double crystal monochromator in which X-ray energy selectivity derives from Bragg diffraction. A monochromator crystal is comprised of a near-perfect single crystal, with Bragg diffraction governing the relationship between wavelength λ and angle of incidence θ_B

https://doi.org/10.1515/9783110570441-003

of the X-ray beam on the crystal. Bragg's law eq. (3.1) can be illustrated using the schematic of Figure 3.2:

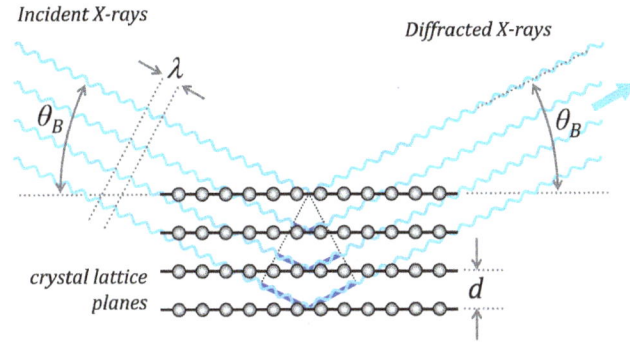

Figure 3.2: Bragg diffraction with Bragg angle θ_B from a crystalline material. The crystal lattice planes are separated by the lattice spacing d, with the X-ray path differences from successive crystal lattice planes being a whole number of wavelengths λ.

$$n\lambda = 2d\sin\theta_B \tag{3.1}$$

$$E = n\frac{hc}{2d\sin\theta_B} \tag{3.2}$$

where n is the order of the diffraction, d is the lattice spacing and θ_B is the angle subtended by both the incident X-ray beam and the diffracted X-ray beam to the lattice plane, called the Bragg angle. Bragg diffraction occurs when the X-ray path differences from successive planes within the crystal correspond to a whole number of wavelengths (shown in blue in Figure 3.2). Only when this condition is satisfied does constructive interference occur, yielding a diffracted beam. Typically, for an X-ray energy E in electron volts (eV) and wavelength λ in Ångströms (Å) we can convert between wavelength and energy using the relationship: $E = hc/\lambda = 12,398.4193/\lambda$. A monochromator comprised of just a single crystal and an aperture for the diffracted beam will be functional but would change both the directions ($2\theta_B$) of the diffracted X-ray beam as the energy is scanned. Because large angular changes would be practically difficult to accommodate in most experiments, double crystal monochromators are very common on X-ray beamlines. Here, two essentially parallel diffracting crystals are arranged as shown in Figures 3.3 and 3.4, with the monochromatic X-ray beam exiting the monochromator at the same angle as the entering polychromatic X-ray beam.

3.2.1 Monochromator crystals

Most X-ray monochromators use elemental silicon crystals; and on third- and fourth-generation sources these crystals are often cooled using pressurized liquid nitrogen. Germanium, indium antimonide (InSb), and diamond have also been used for crystal monochromators, as have some highly specialized materials such as yttrium boride (YB$_{66}$), which we will discuss below. Silicon has a number of advantages for application in monochromators. Firstly, because of its routine use by the semiconductor industry, it is readily available as large very pure single crystals, which can be cut along different Miller lattice planes to give crystals that are very suitable for use in monochromators. Secondly, for temperatures close to 120 K or below, the thermal expansion of silicon is close to zero and it has an excellent thermal conductivity (600 Wm^{-1} K^{-1} at 125 K compared to Cu with 450 Wm^{-1} K^{-1} at 298 K), and both of these conditions mean that the heat load from intense synchrotron radiation will cause only minimal thermal distortion of the silicon. Various crystal cuts are used to access different energy regimes; common cuts found on XAS beamlines are Si(111), Si(220) and Si(311). Each has specific advantages for different X-ray energy ranges and applications that the XAS experimenter should be aware of, and we will discuss these in a subsequent section of this chapter. All commonly used monochromator materials have a cubic lattice, and consequently the d-spacing for a particular diffraction can be simply calculated from the lattice constants using the Miller indices h, k, l and the appropriate lattice constant a_0 with eq. (3.3); for silicon a_0 is 5.43095 Å at 300 K:

$$d_{hkl} = \frac{a_0}{\sqrt{h^2 + k^2 + l^2}} \tag{3.3}$$

3.2.2 Double-crystal monochromators

The simplest type of double-crystal monochromator is a channel-cut crystal, which is really composed of a single crystal. Such a monochromator was used in the very earliest XAS experiments discussed in Chapter 1. Here, a channel is cut through the middle of a single crystal with the direction of the cut being aligned with the required crystal lattice planes. This type of monochromator behaves very similarly to the double-crystal monochromators comprised of two individual crystals but has the advantage of great mechanical stability, as the two halves of the monochromator remain physically attached. A channel-cut monochromator will always have completely parallel crystals, which may be a disadvantage under some circumstances (see Section 3.2.4). While most modern XAS beamlines do not use channel-cut monochromators they do still find specialized uses in X-ray spectroscopy.

Two basic types of double-crystal monochromator are in common use and both rotate about the point where the incoming X-ray beam strikes the first monochromator crystal (Figures 3.3 and 3.4). One type has a variable exit beam height, where the exit beam has a variable offset from the input beam, which is given by eq. (3.4), in which $H(\theta_B)$ is the beam height and D is the separation of the two crystals. With this type of monochromator, the hutch table may be programmed to follow the motions of the crystals with a vertical translation so that the beam height relative to the experiment remains fixed. We note in passing that conventional channel-cut monochromators are necessarily variable exit height devices:

$$H(\theta_B) = 2D \cos \theta_B \qquad (3.4)$$

A second type of double-crystal monochromator has a fixed exit height relative to the beam entrance, which is achieved by translating the second crystal as shown in Figure 3.4, so that the exit height of the beam from the monochromator does not vary with θ_B.

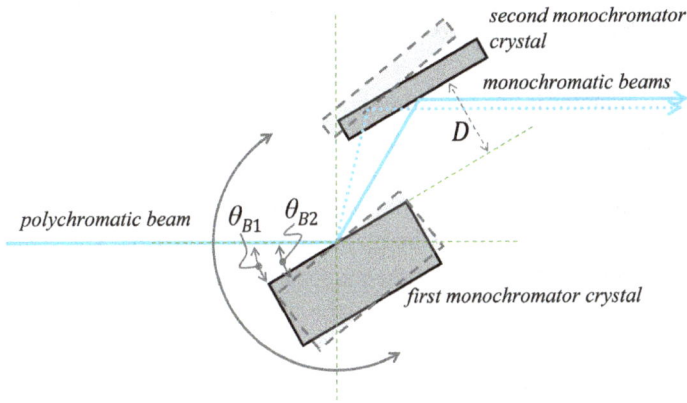

Figure 3.3: Schematic of a variable exit height double-crystal monochromator. The figure shows the side view of a vertically deflecting variable exit height double-crystal monochromator with a crystal separation D. Two crystal positions are shown, corresponding to the Bragg angles θ_{B1} and θ_{B2}, resulting in a vertical displacement of the monochromatic beam (solid blue to dotted blue arrows) on exiting the monochromator. The geometry shown is an upward deflecting double-crystal monochromator, with the thicker first crystal being cooled (often by pressurized liquid nitrogen). The alternative downward deflecting arrangement is also common on many beamlines (e.g. as shown schematically in Figure 3.1).

In many cases, modern monochromators can operate in either of these two modes; variable exit height or fixed exit height, allowing the user to select between the two modes of operation. The advantage of the variable exit height mode is that this can show greater stability on scanning than using the fixed exit height mode, since the latter requires scanning of the translation coordinated with the rotation. If a focusing mirror is used, then, as we will discuss below, the mirror may be positioned such that on scanning the energy with a variable exit height the beam moves across the surface of the mirror with the final position at the sample being invariant. With this mode of

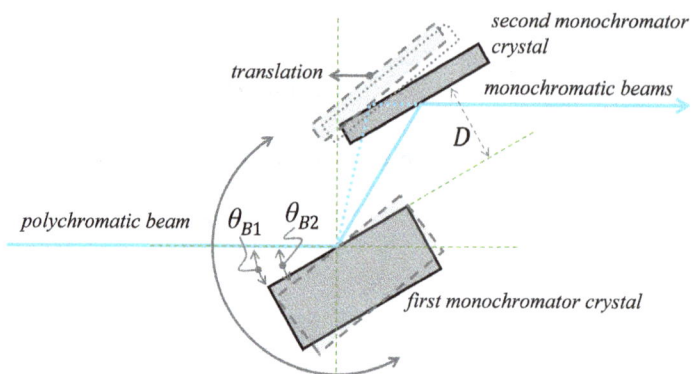

Figure 3.4: Schematic of a fixed exit height double-crystal monochromator. The schematic shows crystal separation D, with two crystal positions corresponding to the Bragg angles θ_{B1} and $\theta_{B2;}$ the monochromatic beam has a variable vertical displacement on exiting the monochromator. The second crystal is translated as shown (dashed outline) to achieve a fixed exit beam height (blue arrow).

scanning only one physical motor is needed – the rotation of the double-crystal mono-chromator, eliminating possible sources of instability and enabling such experimental modalities as rapid scanning.

In most cases, the monochromator crystals are positioned with the displacement in the vertical dimension perpendicular to the plane of the storage ring. For some beamlines the designers have elected to rotate the monochromator by 90°, so that the displacement of the crystals lies in the horizontal plane. The reason for doing this is that the source size is typically larger in the horizontal than in the vertical, with the consequence that the monochromator will be less sensitive to vibrations or other in-stabilities. Such horizontal deflecting monochromators can have poorer energy reso-lution compared to vertically deflecting monochromators, and when θ_B is large, there will be a small relative decrease in the diffracted intensity for X-rays that are polar-ized with the e-vector in the horizontal plane, as is the case with conventional bend magnet sources and with plane wigglers and undulators. For example, with an Si(111) crystal cut, operating at an X-ray energy of 5 keV ($\theta_B = 23.3°$), polarization effects cause the intensity to decrease by about 10% with a horizontal deflecting monochromator.

3.2.3 Crystal monochromator energy resolution and range

In many XAS experiments, good energy resolution for the beamline is required. Sev-eral factors combine to give the overall energy resolution. For a vertically deflecting monochromator these include: the vertical divergence of the source; the divergence afforded by the entrance slit to the experiment; and the natural reflection width of the monochromator crystals, known as the Darwin width ω_D, which decreases with higher order diffraction. Neglecting polarization and asymmetry effects, the Darwin width is given as follows:

$$\omega_D = 2 \frac{e^2}{m_e c^2} \frac{\lambda^2}{\pi V \sin 2\theta_B} |F(hkl)| \qquad (3.5)$$

where e is the electronic charge, m_e is the electron rest mass, c the velocity of light, V the crystallographic unit cell volume and $|F(hkl)|$ is the structure factor of the diffraction. The approximate best-case energy resolution is related to ω_D by eq. (3.6), in which $\Delta\psi$ is the vertical angular divergence of the source (see Chapter 2) for a vertically diffracting monochromator:

$$\frac{\Delta E}{E} = (\omega_D^2 + \Delta\psi^2)^{\frac{1}{2}} \cot\theta_B \qquad (3.6)$$

Table 3.1 shows some common monochromator crystal cuts, together with approximate Darwin widths and calculated best-case energy resolutions ΔE, for $\Delta\psi \rightarrow 0$.

Table 3.1: Parameters for some common monochromator crystal cuts.

Crystal cut	d (Å)	F_{khl}	Energy (eV)	θ_B (°)	ω_D (µrad)	ΔE (eV)
Si(111)	3.13560	58.73	2,500	52.26	167.1	0.32
			5,000	23.29	55.7	0.65
			10,000	11.40	26.1	1.29
			20,000	5.67	12.8	2.59
Si(220)	1.92016	69.33	5,000	40.22	48.4	0.29
			10,000	18.84	19.5	0.57
			20,000	9.29	9.3	1.14
Si(311)	1.63751	46.15	5,000	49.21	32.1	0.14
			10,000	22.25	11.3	0.28
			20,000	10.91	5.3	0.55
Si(400)	1.35775	60.21	5,000	65.94	55.7	0.12
			10,000	27.17	12.8	0.25
			20,000	13.20	5.8	0.50
Ge(111)	3.26655	154.81	5,000	38.17	381.7	0.82
			10,000	13.43	134.3	1.64
			20,000	6.33	63.3	3.27

3.2.4 The harmonic rejection problem and crystal detuning

If the two monochromator crystals of a double-crystal monochromator are completely parallel, then what are called harmonics may be important. Here, according to eq. (3.2), $E = n(hc/2d\sin\theta_B)$; with $n = 1$ we have the required X-ray energy for our XAS experiment, which is called the **fundamental**. However, higher values of n are also possible, which means that any allowed higher order reflection might enter the experiment. Not all Bragg diffraction is "allowed", as some cancel due to symmetrical atomic arrangement and are rigorously not present (a **systematic absence**). For example, for an Si(111) monochromator higher order harmonics will be Si(222), $(n = 2)$; Si(333), $n = 3$; Si(444), $n = 4$; etc. Of these, the Si(222) diffraction is systematically absent, and the first allowed harmonic is therefore $n = 3$, the Si(333) Bragg diffraction. There are simple rules to determine whether a reflection is systematically absent; for cubic systems such as those in Table 3.1 diffraction will be systematically absent if h, k and l have mixed odd and even values, or if they are all even and $h + k + l \neq 4m$, where m is an integer. This means that the first allowed higher harmonic is $n = 3$ for Si(111), Ge(111) and Si(311) cuts and $n = 2$ for Si(220) and Si(400).

It is very good practice to remove higher harmonics from the incident X-ray beam since their presence can be undesirable for a number of reasons. As an example of why this is, we consider a simple XAS experiment in which one measures the absorbance $A = \log I_0/I_1 = \mu x$, where I_0 and I_1 might be readings from gas ionization chambers (see Chapter 5). Higher harmonics will be more effective in penetrating the sample than the fundamental as they fall to higher energies, and also do not exhibit and absorption edge as a function of energy, and hence would give incorrect readings for I_0 and I_1.

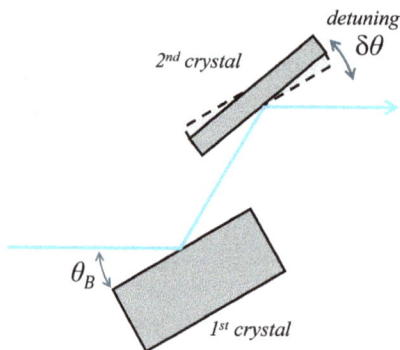

Figure 3.5: Detuning the second monochromator crystal through a displacement angle $\delta\theta$.

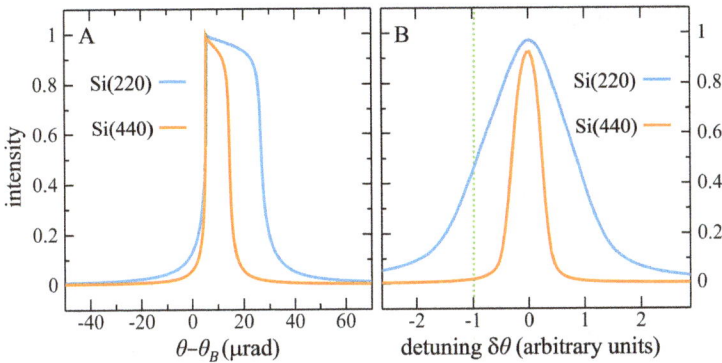

Figure 3.6: (A) Calculated crystal diffraction curves for a 1-cm-thick Si(220) crystal using an X-ray energy of 10,000 eV, showing the fundamental Si(220) reflection with $n = 1$ (blue), and the next allowed harmonic Si(440) with $n = 2$ (orange). The Darwin width ω_D corresponds to the width of the plateau-like region. (B) Measured detuning curves for a Si(220) double-crystal monochromator using the same fundamental X-ray energy of 10,000 eV. In both (A) and (B) intensity curves have been normalized to facilitate visualization. The green broken line in (B) shows the angle of the 50% detuned position, which would permit substantial fundamental intensity and minimal $n = 2$ harmonic to enter the experiment.

One method for rejecting harmonics is known as *detuning* – here, one moves the second monochromator crystal using some means of fine adjustment[1] such that the crystals are slightly offset from the parallel condition (Figure 3.5). This method exploits the fact that the rocking curve (intensity versus small angular offset of the crystal) for higher harmonics is sharper than that for the fundamental. Figure 3.6 compares both calculated rocking curves (A) and experimental detuning (B) for an Si(220) monochromator. The obvious disadvantage of monochromator detuning is that the intensity of the X-ray beam is substantially reduced, and for this reason, other means of rejecting harmonics may be preferred. Some crystal cuts, such as Si(111) or Si(311) have an $n = 2$ harmonic that is systematically absent, so that the first harmonic occurs for $n = 3$, at three times the energy of the fundamental, rather than at $n = 2$, at twice the energy of the fundamental. In such $n = 3$ cases removing harmonics from the beam using X-ray mirrors can be simpler.

---

1 Adjusting the position of the second crystal requires very fine control; older monochromator designs use a piezoelectric transducer, together with an adjustable voltage, and more modern designs might use a piezolinear actuator. It is instructive to note that the angular adjustments needed are measured in microradians. Just how big is a microradian? Consider tilting a kilometre-long straight line; an angular tilt of one microradian corresponds to 1 mm at a distance of 1 km.

3.2.5 Monochromator crystal glitches

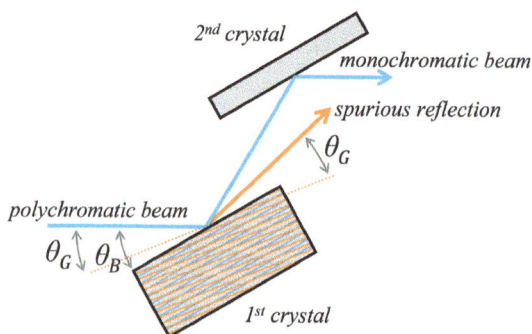

Figure 3.7: Origin of monochromator crystal glitches. A schematic diagram of a spurious reflection from an alternate set of lattice planes (orange planes, not parallel to the crystal surface) within the first monochromator crystal, giving rise to spurious reflection with Bragg angle θ_G (orange ray). This spurious reflection gives rise to a monochromator crystal glitch by detracting from the intensity of the principal monochromatic beam, which has Bragg angle θ_B (blue rays) and arises from the blue lattice planes parallel to the crystal face.

The word "glitch" sometimes tends to be misused by novice users, as a random spike in data that is not due to a particularly reproducible phenomenon. Crystal glitches occur at a reproducible energy when an alternative set of lattice planes diffracts the incident beam via a spurious reflection, detracting from the intensity of the monochromatic X-ray beam emerging from the monochromator. Figure 3.7 illustrates this process. The crystal glitches typically appear as a dip in the intensity of the beam emerging from the monochromator and tend to have some structure, in part because refraction plays a role and also because the harmonic composition of the beam changes as the monochromator energy scans through a crystal glitch. Under ideal circumstances the presence of crystal glitches should not matter, since the signal is always normalized to the incident intensity. However, glitches frequently ratio less than perfectly in part because the harmonic composition of the beam changes as the energy is scanned though the glitch with the detectors in the experiment perhaps responding differently to higher harmonics (e.g. ion chambers and solid-state fluorescence detectors). The presence of crystal glitches is to some extent inevitable, with many modern XAS beamlines having two different orientations of the same crystal cut, for example, both might be Si(220) crystal pairs, with one $\phi = 0°$ and the other $\phi = 90°$ (ϕ rotates about the normal to the diffracting crystal surface). The best crystal for a particular XAS experiment can thus be selected prior to the experiment in order to eliminate problematic crystal glitches.

Figure 3.8 shows monochromator crystal glitches from a ϕ = 90° Si(220) double-crystal monochromator in the vicinity of the Zn K-edge. Note that the glitches are not simple dips in intensity but show the aforementioned structure.

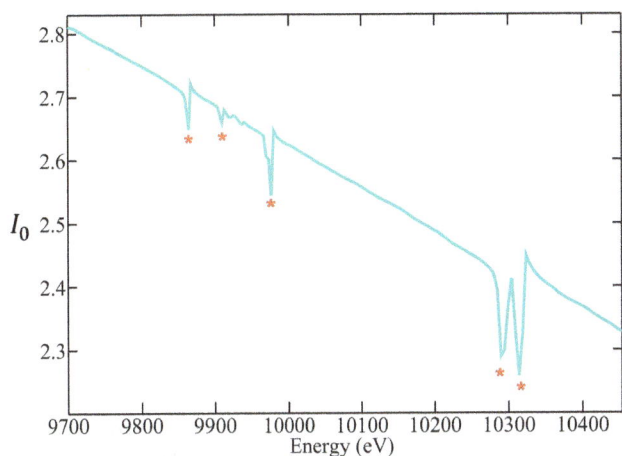

Figure 3.8: Spectrum of monochromator crystal glitches. The spectrum shows the output of an Si(220) double-crystal monochromator ($\phi = 90°$) measured using a gas ionization ion chamber in the experimental hutch. Features marked (*) show significant crystal glitches that are likely to prove problematic in subsequent analysis of XAS data. The large, broad glitches at ~10,300 eV are likely to be especially problematic.

3.2.6 Monochromator designs

As mentioned above, many modern monochromator systems have the ability to swap between monochromator crystals when needed; for example, a monochromator containing two pairs of crystals, might contain both Si(111) and Si(311) double crystal sets to give the users access to a particularly wide total energy range. A modern XAS insertion device beamline on a third- or fourth-generation storage ring will typically require pressurized liquid nitrogen cooling of at least the first monochromator crystal to counteract heating by the intense, polychromatic incident X-ray beam. The working range of a particular double-crystal monochromator will be an integral part of the monochromator design. Typically, the working range will be limited at the low energy by occlusion of the beam by the second crystal at steep Bragg angles, although this will depend upon several design features such as the length of the crystals and their separation D. Likewise, the high energy will be limited by how shallow an angle the monochromator can accept. In most cases a design will be able to range between a little more than 60° at the low X-ray energy to close to 5° at the high X-ray energy. The most common crystal choice for low X-ray energies is Si(111), which, using the above angular range, can access the sulphur K-edge at ~2,470 eV (θ_B = 53.2°). The range of a Si(111) double-crystal monochromator can be extended to lower energies up to about θ_B = 70° by sacrificing the upper end of the energy range by using a second crystal that is shorter on the upstream end, and the K-edge of phosphorus at ~2,140 eV can be accessed (θ_B = 67.5°). If access to the phosphorus K-edge is desired, then an alternative, albeit with poorer energy resolution, is to use a Ge(111) double-crystal monochromator (2,140 eV, θ_B = 62.5°). Table 3.2 provides a practical checklist for different monochromator crystal choices.

Table 3.2: Choice of monochromator crystals.

Crystal cut	Working energy limits (keV)[†]		First higher harmonic	Photon flux, % of Si(111)	Resolution ($\Delta E/E \times 10^4$)
	minimum	maximum			
Ge(111)	2.1	22	$n = 3$	215	32.6
Si(111)	2.3	23	$n = 3$	100	14.1
Si(220)	3.7	37	$n = 2$	74	6.04
Si(311)	4.4	43	$n = 3$	42	2.90
Si(400)	5.2	50	$n = 2$	48	2.53

†Other beamline-specific factors may limit the energy range, such as absorption by beryllium windows effectively limiting the minimum and mirror cut-off limiting the maximum.

Table 3.2 is not intended as a comprehensive list, but as a general guide to some common monochromator crystal choices. The beamline upper limit can be extended by using different crystal cuts; for example, there are reports of XAS at the Pt and Au K-edges (78–71 keV) using a Si(511) double-crystal monochromator, and higher energies still could be reached by employing other crystal cuts, for example, Si(531).

Alternative monochromator designs for multiple crystal sets exist that can swap sets either through a translation or through a rotation. Figure 3.9 shows both internal and external views of one successful monochromator design, which uses a rotation to exchange crystal sets. The advantage of this arrangement is that the monochromator can be made physically narrower than if a translation is used, allowing it to fit within a relatively tight space.

3.2.7 Some corrections to the simple picture

Our discussion thus far has neglected some corrections to the overall picture. The first of these is related to refraction of the X-ray beam as it enters the crystal. If we consider the Bragg angles external to (θ_{out}) and internal to (θ_{in}) the crystal, we can write:

Figure 3.9: The double-crystal monochromator. (A) The inner assembly of the SSRL double-crystal monochromator, which accommodates two sets of crystals (labelled 1 & 2 and 1' & 2') with the blue line showing the X-ray beam path through the assembly (photograph courtesy of SSRL). This design is cooled with pressurized liquid nitrogen. (B) The exterior of a fully assembled monochromator of the same design in its place on the BioXAS-side beamline at the Canadian Light Source (CLS).

$$\frac{\cos\theta_{out}}{\cos\theta_{in}} = \frac{\lambda_{out}}{\lambda_{in}} = \eta \tag{3.7}$$

where η is the refractive index of the crystal, which is an energy-dependent complex quantity, with respective real and imaginary parts $\delta(E)$ and $\beta(E)$, where $\eta(E) = 1 - \delta(E) - i\beta(E)$. However, for silicon monochromator crystals when the X-ray energy is far from the absorption K-edge, the energy dependence simplifies and we can instead write $\eta = 1 - \delta$, where δ is given by eq. (3.8), and is a small quantity:

$$\delta = \frac{e^2\lambda^2}{2\pi m_e c^2} Nf(0) \tag{3.8}$$

Here N is the number of atoms per unit volume, e is the electronic charge, m_e is the electron rest mass, c the velocity of light and $f(0)$ is the atomic form factor for radiation scattered in the forward direction, which is weakly dependent on energy. We can allow for the effects of refraction by using an adjusted lattice spacing d_n:

$$d_n = d\left(1 - \frac{4d^2}{n^2}\frac{\delta}{\lambda^2}\right) \tag{3.9}$$

In most cases the ratio δ/λ^2 will be nearly independent of λ, and can be considered a constant $\delta/\lambda^2 \approx 3.22 \pm 0.1 \times 10^{-6}$ Å^2, and hence d_n depends only on the Bragg order n.

If the crystal is heated by the X-ray beam causing a temperature change ΔT then this may result in a perturbed lattice parameter $a_0(T + \Delta T)$ compared with the ambient $a_0(T)$, and this may be accounted for by using the thermal coefficient of the monochromator crystal material, a_C, as follows:

$$a_0(T + \Delta T) = a_0(T)(1 + a_C \Delta T) \tag{3.10}$$

3.2.8 Specialized crystal monochromators

Some applications require very high resolution beams, for which more than one pair of crystals may be used. A four-crystal monochromators arrangement with either upward/downward or downward/upward deflection has the advantage of providing a fixed beam exit height with no translation of crystals required. Here, channel cut crystals are often needed to provide the greater stability required by such an arrangement. Other specialized types include monochromators with bent crystals to provide a degree of X-ray beam focusing, for which a sagittally bent second crystal (as the first has more complex cooling) might be used to focus the X-ray beam onto the sample. Such designs can increase the flux on the sample by a factor of five, but with some cost in terms of energy resolution.

Specialized crystal materials can be used to access specific energy ranges. Prominent among these are indium antimonide (InSb) and yttrium boride (YB$_{66}$), both of which have large lattice constants that allow double-crystal monochromators to access lower X-ray energies than typically can be achieved using Si(111) or Ge(111). InSb has a_0 of 6.47931 Å at 300 K, which gives access below 2 keV (upwards from about 1.7 keV), and YB$_{66}$ has the unusually large a_0 of 23.44 Å, which gives access to the more challenging 1–2 keV region. Other crystal choices with appropriate d-spacing such as beryl or α-quartz suffered from rapid radiation damage, usually lasting only a matter of hours, and had absorption edges in the range of interest (i.e. the Si K-edge). Because of this and because the technology to create grating monochromators that could operate in this range did not exist, YB$_{66}$ was originally very important despite considerable difficulties in the synthesis of monochromator crystals. We cannot resist a comment on the beautifully complex structure of YB$_{66}$; the boron atoms are contained within B$_{12}$ icosahedral units, with 12 of these icosahedra arranged in a super-icosahedron around a thirteenth B$_{12}$ icosahedron. There are 8 super-icosahedra and 1,248 boron atoms per unit cell. Yttrium is contained within this complex cage-like structure, coordinated by twelve boron atoms from four icosahedral faces. Modern grating monochromators can now easily access the 1–2 keV region, and hence interest in YB$_{66}$ as a monochromator material has waned. While InSb can still be found on a substantial number of modern monochromator systems, we know of no YB$_{66}$ monochromator systems that are currently in routine operation.

3.3 Soft X-ray monochromators

Double-crystal monochromators are used for XAS experiments in what has been called the hard X-ray regime (~5 keV and above) and also for what is known as the tender X-ray regime, which runs from approximately 2–5 keV. In the soft X-ray regime various types of diffraction grating-based monochromators are available. A diffraction grating is a periodic artificial construct consisting of regularly spaced "rulings" or grooves with a periodicity d, which is wider than the wavelength λ of the X-rays for which the monochromator is designed. The operation of a grating is similar in many ways to the diffracting crystal. When illuminated by an incident X-ray beam the grating rulings diffract, with intensity observed only at angles in which the incoming and outgoing rays differ by a whole number of wavelengths; at all other angles destructive interference will occur and no intensity will be observed. Gratings are governed by the well-known grating equation, eq. (3.11) in which α and β are the angles of the beam on either side of the grating normal, and d is groove density of the diffraction grating (Figure 3.10):

$$n\lambda = d(\sin\alpha + \sin\beta) \tag{3.11}$$

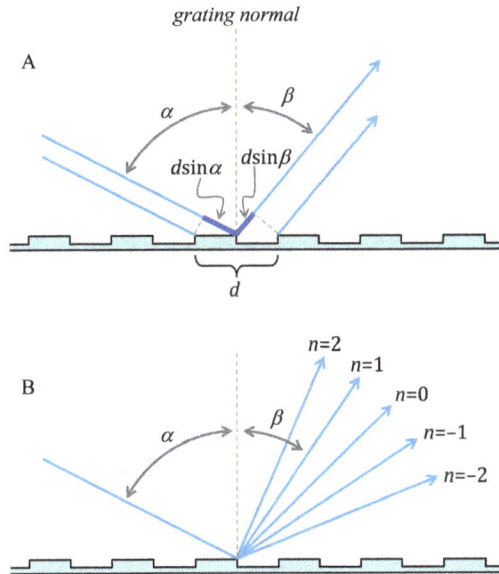

Figure 3.10: Side-view schematic diagram of a diffraction grating. (A) A schematic diagram of a laminar diffraction grating showing the angles α and β relative to the grating normal and satisfying eq. (3.9). (B) Different order diffracted beams from the grating, with the angle β shown relative to the $n = 1$.

The sign convention for β in eq. (3.11) is that β is positive when it is on the same side of the grating normal as α, and negative when it is on the opposite side (all values of β shown in Figure 3.10 would be negative). We have written eq. (3.11) in what may be to some an unfamiliar manner, for consistency with the crystal diffractometers discussed in earlier sections of this chapter. Often, instead of the grating period d, the grating line density N is used, where $N = 1/d$, and this is specified in lines per mm. Moreover, rather than using n for the diffractive order, the symbol k is sometimes used. For a given grating line spacing d (or line density N) and the incident angles α and β there will be a family of diffractive beams symmetrically disposed about the zeroth-order beam with $n = 0$, at which point the grating acts rather like a mirror, with $n = \pm 1, \pm 2, \pm 3$, etc. (Figure 3.10); those with positive order values ($n = +1, +2, +3$, etc.) are known as *internal orders* while the negative ones ($n = -1, -2, -3$, etc.) are known as *external orders*. Most beamlines using a grating monochromator operate with $n = \pm 1$.

$$\Delta\lambda = \frac{d\cos\beta}{n}\Delta\beta \qquad (3.12)$$

The resolving power of a grating is given by eq. (3.12), where $\Delta\beta$ is defined by the width of the exit aperture. Like their crystal counterparts, grating monochromators also come in various types, and examples of common variants are plane grating monochromators, toroidal grating monochromators and spherical grating monochromators. The toroidal and spherical gratings are gently curved gratings that are respectively shaped like a small section of a torus (donut) or of a sphere. A toroidal grating will typically give better focusing, but the spherical grating gives a superior energy resolution although it lacks sagittal focusing. Grating beamlines typically use one or more X-ray mirrors upstream of the grating to work with the grating to control the deflection angle ($\alpha - \beta$). Figure 3.11 shows a side-view schematic of an arrangement in which a plane mirror is used in combination with a spherical grating in a beamline. The use of upstream mirrors can also substantially reduce the heat load on the grating.

Just as many hard X-ray monochromators designs can accommodate more than one set of diffracting crystals, soft X-ray beamlines often have multiple gratings with different ruling densities, affording a greater working energy range for the experimenter. For example, the CLS SGM beamline has three spherical gratings ruled with 600, 1,100 and 1,700 lines per mm to give a total energy range of 250–2,000 eV, with some overlap between gratings at their limits. We will discuss X-ray mirrors in Section 3.4, but here we will note that this beamline also has considerably greater complexity than we have discussed so far, with three mirrors upstream and two downstream of the gratings. The length of the gratings is typically defined by the entrance slits and by the divergence of the source. Technology for producing long gratings was once limited to ~150 mm, but modern facilities can produce long gratings of up to 600 mm in length.

It has been known for many years that specific diffraction grating profiles can concentrate a large fraction of the incident radiation into one diffraction order. This phenomenon is called blazing, and gratings specifically fabricated to produce it are

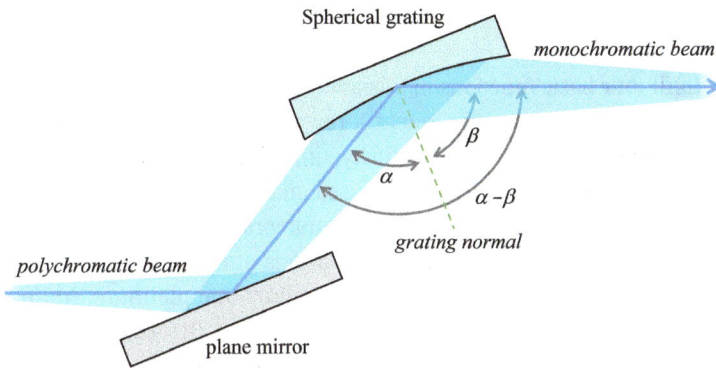

Figure 3.11: Schematic of a spherical grating monochromator combined with a plane mirror. This combination of optics can be arranged to give an equivalent focus at all wavelengths within the range of the grating.

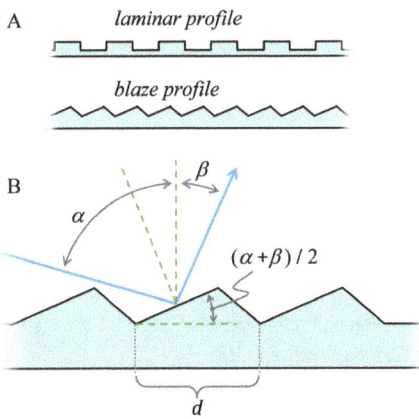

Figure 3.12: Blaze and laminar diffraction gratings: (A) schematic side-view diagrams of blaze and laminar diffraction grating profiles, and (B) details of the blaze angle.

called blazed gratings. In practice, blazed gratings have a diffraction efficiency that is approximately a factor of two higher than other designs, whereas laminar gratings can give higher spectral purity. Figure 3.12 shows a schematic of the profile for both types; the blazed grating can be seen to have a saw-tooth profile. The blaze angle (see Figure 3.12) is selected to be $(\alpha + \beta)/2$ for a given wavelength, so that the diffraction direction coincides with that of the specular reflection from the individual facets of the periodic structure.

3.4 X-ray mirrors

X-ray mirrors may fulfil four basic functions in a beamline: they are used for (i) power filtering to provide heat load relief on subsequent X-ray optics (e.g. the monochroma-tor); (ii) harmonic rejection to improve spectral purity; (iii) collimating the X-ray beam; and (iv) focusing the X-ray beam. While in the past fused quartz with a suitable coating was often used for mirror construction, modern X-ray mirrors almost exclusively are fabricated from a large single crystal of silicon, which also is often coated with a suit-able material, for which common choices are nickel, rhodium and platinum, while a bare silicon surface can be convenient for lower X-ray energies. Before considering X-ray mirrors in detail we will recap Snell's law, given in eq. (3.13) and summarized in Figure 3.13. When light, in our case X-ray light, falls upon the interface between two media, such as the surface of an X-ray mirror, the light can be reflected and refracted. This will depend upon the refractive indices of the two media, which we call η_0 and η:

$$\eta_0 \cos \theta = \eta \cos \theta'$$ (3.13)

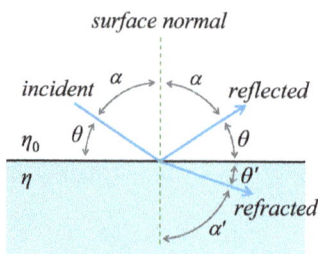

Figure 3.13: Reflection and refraction at an interface between two media of differing refractive index.

We note here that many treatments use the angle α measured relative to the normal of the interfacial surface (i.e. the mirror normal). Here, we will use the angle of attack θ as an incidence angle, which is the angle between the tangent to the mirror surface and the incident beam at X-ray energies $\eta < 1$, and $\eta_0 = 1$ for a vacuum. The critical angle θ_c is the value of θ for which total external reflection occurs, at which $\theta' = 0°$ (Figure 3.13); hence we can write:

$$\cos \theta_c = \eta$$ (3.14)

As θ_c is small, we can use a Taylor series expansion of the cosine and restrict our-selves to the first two terms:

$$\cos \theta_c = 1 - \frac{\theta_c^2}{2} = \eta$$ (3.15)

Remember that the refractive index η is an energy-dependent complex quantity, with real part $\delta(E)$ and imaginary part $\beta(E)$, so that we can write eq. (3.16) (also used above in eqs. (3.4–3.5)):

$$\eta(E) = 1 - \delta(E) - i\beta(E) \tag{3.16}$$

Assuming we are far from an absorption edge, as in the crystal diffraction case we can simplify the expression by neglecting the imaginary term:

$$1 - \frac{\theta_c^2}{2} = 1 - \delta(E) \tag{3.17}$$

Hence,

$$\theta_c = \sqrt{2\delta(E)} \tag{3.18}$$

For X-rays of about 1 Å wavelength (or 12.4 keV), $\delta(E)$ for most materials will be about 10^{-5} or 10^{-6}, and it is immediately obvious that θ_c must be very small, measured in milliradians. Remembering eq. (3.8), which stated that $\delta(E) \propto \lambda^2$ we can surmise that $\theta_c \propto \lambda \propto 1/E$ and that the product $E_c\theta_c$ is close to being a constant.

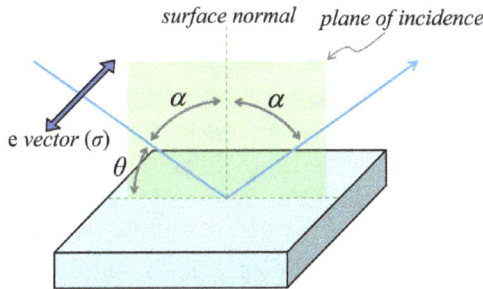

Figure 3.14: Plane of incidence (green rectangle) for an X-ray mirror, showing the **e**-vector for polarization. The vertical green broken line shows the surface normal.

3.4.1 Polarization and mirrors

As we have discussed in Chapter 2, synchrotron radiation from bend magnet, plane wigglers and undulator sources is polarized with the **e**-vector in the plane of the experimental floor. To describe the reflectivity of polarized light we refer to the Fresnel equations. Any polarization state can be resolved using a combination of two orthogonal linear polarizations and following Section 2.7 we refer to these as polarizations as σ, for the component having the **e**-vector perpendicular to the plane of incidence and π for the component whose **e**-vector is parallel to the plane of incidence. The plane of incidence is that containing both the surface normal and the incident and reflected rays and is perpendicular to the surface (Figure 3.14). The Fresnel equations for reflectivity R are given in eqs. (3.19) and (3.20), for σ and π polarizations, respectively:

$$R_\sigma(E) = \frac{\sin\theta - \sqrt{\left(\eta(E)^2 - \cos^2\theta\right)}}{\sin\theta + \sqrt{\left(\eta(E)^2 - \cos^2\theta\right)}} \tag{3.19}$$

$$R_\pi(E) = \frac{\eta(E)^2 \sin\theta - \sqrt{\left(\eta(E)^2 - \cos^2\theta\right)}}{\eta(E)^2 \sin\theta + \sqrt{\left(\eta(E)^2 - \cos^2\theta\right)}} \tag{3.20}$$

Polarization effects tend to be unimportant in the hard X-ray regime and are subtle in the soft X-ray regime, and we will not discuss them further.

3.4.2 X-ray mirror coatings

Returning now to our general consideration of mirror properties, $\delta(E)$ is also directly related to the electronic density ρ_e of the material used to coat the mirror; the higher the atomic number of the coating, the greater ρ_e and the greater the critical angle θ_c. Thus, mirrors coated with high atomic number materials will have larger θ_c values and, for a given value of θ, will provide better reflective surfaces for X-ray mirrors. Figure 3.15 compares the calculated reflectivity for different coatings. The cut-off energy, defined as the energy at which the reflectivity is at 50%, is directly related to the critical angle. Figure 3.15 shows a clear progression of the cut-off to higher energies with increasing atomic number of the mirror coating. Also apparent is the effect of interference, visible as an oscillation on the high energy side of the cut-off energy. This is due to the finite thickness of the mirror coating and approximately manifests as a sinc-type function related to the reciprocal of the X-ray path length, superimposed on the reflectivity curve. For platinum, which otherwise has excellent mirror coating properties, the L_{III}, L_{II} and L_I edges fall directly in the hard X-ray energy range, which is most used for XAS, as does the K-edge of nickel. These absorption edges manifest as sharp dips in the reflectivity, which cause a highly structured beam, which is undesirable for XAS. For this reason, many hard X-ray XAS beamlines use rhodium coatings, which for the most part, do not interfere or have mirrors with multiple longitudinal stripes of different coatings, so that a suitable coating can be moved into the beam path through a translation.

3.4.3 X-ray mirrors in harmonic rejection

As we have mentioned in Section 3.2.4, higher harmonics from the monochromator are unwanted radiation; the mirror cut-off energy provides a practical method to reject harmonics from the X-ray beam. Thus, with an Si(220) double-crystal monochromator for experiments at 10 keV, the Rh mirror of Figure 3.15 would give good rejection of the $n = 2$ harmonic radiation at 20 keV while maintaining good reflectivity at 10 keV, resulting in low harmonic content for the experiment.

The effects of varying the incidence angle θ for the same Rh mirror of Figure 3.15 are shown in Figure 3.16; the cut-off energy decreases with increasing θ, illustrating how a value of θ might be customized for harmonic rejection. We note that a practical consideration for very low values of θ would be that a real mirror would have to be very long and very accurately made.

At some beamlines, such as the CLS BioXAS beamlines, the mirror positioning systems lack the capacity to adjust the mirror angle for harmonic rejection. In this case a pair of flat silicon mirrors set parallel to each other to give a double bounce is used to reject harmonics.

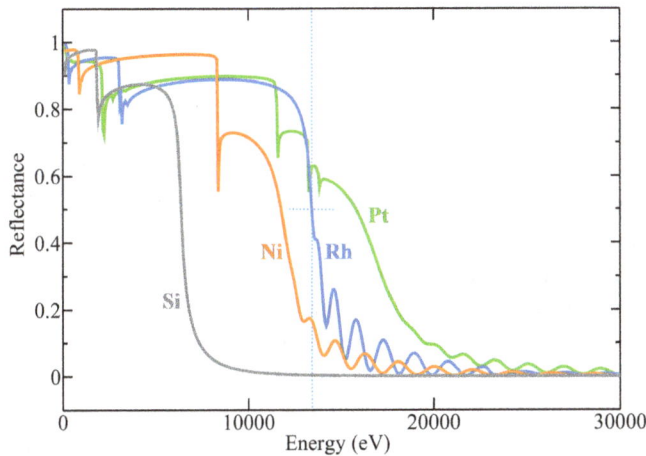

Figure 3.15: Effect of different coatings upon mirror reflectivity. Curves were calculated using an incidence angle θ of 5 mrad with 50-nm-thick coatings of nickel, rhodium and platinum on silicon. The reflectivity of a bare silicon mirror is also shown. The cut-off energy for the Rh coated mirror is shown using the vertical broken blue line.

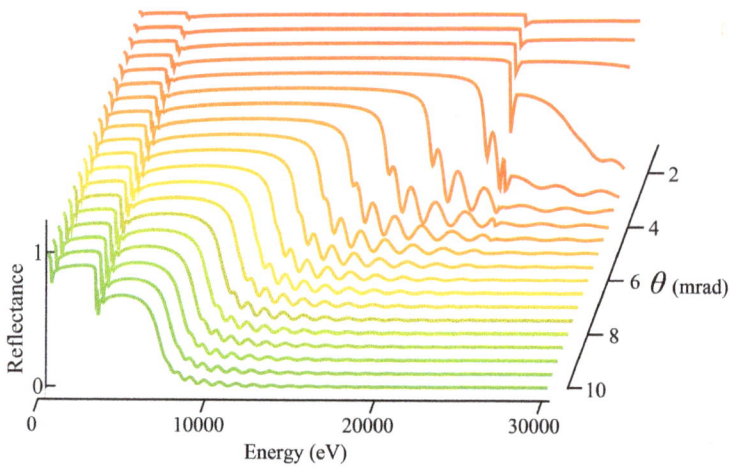

Figure 3.16: Effect of incidence angle θ upon the reflectivity of a 50-nm-thick Rh-coated silicon mirror, calculated assuming zero surface roughness.

3.4.4 X-ray mirror optical figures

A primary reason that X-ray mirrors are used in beamlines is to shape the X-ray beam. While X-ray mirrors can be flat, beamlines frequently contain mirrors with curved surfaces. The nomenclature used is that a curvature along the direction of the beam is called a **meridional curvature**, and a curvature perpendicular to the direction of the beam is called a **sagittal curvature**. With X-ray mirrors meridional radii of curvature need to be large, usually in the range of 2–15 km, while typical values for sagittal radii are 25–100 mm.

We will first consider **collimating mirrors**. As we discussed in Chapter 2, X-rays from the source within a storage ring are diverging. Unless corrected, this divergence will lead to a range of angles incident on the monochromator, which causes broadening of the spectrum of the diffracted beam exiting the monochromator, resulting in a poorer-than-optimal energy resolution. A common solution is a vertically collimating mirror positioned upstream of the monochromator intended to provide a homogenous angle of incidence to the monochromator. Figure 3.17 shows a schematic of how such a mirror functions – the optical figure is known as a bent-flat or cylindrical profile and can be achieved using a flat mirror together with an adjustable meridional mirror bending mechanism. Some examples of collimating mirrors are shown in Figure 3.18.

Apart from various X-ray windows and defining slits, the upstream mirror may be one of the first optical elements in the beamline, in which case the mirror will serve the double function of vertical collimation and power filtering for the downstream optics. Mirrors are well-suited to power-filtering since their glancing angle means the polychromatic beam is distributed over a metre or more of mirror surface (e.g. Figure 3.18), compared with a millimetre or much less in the case of a monochromator crystal. Mir-

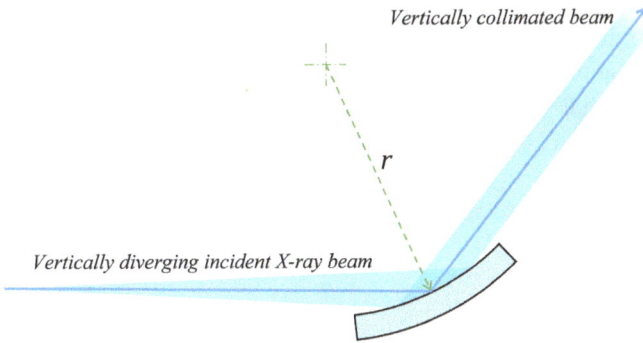

Figure 3.17: Schematic of a vertically collimating X-ray mirror. In reality, the radius of curvature *r* will be several km, as discussed in the text.

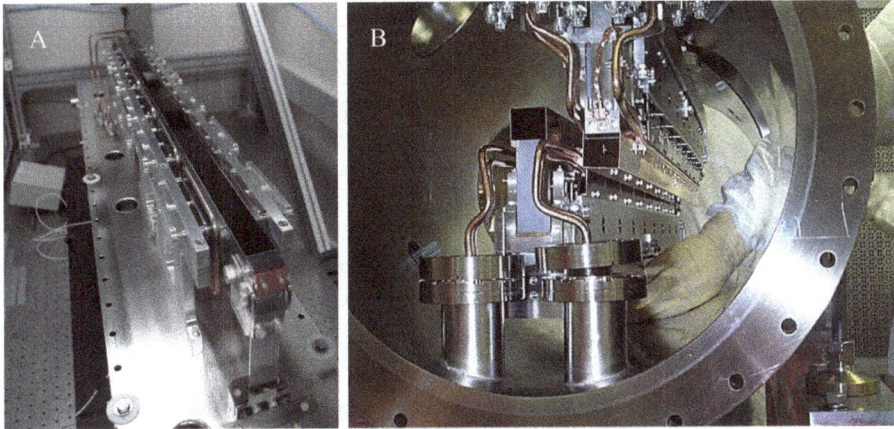

Figure 3.18: Examples of collimating mirrors at different stages of assembly. **(A)** The CLS BioXAS-main vertically collimating mirror, which is fabricated from silicon, 1,050 mm long and 30 mm wide, with a 100-nm-thick rhodium coating, and a working bend radius of 6 km. This mirror has vertical and horizontal beam acceptances of ~0.175 and 1.00 mrad, respectively. **(B)** The collimating mirrors for beamlines 9-2 and 9-3, mounted on their shared mirror tank, while in the process of assembly. Each mirror is surrounded by a rectangular conduit that serves to shield it from scattered radiation from the other mirror. Cooling water pipes are clearly visible for both mirrors.

rors absorb the higher-energy X-rays (Figure 3.15) and so reduce the power load on the first monochromator crystal. For this reason such mirrors require a cooling system, and while some fourth-generation facility beamlines employ liquid nitrogen cooled mirrors, most facilities seek to maintain close to room temperature in specular optics. Given that X-ray mirrors are necessarily long devices, frequently more than 1 m in length, keeping such an extended object cooled can present engineering challenges although the lengthy footprint of the beam on the mirror mitigates the difficulties. For typical heat loads, X-ray mirrors can be cooled by circulating water through heat exchangers that are often positioned at the sides of the mirror. Despite design efforts to provide effective cooling there is inevitably a thermal bump, which will make the mirror less than perfect, and thermal heating can distort X-ray mirrors over prolonged use, so that mirror slope errors accumulate, and the performance of the optics may degrade. Figure 3.19 shows the effects of accumulating slope errors giving degradation over time for an X-ray collimating mirror.

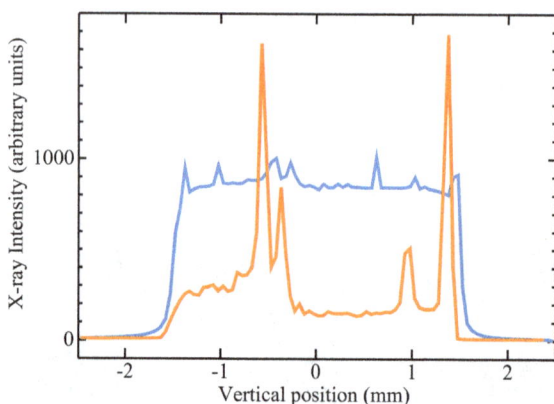

Figure 3.19: Photon flux through a 50-μm vertical aperture for a beamline containing a vertically collimating mirror followed by a double-crystal monochromator. The blue and orange lines are separated by approximately 6 years of mirror operation; the peak-like structure in the orange line is due to mirror slope errors.

The second mirror type that we will consider are **focusing mirrors**. Unlike collimating mirrors, which usually have only meridional bending, a focusing mirror will have both meridional and sagittal bends. An example schematic is shown in Figure 3.20. A common focusing mirror shape is a toroidal section – essentially the optical figure corresponds to the interior profile of a small section of a giant toroid or donut.

The equations describing the focusing behaviour of a toroidal mirror are known as Coddington's equations, and these are given in eqs. (3.21–3.24). Here, the sagittal and meridional radii of curvature are respectively r_s and r_m,[2] with α being the angle of incidence relative to the normal of the mirror surface (Figure 3.14), F_{in} is the source-to-mirror distance and F_{out} is the mirror-to-focus distance.

---

2 We note that some treatments refer to the meridional radius r_m as the **mirror base curve**, or the meridional direction as the **tangential** direction.

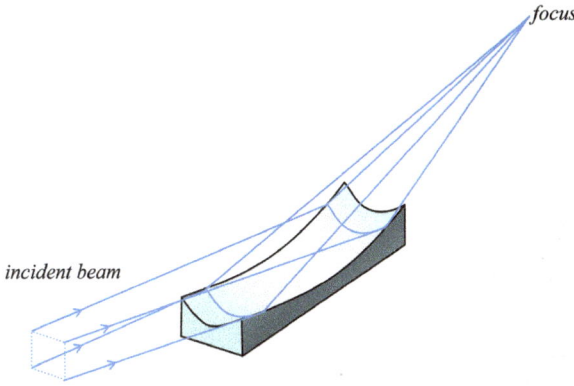

Figure 3.20: Schematic diagram of a focusing mirror accepting a vertically collimated, horizontally diverging incident X-ray beam.

$$\left(\frac{1}{F_{in}} + \frac{1}{F_{out}}\right) = \frac{2}{r_m \cos\alpha} = \frac{2\cos\alpha}{r_s} \tag{3.21}$$

$$r_m = \left(\frac{F_{in}F_{out}}{F_{in} + F_{out}}\right)\frac{2}{\cos\alpha} \qquad r_s = \left(\frac{F_{in}F_{out}}{F_{in} + F_{out}}\right)2\cos\alpha \tag{3.22}$$

with the meridional and sagittal focal lengths f_m and f_s being given by eq. (3.23):

$$f_m = \frac{1}{2}r_m \cos\alpha \qquad f_s = \frac{1}{2}\frac{r_s}{\cos\alpha} \tag{3.23}$$

If $f_m = f_s$ then

$$\frac{r_s}{r_m} = \cos^2\alpha \tag{3.24}$$

As α is close to 90° (θ being measured in milliradians) then we see that $r_m \gg r_s$.

Thus far, aside from Figure 3.19, the figures showing mirror reflectivity in this chapter have all used calculated results. Figure 3.22 shows experimental data from SSRL's beamline 9-3, which receives a side portion of a high-field wiggler fan. The beamline has an optical layout similar to that shown in Figure 3.1, with a bent-flat vertically collimating mirror upstream of the monochromator and a toroid downstream of the monochromator, both Rh-coated, with the mirrors being used to reject the harmonics in the beam. The final focus at the sample position is shown in the inset of Figure 3.22. This has a characteristic bow-tie appearance with an intense central focus together with superimposed frown and smile shaped intensities coming from the upstream and downstream ends of the mirror, respectively.

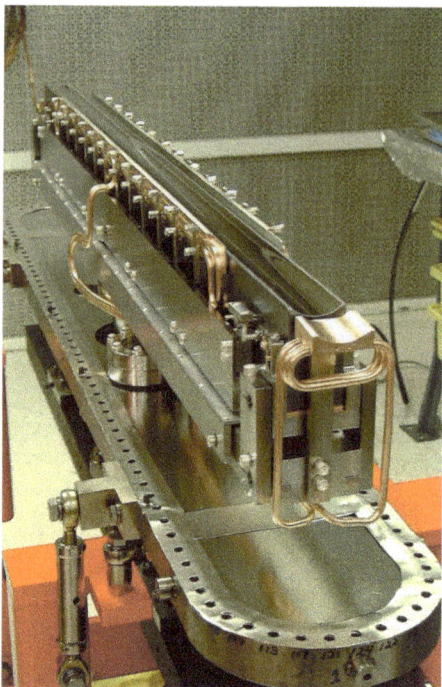

Figure 3.21: Photograph of a focusing mirror mounted in the base of its mirror tank during initial assembly. Note the obvious sagittal curvature of the mirror and the fittings for carrying the cooling water. Photo credit: Dr. Tom Rabedeau, SSRL.

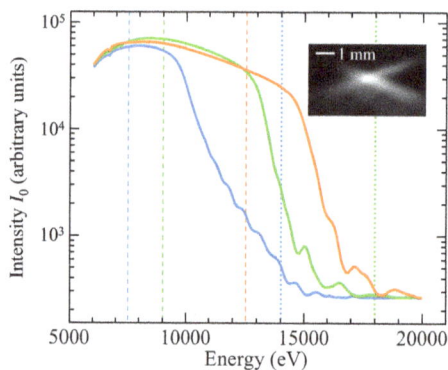

Figure 3.22: Energy scans for different mirror cut-off angles on SSRL beamline 9-3, corresponding to nominal cut-off energies of 10, 13 and 15 keV shown by the blue, green and orange lines, respectively. Note the logarithmic intensity scale. The dashed lines show the energies for the Fe, Cu and Se K-edge XAS (blue, green and orange, respectively) and the dotted lines the corresponding $n = 2$ harmonic energies for the Si(220) monochromator used on beamline 9-3. Data reproduced courtesy of Dr. Matthew Latimer of SSRL. The inset shows the final focus at the sample position recorded shortly after the beamline became operational in 1998, measured by raster scanning a pinhole through the beam while monitoring the X-ray intensity passing through.

3.5 The other pieces: filters, windows, masks and slits

A large number of other components make up a beamline, each with its own degree of sophistication in design. The interface between the storage ring and its integral components (e.g. wigglers and undulators) and the beamline is known as the **front end**. This serves a number of functions, including heat dissipation, safety and vacuum isolation. The front end typically is situated adjacent to the concrete shield wall (Section 3.6) and contains masks to eliminate off-axis radiation, and beam position monitors. In addition, a hard X-ray beamline front-end includes filters and windows, while a soft X-ray beamline may be windowless and with thin or no filters. For hard X-ray beamlines filters are often placed in the beam-path to remove the low energy component of the synchrotron radiation and reduce the heat load on the downstream components of the beamline. Common filter materials are graphite and diamond. Windows are also an important component of the front end and need to be strong, able to dissipate power, and relatively transparent to X-rays in the energy range of the beamline. Typical windows at synchrotron radiation facilities might be made of beryllium, perhaps about 250 μm thick, brazed or otherwise mounted on a conductive flange fabricated from copper or stainless steel, perhaps containing channels for water cooling.

3.6 Radiation shielding and personnel protection

3.6.1 Radiation shielding

There are two types of radiation for which shielding must be in place – these are photons and neutrons. The photons consist of the synchrotron radiation itself and Bremsstrahlung. Bremsstrahlung,[3] from the German *bremesen* (brake) and *strahlung* (radiation), or **braking radiation**, are photons produced when high energy electrons are decelerated by interacting with matter. The matter in question might be residual gas in a vacuum chamber (gas bremsstrahlung), or perhaps the walls of a vacuum chamber. In a synchrotron light source facility bremsstrahlung is typically very high energy photons sharply peaked in the forward direction. Bremsstrahlung shielding is typically tangential to the storage ring, at the end of beamlines and usually consists of lead bricks piled close to the beamline. Neutrons are produced when bremsstrahlung is absorbed by shielding, and because light elements such as hydrogen are good absorbers of neutrons, concrete is used for neutron shielding. Thus, thick concrete walls surround the storage ring for the purposes of

---

3 The reader may hear the Bremsstrahlung referred to as Bremsstrahlung radiation; this is incorrect because radiation is already within the original German, and this usage would translate to "braking radiation radiation".

safeguarding against neutron radiation. At least some of these concrete walls can be moved in order to access the hardware within during repair and maintenance.

3.6.2 Experimental hutch

The final piece of radiation safety equipment that we will consider here is the experimental hutch (also called endstation), which guards against photons and in particular the synchrotron radiation itself and is an integral part of all hard X-ray experiments at synchrotron radiation facilities.[4] The modern hutch is frequently a shielded room (Figure 3.23C), often using lead as a photoabsorber, which might have space for one or more people to work during setup of the experiment. Hutch access is protected by interlocks and a robust search-reset procedure, specific to each synchrotron facility. In general, this requires that the hutch must be searched, along with visible and audible warnings, to ensure that all personnel have vacated the hutch. After the search is complete and the door is closed and interlocked, the beam can be admitted to the hutch to illuminate the experiment.

---

4 Safety considerations and, in particular, radiation safety, are paramount at all synchrotron radiation facilities. Radiation in areas accessed by users and staff is meticulously monitored. Most facilities exhibit sufficient background levels of radiation such that users may experience a greater radiation exposure (albeit still very low) during the flight that brings them to the facility.

Figure 3.23: Evolution of the experimental hutch. (A) The first hutch at SSRL BL1-5, taken in 1974. Sally Hunter, a graduate student of Artie Bienenstock, can be seen loading an XAS sample. This first hutch was reminiscent of a rabbit hutch belonging to the children of one of the early experimenters (thus the name hutch) in size. (B) The experimental hutch for beamline X10-C at the NSLS taken in 1990 when Graham George was spokesperson for X10-C. (C) The interior of a modern hutch, that of the CLS HXMA beamline, taken in 2006, with then-graduate student Limei Zhang (now Professor at University of Nebraska) preparing to start XAS measurements. Unlike the other two hutches the modern hutch has sufficient room for several people inside. Photo credits: (A) Herman Winick, SSRL; (B) and (C) Graham George.

4 Interactions of X-rays with matter

In this chapter, we will discuss various aspects of how X-rays may interact with matter, with an overview of the physical basis for X-ray absorption spectroscopy (XAS). Put simply, as shown in Figure 4.1, X-rays can interact with matter in at least four different ways: they can be **transmitted**, so that they pass through the matter unchanged in energy and in direction vector, but not phase; they can be **elastically scattered**, so that the direction vector but not the energy is altered; they can be **inelastically scattered**, so that both direction vector and energy are altered; and they can be **absorbed** by the matter. It is this last interaction that is central to this book.

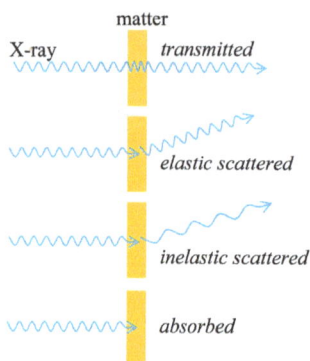

Figure 4.1: X-ray interactions with matter. Photons can be either transmitted, elastically scattered, inelastically scattered or absorbed.

Before focusing upon absorption, we will note that all the above listed modes, including transmission, involve interactions between the photons and the electrons of the matter. As we have discussed in Chapter 2, photons are emitted by accelerated charged particles; almost all the photons around us are derived from this mechanism, although there are other important mechanisms that can give rise to photons, such as some types of radioactive decay. Photons are always in motion, bear no charge, have no rest mass, yet possess momentum. Photons are bosons, and hence they are not subject to the Pauli exclusion principle, and they differ from charged particles, in that energy may be conserved when passing through matter. Before proceeding to discuss the different ways in which photons interact with matter, we will first review some background concepts.

4.1 Background concepts – wave particle duality

The physicist Max Planck, in 1900, first postulated that electromagnetic energy could only occur in quantized form, relating the energy E of a quantum particle, such as a photon, to its frequency v by the now well-known equation:

https://doi.org/10.1515/9783110570441-004

$$E = h\nu \tag{4.1}$$

where h is Planck's constant, having a value of $6.62607004 \times 10^{-34}\ \mathrm{m^2 \cdot kg \cdot s^{-1}}$. The idea of wave-partial duality is credited to de Broglie; considering now a wave, the product of the wavelength λ and frequency ν gives the wave velocity, which for photons is the velocity of light c, eq. (4.2), which in a vacuum is $2.99792458 \times 10^8\ \mathrm{m \cdot s^{-1}}$:

$$\nu\lambda = c \tag{4.2}$$

Einstein's theory of relativity states that

$$E^2 = (pc)^2 + \left(m_0 c^2\right)^2 \tag{4.3}$$

where E is the energy, p is the momentum and m_0 is the particle rest mass. For particles at rest, $p = 0$, and eq. (4.3) gives the familiar expression for rest energy $E_0 = m_0 c^2$. If the particle is non-relativistic ($\beta = v/c \rightarrow 0$), then we obtain the following that contains the familiar expression for kinetic energy, E_K:

$$E = \frac{1}{2}m_0 v^2 + m_0 c^2 = \frac{1}{2}m_0 v^2 + E_0 = E_K + E_0 \tag{4.4}$$

Photons are massless with $m_0 = 0$, so that $E = pc$; we therefore can combine eqs. (4.1) and (4.2) to give

$$p = \frac{h}{\lambda} \tag{4.5}$$

In 1923, de Broglie proposed that a similar relationship held for particles, such as electrons with mass m_e and velocity v, so that the momentum $p = m_e v = h/\lambda_e$, giving what is known as the de Broglie equation:

$$\lambda_e = \frac{h}{m_e v} \tag{4.6}$$

Here, the wavelength λ_e is known as the **de Broglie wavelength**. For non-relativistic electrons, the excess energy over the rest energy is simply the kinetic energy of the electron:

$$E_K = \frac{1}{2}m_e v^2 = \frac{p^2}{2m_e} \tag{4.7}$$

and

$$p = \sqrt{2m_e E_K} \tag{4.8}$$

We can now define a quantity called the wave-vector k, or with photoelectrons, the photoelectron wave vector, given by eq. (4.9), in which $\hbar = h/2\pi$:

$$k = \frac{2\pi}{\lambda_e} = \frac{p}{\hbar} = \sqrt{\frac{2m_e}{\hbar^2} E_K} \tag{4.9}$$

We will return to k later in Section 4.9.

4.2 Background concepts – quantum numbers

Quantum numbers are numerical values that are used to describe different solutions to the quantum mechanical wave equation. There are four quantum numbers that will concern us here, which are the **principal quantum number** n, the **azimuthal quantum number** l, the **magnetic quantum number** m and the **spin quantum number** m_s. Electrons are fermions, and so are subject to the Pauli exclusion principle; hence, when describing electrons in an atom, each electron must possess a unique combination of quantum numbers. These quantum numbers relate to observable physical quantities:

- n, the **principal quantum number**, relates to energy levels;
- l, the **azimuthal quantum number**, relates to the magnitude of the orbital angular momentum, given by $L^2 = \hbar^2 l(l+1)$;
- m, the **magnetic quantum number**, relates to the projection of the orbital angular momentum along an axis, describing the number of orbitals and their orientation, e.g. $L_z = m\hbar$; and
- m_s, the **spin quantum number**, relates to the projection of the electron spin angular momentum along an axis, e.g. $S_z = m_s\hbar$.

Orbital notation specifies n numerically, and l, using what is called the spectroscopic designation,[1] with s, p, d and f orbitals corresponding to $l = 0, 1, 2$ and 3, respectively. The electron spin quantum number specifies the projection of the electron spin angular momentum along an axis, e.g. $S_z = m_s\hbar$. As examples: the most tightly bound electron in an atom would have $n = 1$ and $l = 0$, and is called the 1s level; the chemistry of oxygen often involves the 2p electrons ($n = 2$, $l = 1$); first transition metals show chemistry that involves the 3d electrons ($n = 3$, $l = 2$); and the rare earths show chemistry involving the 4f electrons ($n = 4$, $l = 3$).

Electron orbital and spin angular momenta combine to form a total angular momentum J with a quantum number $j = l + s$, where s can have values m_s of $\pm 1/2$. We add j to our orbital notation as a subscript, so that $3d_{5/2}$ is a state with $n = 3$, $l = 2$ and $j = 5/2$ (with s of +1/2) and $2p_{1/2}$ is a state with $n = 2$, $l = 1$ and $j = 1/2$ (with s of –1/2).

---

1 While the origins of the notation are frequently forgotten, in spectroscopic orbital notation, s stands for sharp, p for principal, d diffuse, and f fundamental.

4.3 X-ray scattering

In Figure 4.1, we highlighted four ways in which X-rays can interact with matter (Figure 4.1).[2] It is well-known that light effectively slows down when it passes through matter, and that the degree to which it slows is related to the refractive index of the matter. In fact, a consequence of the massless nature of photons is that they can only travel at the speed of light, and the aforementioned effective slowing of their passage through matter is a result of interactions with the matter. One can visualize this process as being due to the electromagnetic nature of the incident photons, which causes a disturbance in the electrons of the matter, which in turn causes emission of a new photon. This interaction takes time, hence the slowing of the light to a velocity given by $c' = c/\eta$, where η is the refractive index of the matter.

With elastic X-ray scattering, the energy of the scattered radiation does not change, but the direction of the X-ray does, and with inelastic X-ray scattering, both direction and energy change. Elastic X-ray scattering is also known as **Thompson scattering**, and inelastic X-ray scattering as **Compton scattering**.

4.3.1 Elastic X-ray scattering

To understand elastic X-ray scattering, we consider the incoming X-ray light as an electromagnetic wave, which will accelerate the electrons of the matter in the direction of the X-ray electric vector in an oscillatory manner. As we saw in Chapter 2, this will generate light, and because most of the electrons in matter are not moving at relativistic velocities, the emitted light will have a donut-shaped profile, with intensity proportional to the square of the electric field amplitude, or proportional to $\cos^2\vartheta$, where ϑ is the angle between the X-ray direction vector and the observer. The nature of this elastic X-ray scattering gives rise to yet another name – **dipole radiation**. This is important when X-ray fluorescence detection (see Section 4.6) is used because, as we will discuss below, the fluorescence is isotropic over 4π steradians; if background due to scatter is to be minimized, then the best geometry is when $\vartheta = 90°$. The capacity of an isolated electron to scatter X-rays is given by an X-ray scattering length, which is called the **Thompson scattering length**, which again is known by different names,

---

2 For the sake of completeness, we should emphasize that we have neglected phenomena such as pair-production, where a high energy photon (or any uncharged boson) travelling through matter will cause the genesis of a particle-antiparticle pair, most often an electron and a positron. This will not occur at normal X-ray energies, as the rest mass of each of the two particles corresponds to an energy of 511 keV, hence photons at twice this energy, i.e. 1,022 keV, would be needed for electron-positron pair production.

specifically the **Lorentz radius** and the **classical electron radius**, r_0. It is given by eq. (4.10)[3] in which e is the electronic charge and ϵ_0 is the permittivity of a vacuum:

$$r_0 = \frac{1}{4\pi\epsilon_0}\frac{e^2}{m_e c^2} \tag{4.10}$$

As might be expected, due to the small size of electrons, r_0 is also small, with a value of 2.8179×10^{-5} Å. The total Thompson cross-section σ_T can be shown to be given by $\sigma_T = r_0^2 8\pi/3$. Thompson scattering describes the scattering from an isolated electron, but when the electron is bound to an atom, what is known as **Rayleigh scattering** occurs. This is directly proportional to σ_T with additional terms, including a wavelength dependency. For an atom, the intensity of the elastic scattering is related to both σ_T and the square of the number of electrons in the atom, or Z^2. Bragg diffraction, which we discuss elsewhere in this book (e.g. Section 3.2), is the result of elastic scattering from the array of atoms arranged in a periodic crystal lattice.

4.3.2 Inelastic X-ray scattering

Compton scattering occurs when some of the kinetic energy of the incoming photon is imparted to an electron, which recoils, giving a photon of lower energy (longer wavelength, λ_c). Compton's formula, given in eq. (4.11), expresses the shift in wavelength ($\Delta\lambda$) as a function of the photon scattering angle θ (the angle between the incoming and outgoing photon direction vectors). The related form, in terms of incident X-ray energy (E) and eq. (4.11) rearranged to give the Compton scattered X-ray energy (E_c), is given in eq. (4.12):

$$\Delta\lambda = \lambda_c - \lambda = \frac{h}{m_e c}(1 - \cos\theta) \tag{4.11}$$

$$E_c = \frac{m_e c^2 E}{m_e c^2 + E(1 - \cos\theta)} \tag{4.12}$$

The Compton scattering cross-section σ_c is described by eq. (4.13), the Klein-Nishina formula, which provides the differential cross-section with solid angle Ω, which we give, for the sake of completeness, in the simpler form using wavelength rather than energy:

---

3 r_0 is one of three associated lengths, the others being the Bohr radius a_0 and the reduced electron Compton wavelength λ_e which can be related by the fine structure constant $\alpha = e^2/(4\pi\epsilon_0\hbar c)$, so that $r_0 = \lambda_e\alpha = a_0\alpha^2$.

$$\frac{d\sigma_C}{d\Omega} = \frac{r_0^2}{2}\left(\frac{\lambda}{\lambda_c}\right)^2\left(\frac{\lambda}{\lambda_c} + \frac{\lambda_c}{\lambda} - 2\sin^2\theta\cos^2\varphi\right) \tag{4.13}$$

Here, φ is the angle between the direction vector of the scattered photon and the *e*-vector of the incident X-ray.

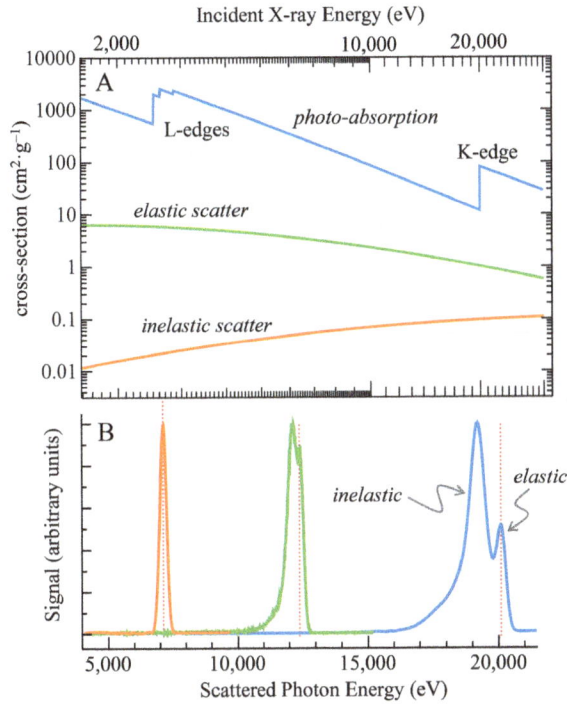

Figure 4.2: Elastic and inelastic scattering. (A) The relative cross-sections for photoabsorption, elastic scatter and inelastic scatter for molybdenum. (B) X-ray scattering from water, measured experimentally using a germanium solid-state detector. The vertical red broken lines indicate the energy of the incident X-rays.

Figure 4.2A compares the calculated X-ray cross-sections for molybdenum, showing the photoabsorption, which we will discuss below, and both the elastic and inelastic scatter. As we will discuss in Chapter 6, it is conventional to express the X-ray cross-section in the somewhat odd-seeming units of $cm^2{\cdot}g^{-1}$. Figure 4.2B shows X-ray scattering at three different incident X-ray energies 7,100 eV, 12,300 eV and 21,000 eV. These data were measured with a solid-state detector system (see Chapter 5), which under the conditions of measurement has ~250 eV energy resolution. For the highest energy, the elastic and inelastic scatters are easy to distinguish and become somewhat less well separated as the incident energy is decreased. At very high energies, the inelastic scatter becomes completely resolved from the elastic, and its onset at low energies is called the Compton edge.

4.4 X-ray absorption

This brings us to X-ray absorption in our list of the modes of interaction. This is typically described by the X-ray absorption coefficient $\mu(E)$, which depends strongly upon the X-ray energy, and the nature of the matter. With an incident X-ray intensity of I_0, the transmitted X-ray intensity I is given by $I = I_0 \exp(-\mu(E)x)$, where x is the X-ray path length through the matter (e.g. the sample). The absorbance A is simply given as follows:

$$A = \mu(E)x = \log\frac{I_0}{I} \tag{4.14}$$

If we, for the moment, neglect absorption edges, $\mu(E)$ is empirically given as follows:

$$\mu(E) = \frac{\rho Z^4}{ME^3} \tag{4.15}$$

where Z is atomic number, ρ is density, M is atomic mass and E is the X-ray energy, so the penetration of X-rays will approximately increase as a function of the inverse cube of their energy. Nearly all of the content of this book focusses on events surrounding X-ray absorption edges, which comprise relatively sharp discontinuities in $\mu(E)$ arising from excitation of core-level electrons (Figure 4.3).

For most elements, the energy needed for core-level excitation falls into the X-ray regime. The different absorption edges have names that change with the principal quantum number n of the electron being excited, with a K-edge signifying $n = 1$ (i.e. 1s) excitation, an L-edge $n = 2$ excitation (i.e. 2s or 2p) excitation, an M-edge $n = 3$ and so on. Since the K-edge excites the most tightly bound core electrons, this is always the highest energy edge. Figure 4.3a shows a schematic of a K-edge 1s core excitation, with the K-edge X-ray absorption spectrum of molybdenum metal foil in Figure 4.3b.

X-ray absorption by the electrons belonging to different core levels gives rise to different X-ray absorption edges. For L_{II} and L_{III} edges, the 2p electron ($n = 2$, $l = 1$) is excited, giving a 2p core hole. The presence of this hole means that for this level, the total angular momentum is no longer zero, so the aforementioned coupling between orbital and spin angular momenta (spin-orbit coupling) splits the 2p level into $j = l - 1/2$ and $j = l + 1/2$ levels, which are called the $2p_{1/2}$ and $2p_{3/2}$ levels, respectively. These give rise to the L_{II} and L_{III} absorption edges, respectively. For the p-orbitals, we have three possibilities, corresponding to $m_l = +1$, 0 and –1, with $m_s = +1/2$ or –1/2 to make $3 \times 2 = 6$ different states; a degeneracy of 4 for the $2p_{3/2}$ arises from $m_l = +1$ or –1 and $m_s = +1/2$ or –1/2, while $m_l = 0$ and $m_s = +1/2$ or –1/2 gives a degeneracy of 2 for the $2p_{1/2}$. Similar spin-orbit coupling applies to the M edges for 3p and 3d levels, and to the N edges for

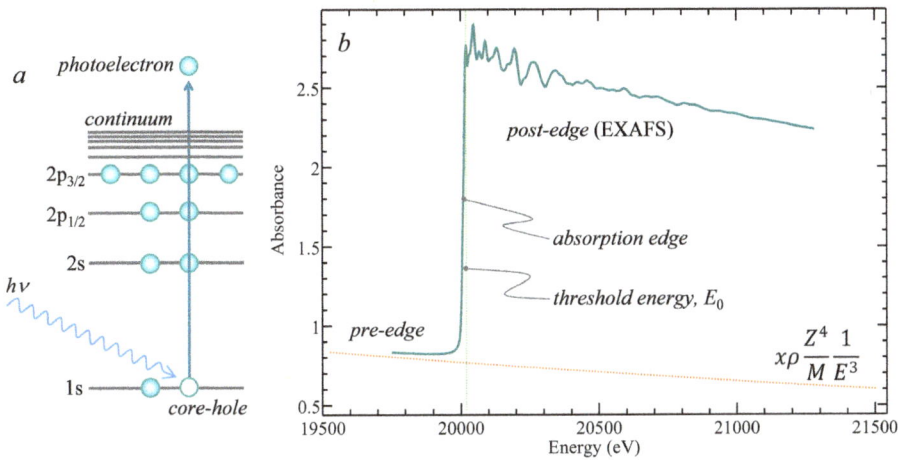

Figure 4.3: The X-ray absorption edge. (a) A schematic diagram of the excitation of a core electron by absorption of an X-ray photon. (b) The experimental K-edge X-ray absorption spectrum of a molybdenum metal foil (the empirical background absorption is shown by the broken orange line). In the pre-edge region, the X-ray photon has insufficient energy to excite the core electron, in this case a 1s electron, whereas at and above the threshold energy E_0 (the post-edge region), there is sufficient energy to excite the core electron, which leaves that atom as a photoelectron. The sharp rise in absorption coefficient is known as the absorption edge.

the 4p, 4d and 4f, and so on. This is summarized in Table 4.1, with a schematic diagram of the different core-electron excitations giving rise to K, L and M X-ray absorption edges, shown in Figure 4.4.

The various absorption edges of sodium molybdate and sodium tungstate are shown in Figure 4.5. Because it depends upon atomic physics, XAS will always see all occurrences of an element within a sample, whereas other methods, such as electron paramagnetic resonance (EPR) spectroscopy, may only allow the investigator to observe a fraction of a particular element (e.g. EPR will only detect metal ions that are isolated and paramagnetic). This all-seeing nature is both a strength and a weakness of the method, a strength because no occurrences of an element of interest can remain undetected, and a weakness because data from complex mixtures may be complicated to analyse, and in some cases, this may be intractable.

To a first approximation, the X-ray absorption edge energies scale as the square of the atomic number, as shown in Figures 4.6 and 4.7 for the first transition metals. This makes X-ray absorption element-specific, with no confusion with nearby elements because each element has its own unique X-ray absorption edge energies. Having made this bold statement, there are of course a very small number of near coincidences, as is the case with the Tl L_{III}-edge with the Se K-edge.

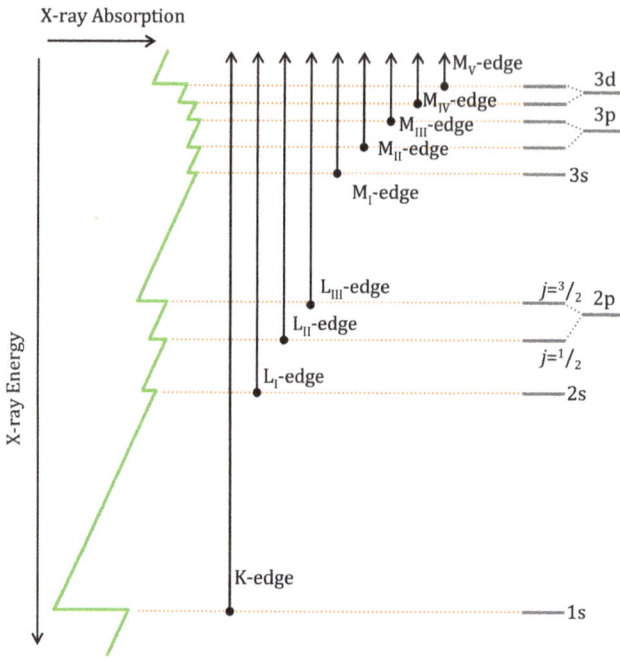

Figure 4.4: Schematic diagram of the different X-ray excitations giving rise to K, L and M edges. We note that the X-ray absorption would actually be much higher at the lowest X-ray energies.

Table 4.1: X-ray absorption edges.

Series	Number of edges	Core hole	Edge	Degeneracy
K	1	1s	K	2
L	3	2s	L_I	2
		$2p_{1/2}$	L_{II}	2
		$2p_{3/2}$	L_{III}	4
M	5	3s	M_I	2
		$3p_{1/2}$	M_{II}	2
		$3p_{3/2}$	M_{III}	4
		$3d_{3/2}$	M_{IV}	4
		$3d_{5/2}$	M_V	6

Table 4.1 (continued)

Series	Number of edges	Core hole	Edge	Degeneracy
N	7	$4s$	N_I	2
		$4p_{1/2}$	N_{II}	2
		$4p_{3/2}$	N_{III}	4
		$4d_{3/2}$	N_{IV}	4
		$4d_{5/2}$	N_V	6
		$4f_{5/2}$	N_{VI}	6
		$4f_{7/2}$	N_{VII}	8

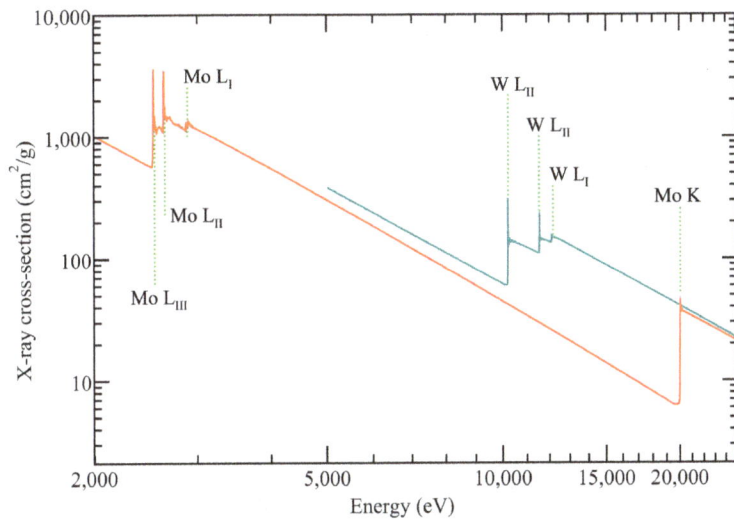

Figure 4.5: X-ray absorption spectra of Na_2WO_4 (blue line) and Na_2MoO_4 (orange line), showing the accessible X-ray absorption edges between 2.4 and 20 keV.

4.5 Selection rules and photoexcitation

Selection rules specify which electronic transitions are likely to occur within a physical system. Selection rules may facilitate the identification of "allowed transitions", which would be those with a high probability of occurring, or "forbidden transitions", which would be those with a small or no probability. We will discuss selection rules in more detail later in this book. Here, we begin with the property that photons carry angular momentum; when a photon is absorbed by an electron, this angular momentum is conserved by adding to or subtracting from the electron angular momentum, which is specified by the quantum number l. This gives rise to the selection rule of

$$\Delta l = \pm 1 \tag{4.16}$$

For a K-edge, the 1s electron is excited. This has no angular momentum ($l = 0$), thus $\Delta l =$ +1 and the resulting photoelectron has $l = 1$. The outgoing photoelectron de Broglie wave, therefore, has p symmetry, resembling a p-orbital, and will "fit" into unoccupied states with a lot of p character. The situation is very similar for an L_I edge with 2s excitation. For an L_{II}- or L_{III}-edge, however, the initial state electron is 2p with $l = 1$ and the photoelectron de Broglie wave thus has either $l = 2$ (d symmetry) or $l = 0$ (s symmetry). The $l = 2$ final state is more probable than the $l = 0$ by about a factor of 50, and because of this, the $l = 0$ final state often has been neglected in theoretical treatments.

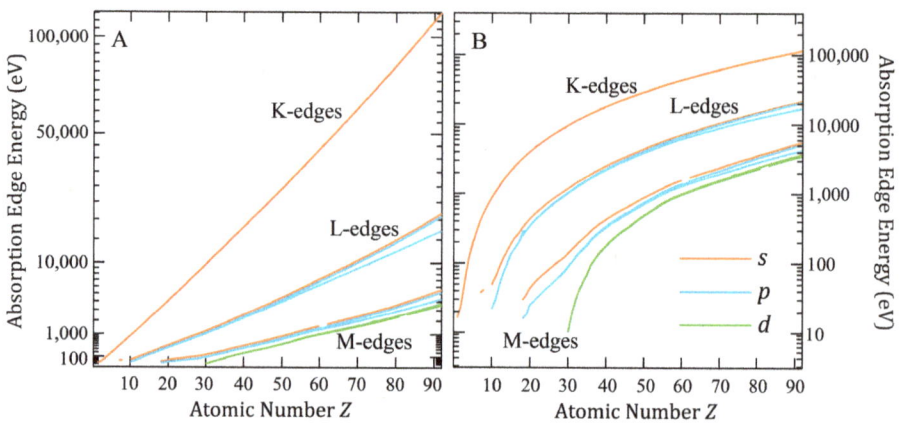

Figure 4.6: X-ray absorption edge energies (E_0) versus atomic number, showing the K, L and M-series. Different line colours are used to signify different core-level types, with orange for 1s, 2s and 3s excitation (K, L_I and M_I), blue for 2p and 3p excitation (L_{II}, L_{III}, M_{II}, M_{III}) and green for 3d excitation (M_{IV}, M_V). A and B are, respectively, plotted with a square root and $\log_{10}$ ordinate scales. A clearly shows the near Z^2 proportionality of E_0.

We have discussed above how the energy at which the X-ray photon is sufficient to eject a photoelectron to the continuum is called the threshold energy, E_0. If the X-ray energy E is steadily increased beyond E_0, the kinetic energy E_K of the emitted photoelectron increases, while its de Broglie wavelength λ_e correspondingly decreases (Figure 4.8). From eq. (4.9) and from $E_K = E - E_0$, we can now write a useful expression for k, as follows:

$$k = \frac{2\pi}{\lambda_e} = \sqrt{\frac{2m_e}{\hbar^2}(E - E_0)} \tag{4.17}$$

which, in practical units relevant to this book, can be written as follows:
$$k[\text{Å}] = \sqrt{0.262467(E[\text{eV}] - E_0[\text{eV}])}$$

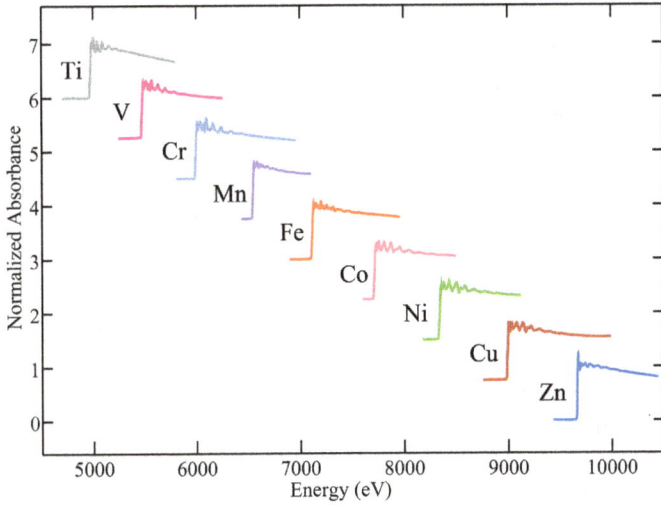

Figure 4.7: X-ray absorption spectra (K-edges) of first transition metals. Spectra are offset vertically for clarity, and are coloured, according to the element, using the CPK colour scheme.

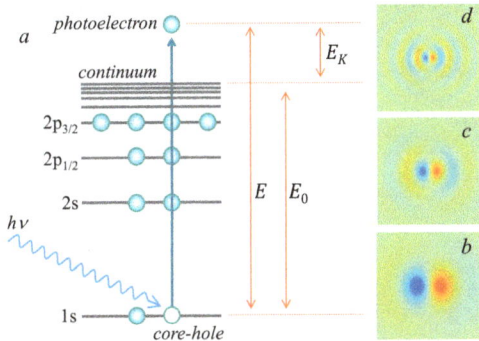

Figure 4.8: Photoexcitation process. (a) A schematic of the photoexcitation process, indicating the threshold energy E_0, kinetic energy of the photoelectron, E_K, and the energy of the incident X-ray photon E, as discussed in the text. (b–d) The photoelectron final state wavefunctions, propagating out from a centrally placed atom for 1s excitation, with the polarization **e** vector oriented along the x axis, and showing clear p-symmetry. Kinetic energy E_K increases from b to d, showing the progressively decreasing de Broglie wavelength.

4.6 X-ray fluorescence and Auger emission

The electron subject to photoexcitation in X-ray absorption is typically very tightly bound (e.g. 1s level for a K-edge). Consequently, the core hole created by X-ray absorption will have a very short lifetime and will be rapidly filled by decay of an outer electron, with the excess energy being carried away through concomitant emission of either an Auger electron, or an X-ray fluorescent photon (Figure 4.9).

With Auger electron emission, the electron decaying typically has the same symmetry as the core hole, with the emitted electron being close in energy to that decaying to fill the core hole. For example, K-edge photoexcitation generates a 1s core hole, with a dominant Auger decay process being 2s decay to fill the 1s core hole with concomitant emission of an Auger electron originating from the 2p levels. Using the nomenclature of Auger electron spectroscopy, such an emission is called a KLL emission, with K signifying the core hole, L the L_I (2s) initial state of the electron filling the core hole and the final L the $L_{II,III}$ (2p) initial state of the emitted Auger electron. Likewise, an L_{III} ($2p_{3/2}$) core hole might give rise to LMM Auger emission, with a 3p electron decaying to fill the 2p hole, and emission of another 3p electron as an Auger electron. The origin of the Auger electron depends upon the strength of electron-electron interactions, which are strongest in levels that are close together. With large atoms, the number of possible Auger emissions increases, including what are known as Coster-Kronig transitions, in which the core hole is filled by an electron from a higher subshell of the same shell. In any case, the final state will be an atom with two core holes that are created by the electron decaying to fill the core hole from the primary photoexcitation event, and that created by the emission of the Auger electron, which carries away the excess energy.

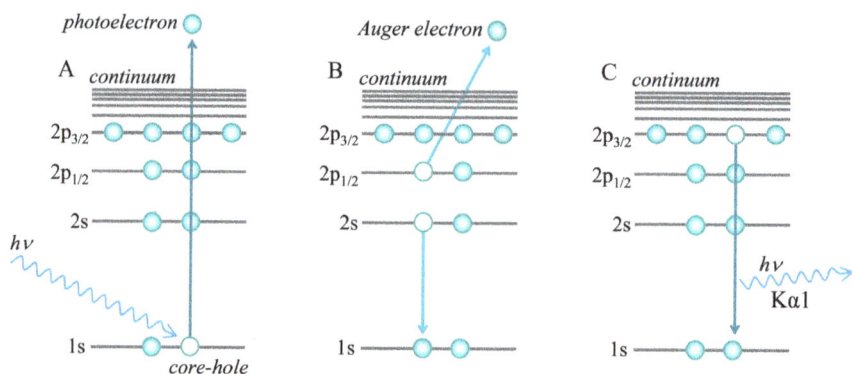

Figure 4.9: X-ray Auger and fluorescence emission. (A) The initial photoexcitation event, generating a 1s core hole. (B) KLL Auger electron emission, with decay of a 2s electron to fill the 1s core hole, with excess energy being carried away by emission of a $2p_{1/2}$ electron as an Auger electron. (C) The X-ray fluorescence process with decay of a $2p_{3/2}$ electron to fill the 1s core hole with emission of a $K\alpha1$ X-ray fluorescent photon.

X-ray fluorescence events are governed by the same $\Delta l = \pm 1$ selection rules of eq. (4.16). Thus, with a K-edge, the 1s core hole might be filled by decay of 2p electron with emission of a $K\alpha$ photon, with $K\alpha1$ emission arising from $2p_{3/2} \rightarrow 1s$ transitions, and $K\alpha2$ emission arising from $2p_{1/2} \rightarrow 1s$ transitions. In accordance with normal XAS practice, in this book, we will employ what is called the Siegbahn notation for X-ray

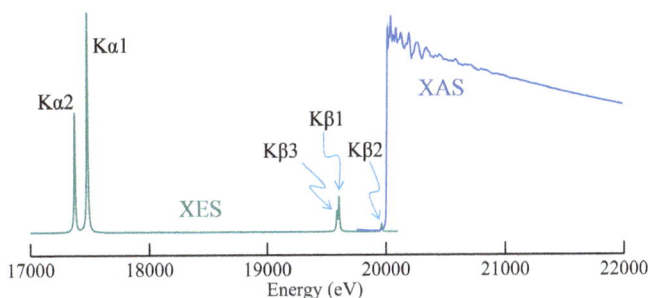

Figure 4.10: Comparison of the X-ray absorption (XAS) and X-ray fluorescence emission spectra (XES) for molybdenum metal foil. The spectra are arbitrarily scaled in the ordinate.

fluorescence lines (e.g. Kα1 denoting $2p_{3/2} \rightarrow 1s$), rather than the international union of pure and applied chemistry (IUPAC) notation [1], which employs the X-ray notation for the levels involved, so Kα1 would be a K-L3 or K-L$_{III}$ fluorescence line. The Siegbahn notation is an evolved nomenclature, based upon the historical order of discoveries. The X-ray notation of Barkla (e.g. K, L and M) dates from 1911, while in 1913, Moseley discovered that there were two types of X-ray emission lines for each element, and named them α and β; thus Kα and Kβ. In 1916, Siegbahn discovered that the α and β lines were split, thus the notation Kα1, Kα2. Correspondingly, the major fluorescence line for a K-edge is the Kα1 ($2p_{3/2} \rightarrow 1s$; K-L3), with minor lines Kα2 ($2p_{1/2} \rightarrow 1s$; K-L2) and Kβ1 ($3p_{3/2} \rightarrow 1s$; K-M3) – thus far, the order seems systematic, but in a departure from this, the $3p_{1/2} \rightarrow 1s$ transition is denoted Kβ3, while the Kβ2 is a $4p_{3/2} \rightarrow 1s$ transition. Likewise, while the major line for an L$_{III}$ edge is the Lα1 ($3d_{5/2} \rightarrow 2p_{3/2}$), that for the L$_{II}$ is the Lβ1($3d_{5/2} \rightarrow 2p_{1/2}$) and for the L$_I$, the major line is the Lβ3 ($3p_{3/2} \rightarrow 2s$), with the Lβ2 originating from a $4d_{5/2} \rightarrow 2p_{3/2}$ transition (and a minor line from the L$_{III}$-edge). Figure 4.10 compares the K X-ray fluorescence emission and absorption spectra for molybdenum metal. Figure 4.11 provides a graphical guide to the major and minor K-edge and L-edge fluorescence lines, respectively. Table 4.2 lists the strongest lines for each absorption edge.

The partition between Auger emission and fluorescence depends upon the absorption edge and the size of the atom. In general, fluorescence yield increases with increasing absorption edge energy and, therefore, with atomic number, while L-edges have lower fluorescence yield than do K-edges for a given element. Auger emission tends to be the dominant mechanism of core-hole decay in small atoms, whereas with larger atoms, X-ray fluorescence dominates. Figure 4.12 shows the fluorescence yields for K- and L-edges for a range of the chemical elements.

Similar to the X-ray absorption edges (Figure 4.6), the fluorescence energies change systematically with element and edge, also increasing approximately proportional to E^2. Figure 4.13 shows the energy dependence of the major fluorescence lines for the K- and L-edges. Like the absorption edge energies, the fluorescence line ener-

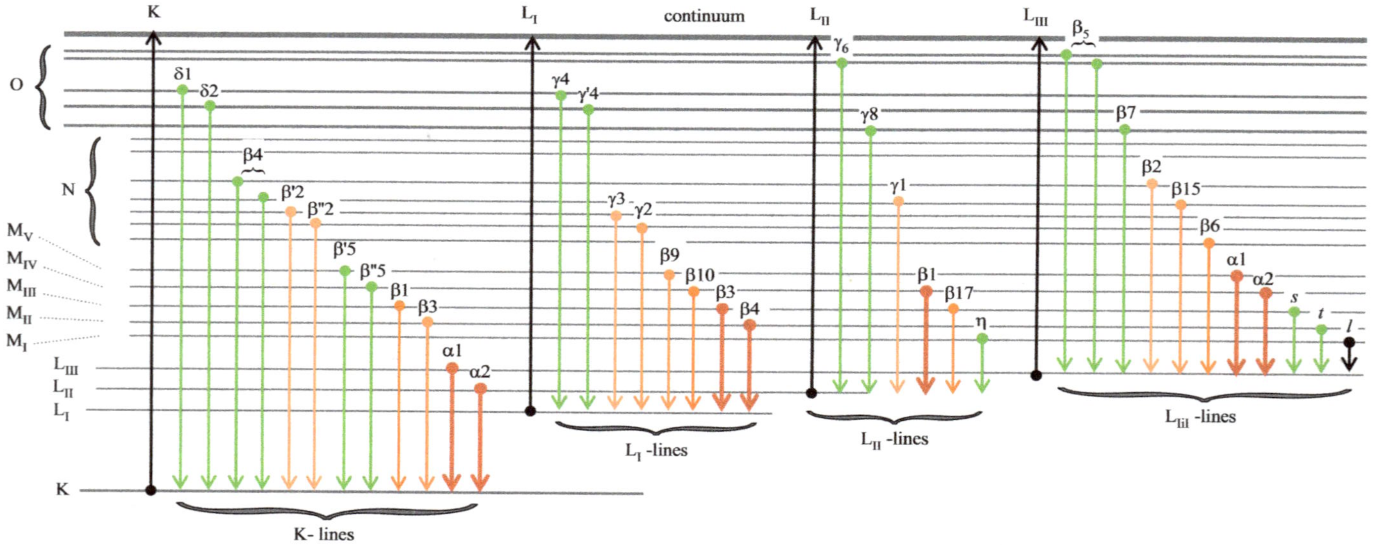

Figure 4.11: X-ray fluorescence lines using the Siegbahn notation. Transitions are coloured by intensity (red/bold-orange-green: strongest to weakest), independently for each of the K, L_I, L_{II} and L_{III} lines.

Table 4.2: Major X-ray fluorescence lines*.

Edge	Siegbahn	Transition	IUPAC	Relative intensity	Energy (Mo), eV
K	Kβ1	$3p_{3/2}{\to}1s$	K-M3	15	19,608.3
	Kβ3	$3p_{1/2}{\to}1s$	K-M2	8	19,590.3
	Kα1	$2p_{3/2}{\to}1s$	K-L3	100	17,479.3
	Kα2	$2p_{1/2}{\to}1s$	K-L2	52	17,374.3
L_I	Lγ3	$4p_{3/2}{\to}2s$	L1-N3	~5	2,832.2
	Lβ3	$3p_{3/2}{\to}2s$	L1-M3	35	2,473.1
	Lβ4	$3p_{1/2}{\to}2s$	L1-M2	20	2,455.7
L_{II}	Lγ1	$4d_{3/2}{\to}2p_{1/2}$	L2-N4	3	2,623.5
	Lβ1	$3d_{3/2}{\to}2p_{1/2}$	L2-M4	53	2,394.8
L_{III}	Lβ2	$4d_{5/2}{\to}2p_{3/2}$	L3-N5	5	2,518.3
	Lα1	$3d_{5/2}{\to}2p_{3/2}$	L3-M5	100	2,293.2
	Lα2	$3d_{3/2}{\to}2p_{3/2}$	L3-M4	11	2,289.8
	Lℓ	$3s{\to}2p_{3/2}$	L3-M1	5	2,015.7

*Fluorescence line energies for Mo are given as an example, with approximate relative intensities normalized relative to the most intense (100) line for the K- and L-series lines.

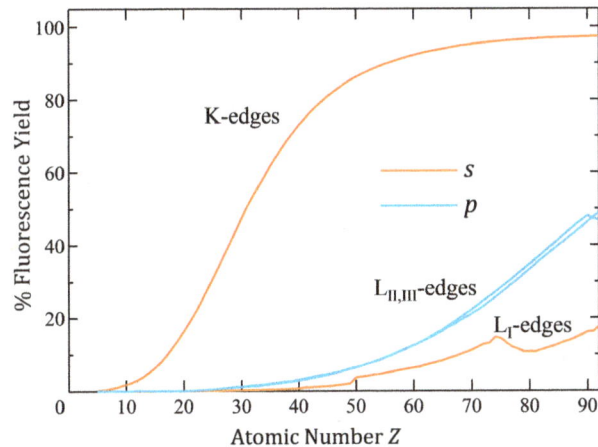

Figure 4.12: X-ray fluorescence yields for the K and L X-ray absorption edges. The Auger electron yields are approximately inverted on the ordinate – 100% minus the fluorescence yield.

gies are also essentially unique and, in most cases, clearly identify the element. Also like the absorption edges, there are near coincidences that potentially can cause confusion, for example the Pb Lα1 and As Kα1 lines fall at 10,543.7 and 10,551.5 eV, respectively. While these clearly differ, solid-state detectors have a best energy resolution of ~150 eV, hence these fluorescence lines could be potentially confused, although the absorption edges themselves are at quite different energies (Pb L_{III} and As K at 13,035 and 11,867 eV, respectively). One other example is that of the Sb Lα1 at 3,604.7 eV, which has been confused with the Ca Kα1 at 3,691.7 eV. The use of polyethylene terephthalate (PET) cover slips as supports for biological samples have occasionally caused confusion as production of PET uses antimony compounds to catalyse polymerization (e.g. Sb_2O_3, $Sb(CH_3CO_2)_3$ or $Sb_2(OC_2H_4O)_3$), and commercial PET usually contains low levels of antimony, which has been mistaken for calcium in the sample. These rare confusions can be resolved experimentally by the simple process of driving the monochromator either side of an absorption edge, as the combination of fluorescence and absorption edge is genuinely unique.

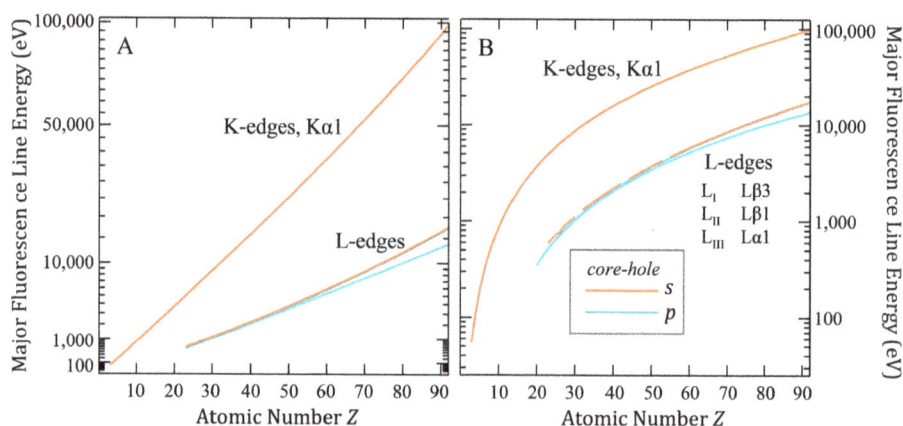

Figure 4.13: X-ray fluorescence emission energies versus atomic number. The most intense fluorescence lines for each of K, L_I, L_{II} and L_{II} edges are shown for a range of the chemical elements. A and B are, respectively, plotted with a square root and log_{10} ordinate scales. A clearly shows the near Z^2 proportionality of the fluorescence line energies.

4.7 X-ray photoabsorption timescales

It is instructive to ask about the time needed for the process of X-ray photoabsorption. X-ray absorption is a very fast process by almost any standards. The core levels containing the electrons to be excited are spatially very compact; for example, for Mo K-edge XAS, examination of the radial component of the Mo 1s wavefunction indicates that the electron will reside predominantly within ~0.03 Å (3×10^{-12} m) of the nucleus. Given the speed of light of ~3×10^8 m·s^{-1}, photoabsorption must occur within the time that a photon will take to cross the 1s orbital, which will be approximately 10^{-20} s or 10 zeptoseconds (zs). Lower energy X-ray absorption edges will have somewhat more diffuse core levels, and the time of photoabsorption is thus correspondingly longer, but irrespective of this, the actual photoabsorption event behind XAS is very fast.

In comparison to the timescale of photoabsorption, the core hole that is created by the X-ray photoabsorption process is relatively long-lived. Again, using the Mo K-edge as an example, the 1s core-hole lifetime is 1.46×10^{-14} s or 0.146 femtoseconds (fs). Other elements with lower absorption edge energies have longer core-hole lifetimes; for example the V 1s core-hole lifetime is 0.652 fs. These lifetimes, Δt, are critically important for XAS spectroscopic resolution, as they determine the width of near-edge spectral features via the Heisenberg uncertainty principal. Specifically, the spectroscopic resolution has a core-hole lifetime contribution ΔE_L that is given by $\Delta E_L \approx (\hbar/2)/\Delta t$, which contributes a Lorentzian lineshape[4] to the overall spectroscopic broadening. This core-hole lifetime broadening is convoluted with a Gaussian contribution from the beamline optics, resulting in the observed width of measured spectra. With hard X-ray measurements, the core-hole lifetime broadening typically dominates the observed broadening of measured spectra, while with soft X-ray measurements, the optical broadening from the beamline is usually limiting.

4.8 Nomenclature

Before we turn from some essentially atomic phenomena to consider how the X-ray absorption of molecular species differ, we will briefly discuss the nomenclature for the different parts of the X-ray spectrum. Figure 4.14 shows an example XAS spectrum with the different spectroscopic regions indicated. Unfortunately, there is confusion in the literature about nomenclature, especially about the near-edge portion of the spectrum. The entire spectrum is frequently referred to as the XAS (X-ray absorption spectrum), but also as the X-ray absorption fine structure or XAFS; EXAFS without the "E". The near-edge portion of the spectrum is most often referred to as the X-ray absorption near-edge structure or XANES, but other terms are common. These include NEXAFS (near-edge X-ray absorption fine structure), particularly in the soft X-ray regime or simply the XAS, for example in studies of first transition metal L-edge spectroscopy, when the EXAFS is rarely studied. More obscure terms include some that have sample or experimental specificity incorporated such as XAMES (X-ray absorption metal-edge structure), FYNES (fluorescent-yield near-edge structure) [2] or SPXANES – spin-polarized XANES [3]. Pre-edge is sometimes used to refer to low-lying bound-state transitions, which form part of the near-edge spectrum, and XANES is sometimes used to refer to the early post-edge region. The EXAFS does not suffer as much from multiple acronyms, although XAFS (no "E") has been used to refer to the EXAFS oscillations, and there are technique-specific variants such as SPEXAFS (spin-polarized EXAFS) [4]. We do not criticize any of these

---

4 The Lorentzian or Cauchy line shape in the energy domain arises from the exponential nature of the decay in time (e.g. see Appendix B).

acronyms. Some such as SPEXAFS add information to the picture, and indeed we have been guilty of adding to the list ourselves [5], but in this text, we will refer to the whole spectrum as the XAS.

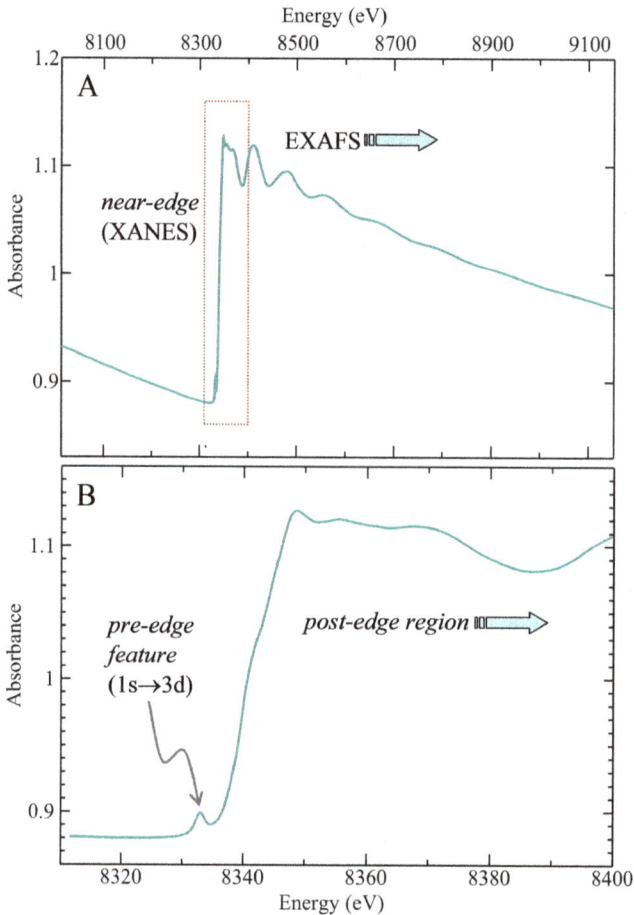

Figure 4.14: Regions of the X-ray absorption spectrum (A) and the near-edge spectrum or X-ray absorption near-edge structure (XANES) (B).

4.9 X-ray absorption by molecules and solids

Up to this point, we have dealt with what are essentially atomic phenomena occurring in isolated atoms. We will now begin to consider X-ray absorption of molecular or extended solid systems. Figure 4.15 compares the X-ray absorption spectra of elemental sulfur, which in the α-allotrope exists as S_8 rings, and argon gas, which is a monatomic entity. These data were selected because the elements are close to each other in the periodic table. There are some notable and interesting differences between the spectra. Most obviously, the portion of the spectra near to the absorption edge (which we will call the near-edge spectrum) is very different, as highlighted in Figure 4.16. In addition, the extended region on the high-energy side of the absorption edge shows significant difference, which the inset in Figure 4.15 shows in more detail – the sulfur spectrum can be seen to oscillate about the (aligned) relatively featureless argon spectrum. These oscillations are a phenomenon that is found in molecular species or extended solids, and are known as the extended X-ray absorption fine structure (EXAFS). We will discuss the EXAFS in detail below, but in brief, these are modulations in the X-ray absorption, caused by photoelectron backscattering by neighbouring atoms; as argon is a monatomic gas with no neighbour atoms, there is no EXAFS, whereas the atoms in elemental sulfur have two sulfur neighbours, each separated by 2.037 Å.

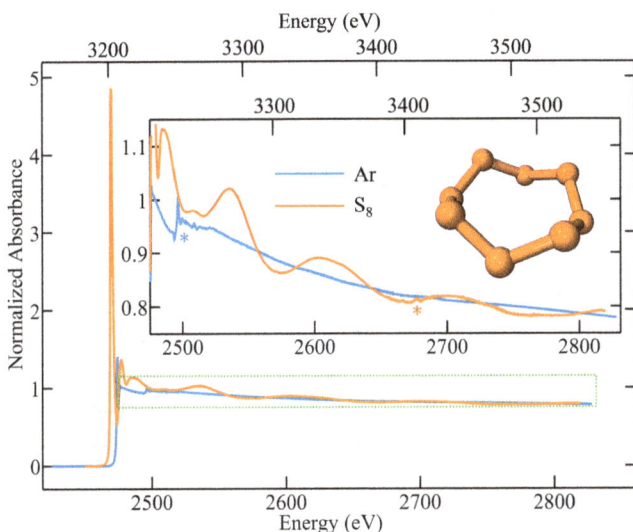

Figure 4.15: Comparison of the K-edge X-ray absorption spectra of gaseous argon and elemental sulfur. The spectra have been offset horizontally to align the different absorption edge energies, with the upper abscissae scales corresponding to the argon spectra and the lower to sulfur. The green broken line corresponds to the region enlarged in the inset, which also shows the structure of elemental α-sulfur, comprised of S_8 rings, with all S atoms coordinated by two others. The features marked (*) are multiple electron excitations discussed in the text and shown in more detail in Figure 4.16.

The most striking difference between the K-edge XAS of S_8 and argon is the intense peak, labelled **A** in the sulfur spectrum, which is absent in the argon spectrum. This is due to the strongly allowed $\Delta l = +1$ transitions from the 1s to the unoccupied orbitals that possess a lot of p-orbital character, which includes S–S σ^* and π^* orbitals. A simulation using molecular orbital theory for S_8 is shown in Figure 4.17. The formal electron

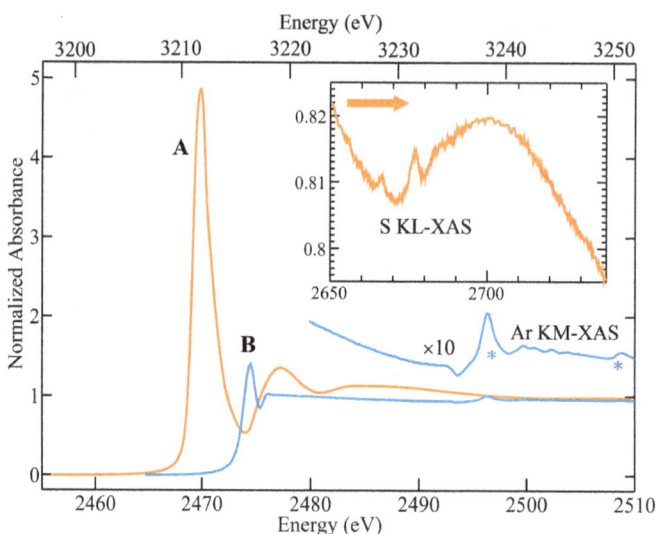

Figure 4.16: Comparison of the near-edge portion of the XAS of sulfur (S_8) and argon. The most intense features in the sulfur and argon spectra are marked **A** and **B**, respectively. The features marked * in the 10× vertically expanded region of the argon spectrum are due to $KM_{II,III}$ and KM_I multiple electron excitations. The inset shows the sulfur $KL_{II,III}$ multiple electron excitation regions, indicated in Figure 4.15.

configuration of zero-valent sulfur has two vacancies in the 3p level, but these are filled in monatomic argon. The smaller feature **B** in the argon spectrum can be attributed to 1s→4p transitions. The structure ~22 eV above the argon edge shown in the expanded region of Figure 4.14 is essentially a $KM_{II,III}$ feature. There is also an even smaller feature ~200 eV above the sulfur edge, which corresponds to a $KL_{II,III}$ feature. These structures arise from one photon, two electron transitions and a two-hole final state. The argon $KM_{II,III}$ involves 1s and 3p excitation, while the sulfur $KL_{II,III}$ involves 1s and 2p excitation. Due to the additional core hole, these multiple excitations are separated from the K-edge by approximately the L-edge or M-edge energy of the element with $Z+1$ (e.g. chlorine for a sulfur KL, or potassium for the argon KM) because the additional core hole has approximately the same energetic effects as an increased nuclear charge. These multiple electron features have been studied in a number of systems [6, 7], but their presence may present problems for EXAFS analysis, although their significant manifestation is controversial in some cases.

An understanding of the origins of the prominent features in near-edge spectra as transitions to bound states suggests that the near-edge spectrum will inform in some detail about the electronic structure of the atom undergoing photoexcitation. We will discuss the near-edge in substantially more detail in Chapter 9.

Returning now to the EXAFS, as shown in Figure 4.15, these are oscillatory modulation of the X-ray absorption coefficient $\mu(E)$ on the high-energy side of the absorption edge. The first physical interpretation of EXAFS was put forward in the twentieth

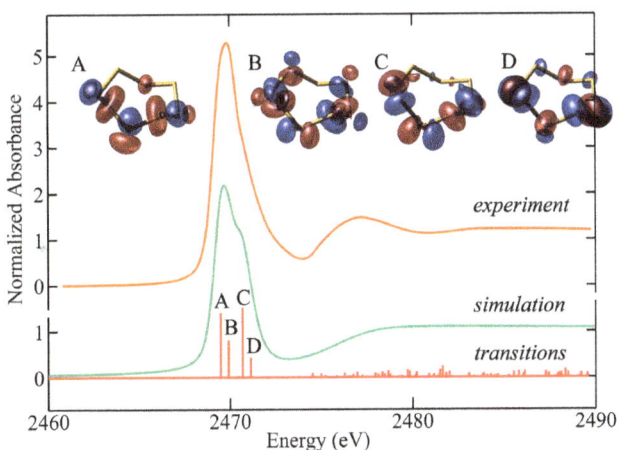

Figure 4.17: Density functional theory simulation of the near-edge portion of the XAS of S_8. The simulation is shown by the green line, with the transition intensities indicated by the red line (stick spectrum). The molecular orbitals isosurfaces (0.05 e$^-$/a.u.^{3}) involved in the corresponding excited states are shown above the experimental spectrum (A–D, top), corresponding to transitions A–D in the stick spectrum (red); all have substantial sulfur 3p orbital contribution. The excited sulfur atom is shown lowest in each of the plots, about which the excited state molecular orbitals are symmetrically disposed.

century by German physicist Ralph Kronig (also of the Kramers-Kronig relation, see Chapter 15), with EXAFS originally called Kronig structure. The story of the origins of our understanding of EXAFS is eloquently described by Lytle [8], and we will not reproduce the history here, except to say that the origins of the EXAFS were unclear until the 1970s, when Dale Sayers, Ed Stern and Farrel Lytle finally demonstrated how it could be used as a quantitative structural probe [9].

The nature of the photoelectron de Broglie wave has already been discussed in Sections 4.1 and 4.5, and a depiction of the final state photoelectron wavefunction is shown in Figure 4.8 for three different incident X-ray energies. A highly schematic diagram of X-ray absorption when more than one atom is involved is shown in Figure 4.18, in which the excited photoelectron is backscattered by the neighbour atom.

As described in the caption to Figure 4.18, when the backscattered photoelectron de Broglie wave is in-phase with the outgoing wave, then the absorption shows a maximum. This can be viewed as constructive interference between outgoing and backscattered waves. When the backscattered wave is out of phase with the outgoing wave, then the absorption shows a minimum, and this can be viewed as destructive interference. As E is scanned through the absorption edge (E_0) to increasingly higher energies, the photoelectron kinetic energy E_K will increase and its de Broglie wavelength λ_e will decrease, giving rise to successive cycles of constructive and destructive interference that in turn will give rise to an oscillatory behaviour in the absorption coefficient. These are the EXAFS oscillations. Despite being very approximate, this em-

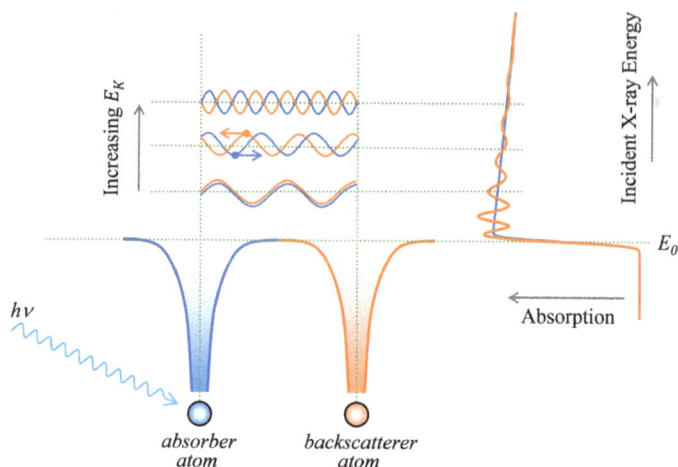

Figure 4.18: X-ray absorption for an absorber-backscatterer pair. Arrival of an X-ray photon with energy $E = h\nu > E_0$ results in X-ray absorption via photoexcitation of a core electron. The resulting photoelectron has a kinetic energy and a de Broglie wavelength λ_e that depend upon the excess energy $E - E_0$ (eq. (4.17)). The photoelectron may be scattered by the electrons around a neighbouring atom (backscatterer atom) with the phase of the backscattered photoelectron relative to the outgoing wave depending upon the de Broglie wavelength. The X-ray absorption is modulated according to the phase of the backscattered electron at the nucleus of the absorber atom, with in-phase corresponding to an absorption maximum and 180° out of phase corresponding to an absorption minimum (shown on the right).

pirical understanding of some aspects of the origins of the EXAFS allows us to predict some of the factors to which the EXAFS will be sensitive:

- **The distance between absorber atom and backscatterer atom.** Moving the backscatterer relative to the absorber in Figure 4.18 will modify the frequency of the EXAFS.
- **The number of electrons around the backscatterer.** The more electrons that the backscatterer atom has, the more effective at photoelectron backscatterering it will be. Thus, we can predict that the amplitude of the EXAFS oscillations will generally increase with the atomic number of a backscatterer.
- **The number of backscatterers of a particular type at a particular distance.** The more backscatterers there are at a given distance, the bigger we expect the amplitude of the EXAFS oscillations to be. For N equivalent backscatterers at the same distance we expect the EXAFS to scale in proportion to N.

As originally noted by Sayers, Stern and Lytle [9], the oscillatory nature of EXAFS means that Fourier analysis can be very useful. Readers who are unfamiliar with Fourier methods and in particular, the Fourier transform, are referred to Appendix B of this book.

To treat the EXAFS more quantitatively, we now define the absorption coefficient $\mu(E)$ as being composed of a non-oscillatory part $\mu_0(E)$ and an oscillatory part $\chi(E)$, as follows:

$$\mu(E) = \mu_0(E)(1 + \chi(E)) \tag{4.18}$$

So, we can write the following equation:

$$\chi(E) = \frac{\mu(E) - \mu_0(E)}{\mu_0(E)} \tag{4.19}$$

The EXAFS $\chi(E)$ is thus a dimensionless quantity. It is normal practice to express the EXAFS χ as a function of the photoelectron wave vector k, eq. (4.17), $\chi(k)$, and as we will discuss in Chapter 8, in most cases, to weight $\chi(k)$ by k^3, so that the highly damped high-k structure is readily visible. In Chapter 8, we will also discuss the extraction and interpretation of the EXAFS $\chi(k)$ in some detail, but for now, we will briefly give an example, the Mo K-edge EXAFS of $[Cl_2FeS_2MoS_2FeCl_2][PPh_4]_2$, which is shown in Figure 4.19. The Mo K-edge EXAFS Fourier transform of this data is shown in Figure 4.20, together with the crystal structure of the anionic $[Cl_2FeS_2MoS_2FeCl_2]^{2-}$ core. This species has a central Mo(VI) atom, coordinated by four essentially equivalent sulfides with approximately tetrahedral local geometry. The sulfides are also coordinated to two ferrous irons, which also possess approximate tetrahedral coordination geometry, each with two chlorides completing their coordination; the multi-metallic core has a linear Fe···Mo···Fe geometry. The molybdenum EXAFS is thus expected to show the presence of two different backscatters, the four directly coordinated sulfides, and the two longer-range Mo···Fe. The isolated EXAFS (Figure 4.19) shows a clear beat at k~11.5 Å^{-1} where the amplitude dips towards zero, derived from the different EXAFS frequencies from the two different backscatterer types, with the Fourier transform (Figure 4.20) showing two clear peaks corresponding to the Mo–S and Mo···Fe EXAFS.

The process of extracting structural information from the EXAFS is one of **model building**. We will deal with this in some detail in Chapter 10, but for now, we can summarize the basic methods: postulate a structural model, use this model to compute the EXAFS using *ab initio* theory, and then refine structural parameters by minimizing the difference between the experimental data and the computed EXAFS. This process can be done either in k space or in R space (Section 10.8), with each method having particular advantages and limitations, and we will discuss these in a subsequent chapter. Such curve-fitting analyses are the mainstay of modern EXAFS analysis.

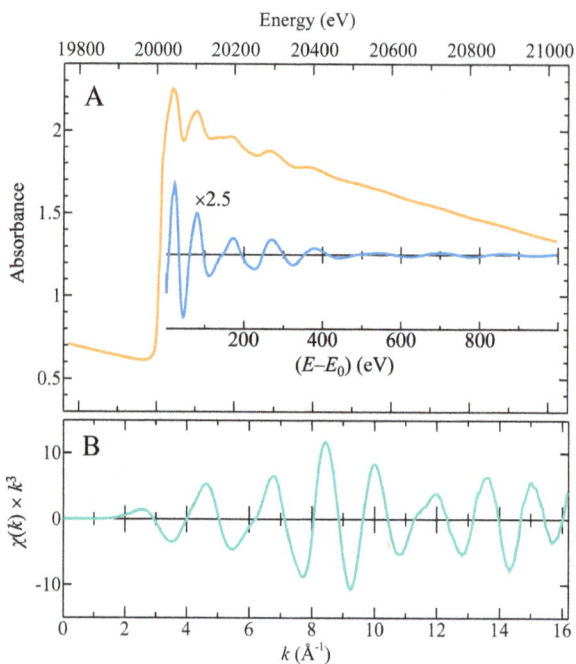

Figure 4.19: A shows the raw Mo K-edge XAS data for $[Cl_2FeS_2MoS_2FeCl_2][PPh_4]_2$, (orange line), with the inset (blue line) showing the background extracted EXAFS oscillations, eq. (4.15), with an ordinate scale 2.5× relative to the raw data and plotted against the photoelectron kinetic energy $(E - E_0)$. B shows the same EXAFS oscillations plotted against the photoelectron wave vector k, eq. (4.13) and weighted by k^3.

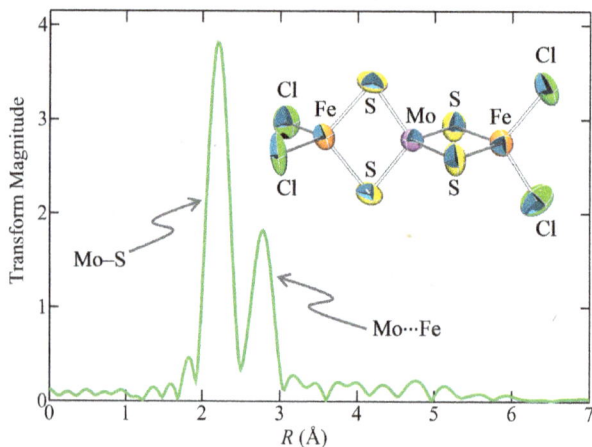

Figure 4.20: The EXAFS Fourier transform of the EXAFS data from Figure 4.19 (Mo–S phase-corrected) showing clear transform peaks due to Mo–S and Mo···Fe backscattering. The inset displays the crystal structure of the multi-metallic core, showing 75% thermal ellipsoids.

References

[1]   Jenkens, R.; Manne, R.; Robin, R.; Senemaud, C. Nomenclature system for X-ray spectroscopy. *Pure & Appl. Chem*. **1991**, *63*, 735–746.

[2]   Dwyer, D. J.; Rausenberger, B.; Lu, J. P.; Bernasek, S. L.; Fischer, D. A.; Cameron, S. D.; Parker, D. H.; Gland, J. L. Fluorescence yield near edge spectroscopy of π-bonded CO on Fe(100). *Surf. Sci*. **1989**, *224*, 375–385.

[3]   Sugiyama, A.; Okamato, K.; Nagamatsu, S.; Kawata, M.; Hayashi, H.; Udagawa, Y.; Fujikawa, T. Spin-polarized Mn K-edge XANES of Mn oxides. *J. Magn. Magn. Mater*. **2007**, *310*, e953–e955.

[4]   Schütz, G.; Frahm, R.; Mautner, P.; Wienke, R.; Wagner, W.; Wilhelm, W.; Kienle, P. Spin-dependent extended X-ray-absorption fine structure: Probing magnetic short-range order. *Phys. Rev. Lett*. **1989**, *62*, 2620–2623.

[5]   Pickering, I. J.; Sansone, M.; Marsch, J.; George, G. N. Site-specific X-ray absorption spectroscopy using DIFFRAXAFS. *Jpn. J. Appl. Phys*. **1993**, *32*, 206–208.

[6]   Filipponi, A.; Tyson, T. A.; Hodgson, K. O.; Mobilio, S. KL edges in X-ray-absorption spectra of third-period atoms: Silicon, phosphorus, sulfur, and chlorine. *Phys. Rev. A*. **1993**, *48*, 1328–1338.

[7]   Kodre, A.; Arčon, I.; Padežnik Gomilšek, J.; Prešeren, R.; Frahm, R. Multielectron excitations in X-ray absorption spectra of Rb and Kr. *J. Phys. B: At. Mol. Opt. Phys*. **2002**, *35*, 3497–3513.

[8]   Lytle, F. W. The EXAFS family tree: A personal history of the development of extended X-ray absorption fine structure. *J. Synchrotron. Radiat*. **1999**, *6*, 123–134.

[9]   Sayers, D. E.; Stern, E. A.; Lytle, F. W. New technique for investigating noncrystalline structures: Fourier analysis of the extended X-ray absorption fine structure. *Phys. Rev. Lett*. **1971**, *27*, 1204–1207.

5 X-ray detectors and detector systems

5.1 Introduction

The topic of X-ray detectors is a particularly practical one. In this chapter we will discuss aspects of how X-rays are experimentally detected and will review various kinds of detectors that are currently in common use for X-ray absorption spectroscopy experiments. In Chapter 1, we reviewed one method of detection, by direct measurement of the X-ray absorption spectrum. In Chapter 4 we reviewed some of the various processes by which a core-hole from the primary X-ray photoexcitation event can decay, through emission of an X-ray fluorescent photon or emission of an Auger electron. XAS is commonly detected via all three processes, by monitoring the absorption of X-rays directly, by measuring the X-ray fluorescent yield, and by measuring the electron yield. In all cases, the intensity of the beam incident on the sample also needs to be monitored simultaneously with each of the above measurements, since this intensity is variable with energy, time and other factors. An important factor in the success of any experiment is the correct choice of experimental approach, including the detector system for a particular measurement. Table 5.1 shows a summary.

Table 5.1: Experimental approach and detection systems.

Detection method	Signal	Concentration range	Comments
Transmittance	$A = \log(I_0/I_1)$	High concentration (ca. > 1 wt%), mainly hard X-rays	Simplest method, least prone to distortions
Fluorescence, non-dispersive	$A \propto I_F/I_0$	Medium to low concentrations (ca. 5 mM aq.)	Beware self-absorption effects with high concentrations
Fluorescence, dispersive	$A \propto I_F/I_0$	Low concentrations (ca. 2 mM–50 μM aq.)	Beware detector linearity and count-rate problems
Electron yield	$A \propto I_E/I_0$	High to medium concentrations, low Z solids mainly soft X-rays	Surface-sensitive (most signal comes from top ca. 100 Å), beware charging problems

5.2 Gas ionization chambers

5.2.1 Ion chamber function

Gas ionization chamber detectors are commonplace at all synchrotron beamlines and are considered standard equipment for hard X-ray XAS. They are frequently used as

https://doi.org/10.1515/9783110570441-005

the upstream detector to measure the incident beam (I_0), as well as downstream of the sample for transmission measurements including of the calibration foil (see Section 6.2).

Typically, a gas ionization chamber, called an ion chamber for short, consists of a gas-tight container closed by X-ray transparent windows containing a pair of parallel metal plates connected to a high voltage power supply, called the sweeping voltage (Figure 5.1). In most cases the chosen gas (e.g. He, N_2 or Ar), or a gas mixture, is gently flowed through the ion chamber, or the chamber is filled and then sealed. The X-ray beam passes through the gas between the plates. A fraction of the incident X-ray flux is absorbed by the gas, yielding photoelectrons and positive ions (ion pairs), which act as charge carriers. The presence of these charge carriers gives rise to a current flow between the plates of the ion chamber, which is proportional to the X-ray flux passing through the ion chamber. The current flow to ground is typically through a current-to-voltage amplifier and is registered as a voltage.

The choice of sweeping voltage V (Figure 5.2) must be sufficiently high to prevent significant ion pair recombination, while not so high as to promote cascade gas amplification of the signal, or (importantly) to induce electrical arcing between the ion chamber plates. A common problem with ion chamber detectors is to use a voltage that is too low, in what is called the recombination region, in which case the ion chamber response will be non-linear and will fail to ratio properly with other detectors. In general, the high voltage must be increased to access what is called the plateau or saturation region (Figure 5.2), where the signal does not increase with increasing voltage. The sweeping voltage must also be low enough that there is no significant chance of arcing between the plates of the ion chamber. Electrical arcing can damage both the ion chambers themselves and the associated electronics; once an arc has discharged, often from the edge of a plate, there can form a defect at the site of the arc, which will lower the voltage required for future electrical arcs. Modern high-flux density beamlines experimenters often need to use a gas that is not prone to arcing, such as helium. Figure 5.2 shows a typical response curve, with the green region corresponding to the operational range of gas ionization chambers.

5.2.2 Proportional counters and Geiger-Muller detectors

Neither proportional counters nor Geiger-Muller detectors are much used in modern X-ray absorption spectroscopy experiments, but in this final section on ion chamber-type detectors we note that the white region to the right of the plot in Figure 5.2 approaches the sweeping voltage range that is used in a type of detector called a proportional counter. The proportional counter is a type of energy-dispersive detector, which has basically the same design as a gas ionization chamber, except that the cathode is a uniform cylinder, and the anode is a very thin and centrally placed wire. The entry of X-rays occurs through a window typically placed at one end of the cathode cylinder; the X-rays ionize the gas (usually a 90% argon, 10% methane mixture) within the detector in the same way as in a conventional ion chamber, with the resulting electrons drifting towards the central wire anode. Within a fraction of a millimetre of the anode wire,[1] the electric field strength is strong enough to cause avalanche amplification of the electron, generating a current pulse that is proportional to the energy of the X-ray photon initiating the initial ionization event. Readout of the proportional counter has much in common with solid-state energy-dispersive detectors discussed in Section 5.3. At higher sweeping voltages still, the detector would operate as a Geiger-Muller detector (which would have a similar arrangement of anode and cathode to a proportional counter), until electrical arcing occurs.

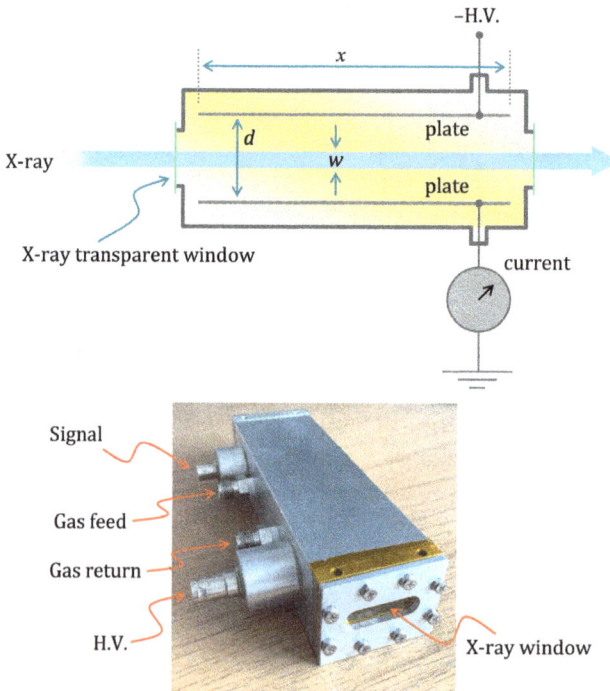

Figure 5.1: Gas ion chamber detectors. A schematic of a gas ionization chamber detector (top), plus a photograph of a standard ion chamber (bottom). Note that this particular design is capable of being pressurized.

---

1 The electric field strength E (not to be confused with X-ray energy) at radial distance r from the central anode wire of diameter a for a proportional counter of cathode inner diameter b is given by $E(r) = V/r\log(b/a)$, where V is the applied sweeping voltage.

5.2.3 Gas ionization chamber readout

The conventional ion chamber used to measure the incident and transmitted beam is typically connected to a current-to-voltage amplifier, and the resulting voltage is proportional to the incident X-ray flux. Two common methods for reading this voltage into a computer are to use an analog-to-digital converter (ADC), or a voltage-to-frequency (VtoF) converter[2] and a counter. Both have advantages and disadvantages. A VtoF averages high-frequency statistical noise and is much easier to gate with a time signal, so that fluorescence detector counts can be measured over essentially the same time slice. However, ADCs can have superior response times. Moreover, typical VtoF converters will not read negative voltages, so that if the offsets of the current-to-voltage amplifier are incorrectly set then the readout will be artefactually zero.

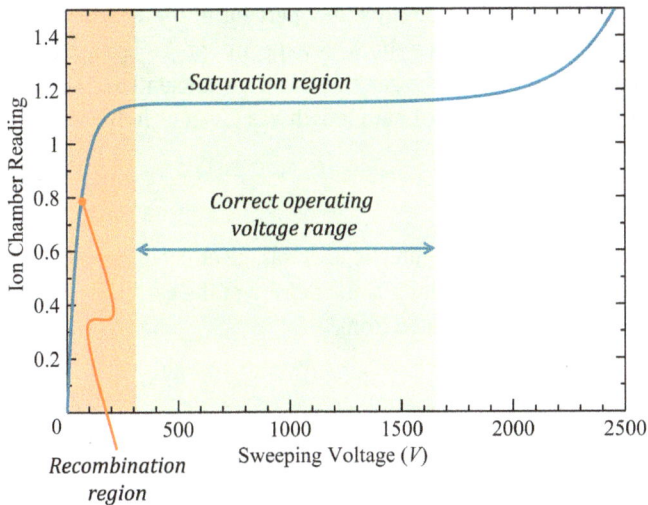

Figure 5.2: Typical response curve versus applied sweeping voltage of a gas ionization chamber detector. The green area – called the saturation region – defines the normal operational range of gas ionization chambers.

---

2 The reader might wonder why these details might be useful. One of us (George) remembers the first night of beamtime of his very first SSRL visit. Somewhere in the middle of the night, another researcher came by our beamline, and while rubbing his hands together he cast hungry eyes over our NIM bins and CAMAC crates, and asked "are you using your VtoF converter?" Fearing that he would remove some vital piece of hardware, I replied "yes" because data was rolling in nicely and I did not want to interrupt that. In truth, I had no idea whether we were using our VtoF converter or not, and in fact we were, but it would have been better to have known.

5.2.4 Calculation of photon flux using a gas ionization detector

The energy needed to produce an electron-ion pair E_{ion} depends upon the ion chamber gas. For N_2 as an ion chamber gas, $E_{ion} \approx 36$ eV,[3] hence absorption of one X-ray photon with an energy E of 9,000 eV by N_2 will produce $E/E_{ion} \approx 9,000/36 = 250$ charge carriers.

The number of charges registered in one second can be calculated from the current flowing in the ion chamber (the voltage divided by the gain) using the elementary charge e (1.602×10^{-19} C). Thus, a single incident X-ray photon at 9,000 eV when absorbed by N_2 will produce a current of $I = e(E/E_{ion}) = (9,000/36) \times 1.602 \times 10^{-19} \approx 4 \times 10^{-17}$ A, and the number of photons registered (absorbed) in the ion chamber can be calculated using

$$N_{abs} = \left(\frac{E_{ion}}{E}\right)\left(\frac{I}{e}\right) \tag{5.1}$$

Ideally, a gas ionization chamber will absorb only a few percent of the incident photons, leaving most of the X-rays available for the XAS experiment. To estimate the photon flux the absorption of the ion chamber gas must also be calculated. The fraction of photons absorbed by an ion chamber of path length x is given as follows:

$$\frac{N_{abs}}{N_{in}} = 1 - e^{-\mu(E)x} \tag{5.2}$$

where $\mu(E)$ is the absorption coefficient of the ion chamber gas, given by $\mu(E) = \sigma(E)_{tot}\rho$ where $\sigma(E)_{tot}$ is the total X-ray cross-section and ρ is the density of the gas. The incident photon flux in photons $\cdot$ second^{-1} can be calculated from

$$N_{in} = \left(\frac{E_{ion}}{E}\right)\left(\frac{I}{e}\right)\frac{1}{(1 - e^{-\mu(E)x})} \tag{5.3}$$

For an accurate calculation of X-ray photon flux, N_{in} in eq. (5.3) should be scaled by the sum of the photoelectric cross-section $\sigma(E)_{pe}$ and the incoherent X-ray cross-section $\sigma(E)_{incoh}$ divided by the total X-ray cross-section $\sigma(E)_{tot}$, $(\sigma(E)_{pe} + \sigma(E)_{incoh})/\sigma(E)_{tot}$.

5.2.5 Choice of ion chamber fill gas

The gases used in ion chambers must be carefully chosen, since the choice depends upon a number of factors, including the length of the ion chamber (x), the separation of the plates (d), the X-ray energy range of the experiment and the flux density of the beamline. Common choices are helium, nitrogen (N_2) and argon, plus (in some cases) mixtures of these. Beamlines providing low X-ray flux densities will often use helium

---

3 Corresponding values of E_{ion} for helium, air and argon are 41.0, 34.4 and 26.0 eV, respectively.

for the tender X-ray regime (e.g. 2–4 keV), N_2 for mid-range hard X-ray experiments (5–12 keV) and argon for higher energies (13–30 keV). Modern beamlines frequently produce X-ray beams with a high X-ray flux density, for which the voltage required to attain the ionization chamber saturation region may not be accessible without arcing. Two solutions to this problem are possible – to use an ion chamber design that allows a gas pressure of more than one atmosphere, or to use a gas that is more reluctant to arc, with a lower X-ray cross-section, so that the saturation region is easier to attain. This second option means that the ion chamber might absorb only a very small fraction of the beam; for example, at 9 keV helium gas at one atmosphere in a standard 15 cm ion chamber will absorb only about 0.01% of the incident X-ray beam with less than ideal signal readout, in comparison with N_2 which would absorb about 8% of the beam. Similarly, N_2 at 20 keV would absorb only 0.6% of the beam, compared to Ar which would absorb 20.5%. Nonetheless, the effective signals from gas ionization chambers are sufficiently high that this is not really a problem. Pressurization of ion chambers requires the use of thicker windows, which can attenuate the beam at low X-ray energies. For hard X-ray applications polyimide (Kapton) adhesive tape is typically used for ion chamber windows, with a practical upper pressure limit of two atmospheres.

5.2.6 Ion chamber Soller slits

Before moving on to discuss other aspects of gas ionization chambers, a brief mention of foil Soller slits is appropriate. In a hard X-ray experiment (Section 6.2), it is common practice to use three ion chambers in series. The first is used to register the incident X-ray flux, I_0, the second the transmitted flux, which is called I_1, and the third to measure an elemental foil or other energy calibration standard and is typically called I_2. The absorbance of the sample is simply given by $A_{samp.} = \log(I_0/I_1)$ while that of the foil (calibration standard) is $A_{foil} = \log(I_1/I_2)$. A problem can arise due to X-ray fluorescence from the foil, which can be very concentrated relative to the sample, and which is isotropically emitted over 4π sr. A small fraction of the foil X-ray fluorescence will enter both I_1 and I_2 ion chambers and will contribute to the signals registered by these detectors. This is not a significant problem for I_2, but as we discuss in Section 12.6, it can be a major issue for I_1, especially if the sample is dilute. The solution to this issue is to use a set of parallel-bladed Soller slits [1] between the foil and I_1, as described by Tse et al. [2], which will effectively block most of the foil fluorescence coming back in the direction of I_1 and facilitate a truer measure of the sample absorbance.

Alternative geometries of Soller slits are frequently used with fluorescence detectors in combination with filters; these are addressed below (Section 5.4.9).

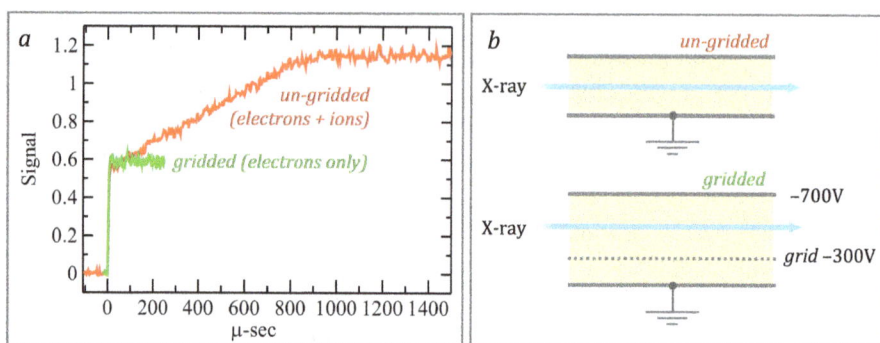

Figure 5.3: Response of gridded versus ungridded gas ionization chamber detectors. Time dependence (a) of conventional (ungridded) and gridded gas ionization chamber detectors, together with a schematic of the detectors (b). Replotted from Müller et al. [3].

5.2.7 Gridded ion chamber designs

Many modern beamlines are designed to scan the incident rapidly in order to do what is often called qXAS (quick-XAS) or QEXAFS (quick-EXAFS). In these cases, the response time of an ion chamber can become important. For an ion pair created by the passage of an X-ray, electrons drift towards one ion chamber plate much more quickly than ions drift towards the other, with a difference in velocity of approximately 1,000. A single ion pair will thus result in a current pulse, which would be biphasic – a very rapid rise in current from the electrons overlaid with a slower rise in current from the ions. For applications that require a fast response time, gridded ion chambers can provide a solution. Here, a grid, often referred to as a Frisch grid,[4] is kept at an intermediate electrical potential and positioned to prevent the ions from inducing charge on the collecting electrode, hence the signal derives only from the much faster electrons. Figure 5.3 compares the responses of gridded and conventional un-gridded ion chambers using a chopper to modulate the incident beam, and clearly shows the biphasic time response of the conventional ion chamber and the superior time response of the gridded design.

---

4 Named after the Austrian-born British Physicist and nuclear pioneer Otto Robert Frisch (1904–1979).

A

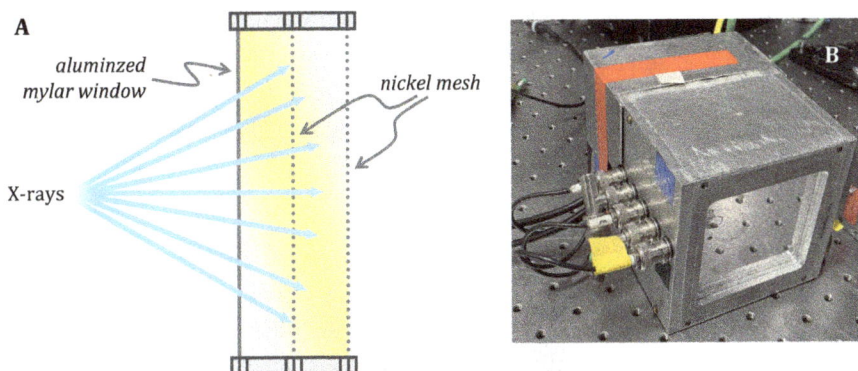

Figure 5.4: Fluorescent ion chamber detectors. (A) Schematic diagram of a simple 3-grid fluorescent ionization chamber detector (side view); (B) a 5-grid X-ray fluorescent ion chamber detector (a Stern-Heald -Lytle detector), with the reflective aluminized Mylar window through which the X-rays enter turned towards the camera.

5.2.8 Fluorescent ion chamber detectors

Almost all experienced XAS users have come across a Stern-Heald-Lytle detector at some point. These beautiful little detectors[5] are effectively ion chambers turned on their sides and are used to measure X-ray fluorescence at 90° to the incident beam. One of the two ion chamber plates is effectively replaced by an electrically conducting 6 μm thick aluminized Mylar window, with two 90% transmitting metallic nickel mesh internally positioned (see Figure 5.4). The two outer electrodes, the aluminized Mylar window and the outer mesh, are typically operated at −45 V, with the signal current collected on the central mesh electrode. Because these detectors are designed to measure X-ray fluorescence, which will never have high flux densities, the plateau region is much easier to attain using comparatively low voltages, relative to an ion chamber measuring the incident beam. The ion chamber fill gas is flowed through the fluorescent ion chamber detector at a low rate. The gas is typically a higher X-ray cross-section gas than would be used for a normal ion chamber; for example, for mid-range hard X-ray measurements one might fill a fluorescent ion chamber with krypton or with (expensive) xenon gas. More efficient hard X-ray fluorescent ion chamber detectors are available with five electrodes, which have double the fill gas path length (absorption distance) relative to the three-electrode version just described, incorporating an additional collector grid and back-plane grid. These detectors outperform many other non-dispersive detectors and are relatively simple to operate. However, they do need a fill gas feed, which if not carefully controlled can lead to stretching of the thin aluminized Mylar window by using too great a gas flow, or through "testing"

5 Available from the EXAFS Company, Pioche, Nevada, USA.

the fill gas flow by placing a finger over the outlet. Once the window has been stretched and is no longer taut, the ion chamber becomes very noisy, and the window must be replaced.

5.3 Photodiodes

A photodiode is a semiconductor device, which registers photons through generation of an electrical current. A photon entering the diode creates an electron–hole pair, which acts as charge carrier, and is swept from the p-n junction by the electric field inherent within the diode depletion region – electrons moving to the cathode and holes to the anode, generating a photocurrent. These devices are typically fabricated from silicon, with a range of different photodiodes available for X-ray detection. In practice, since the photocurrent is fed to a current-to-voltage amplifier, these detectors can be treated in much the same manner as an ion chamber. Indeed, specially thinned photodiodes can be found at some beamlines, designed to pass a fraction of the incident X-ray beam, just like an ion chamber detector. Large-area photodiodes are also commercially available, such as PIPS detectors (passivated implanted planar silicon detectors), which can be used in much the same way as a Stern-Heald-Lytle fluorescent ion chamber detector. Photodiodes have the advantage of not requiring a gas flow but have the disadvantage of responding to visible light, which causes extraneous signals and adds noise to the data. PIPS detectors used for X-ray fluorescence detection are typically fitted with an aluminized Mylar window to exclude visible light, but even with this precaution the hutch must be darkened, windows blanked out, and any sources of light internal to the hutch need to be minimized.

5.4 Energy-dispersive solid-state X-ray detectors

If a diode[6] is operated under reverse bias, meaning an applied voltage with a direction opposed to the normally allowed current flow of the diode, the diode can be used as an energy-dispersive detector. Since such detectors form a mainstay of modern X-ray absorption spectroscopy, we will discuss them in some detail. Figure 5.5 shows a schematic of the operation of these detectors; the X-ray photon entering the diode produces a cloud of electron-hole pairs, with the number produced proportional to the X-

---

6 A diode is one of the simplest types of modern semiconductor devices, consisting of a single p-n junction. The p-type silicon is made by doping electron deficient elements such as boron into the silicon, while the n-type silicon uses doping of electron rich elements such as antimony. The level of doping is usually around 1 dopant atom per 10,000 silicon. A diode typically allows current to flow in only one direction. A reverse-bias diode is a diode in which a voltage is applied in the reverse direction – i.e. against the normal flow of current.

ray energy. The bias voltage causes the holes to drift to the front and the electrons to the rear; the resulting current pulse is proportional to the energy of the incident X-ray photon. We will review various types of energy-dispersive solid-state X-ray detectors, but since they share commonalities in their general properties and use, these will be illustrated in our first example, which is the germanium detector.

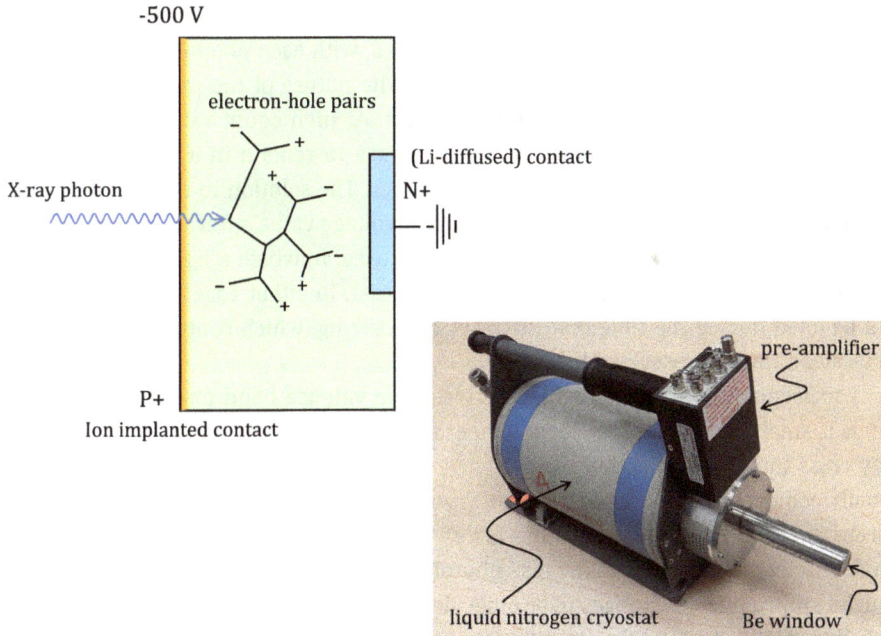

Figure 5.5: Energy-dispersive solid-state detector. Schematic of a reversed bias-based diode germanium energy-dispersive detector, together with a photograph of a single element device. The bulk of the diode is composed of high-purity germanium (HPGe). Photo courtesy of Dr. Ian Coulthard Canadian Light Source.

5.4.1 Germanium detectors

Germanium detectors are workhorse detectors that can be found at many XAS beamlines and are often constructed as arrays of individual diodes [4], or as monolithic devices. Relative to silicon, the higher X-ray absorption cross-section of germanium means that these detectors have a greater efficiency at higher X-ray energies. We will consider the operations of a single diode first, and then proceed to discuss array and multi-pixel monolithic detectors. The germanium detector diode is typically fabricated with a thin ion-implanted p+ contact through which the X-ray photons enter, with the n+ contact at the interior of the device (Figure 5.5). The absorption of the X-ray photon within the diode causes a cloud of electron-hole pairs to form, the number of which is proportional to the X-ray energy, assuming that the X-ray photon and any scattered or fluorescent daughter photons are all absorbed within the diode. The bias voltage causes the electrons and holes to respectively drift towards the front and the rear of the detector, resulting in a current pulse that is proportional to the energy of the X-ray photon.

The charge is transferred to a field-effect transistor (FET), which serves to convert the current into a voltage as part of the pre-amplifier. A pre-amplifier pulse train typical of a germanium detector is shown in Figure 5.6, with each photon causing a sharp rise, followed by a considerably slower decay. The nature of the pulse train will be somewhat different for different detector types. At high count rates the relatively slow decay can be insufficient to allow the voltage to remain in a workable range, which can result in saturation of the pre-amplifier. The solution to this problem is to use an automatic reset, triggered at some preset voltage value, after which the voltage is zeroed. In many cases a pulsed-optical reset is used, in which a light emitting diode triggers a light-sensitive switch, or a transistor-reset. In either case the pre-amplifier will be dead during the time that the reset is occurring, which contributes to the detector dead time, as discussed below.

For germanium, the energy gap between the valence band and the conduction bands is small enough that, at room temperature, thermally excited valence electrons may cross into the conduction band. Hence, these detectors must be cooled, which is usually achieved using a liquid nitrogen cryostat. Germanium detectors have good inherent energy resolution, typically ~150 eV full-width-half-maximum (FWHM), and good efficiency, meaning that most photons entering the detector are measured, which can provide advantages over other detector technologies at higher X-ray energies. In general, most solid state detectors are enclosed by a thin beryllium window,[7] which has low absorption of X-rays in the hard and tender energy range and serves to both exclude visible light and maintain the vacuum needed for cryogenic insulation of the detector. For soft X-ray measurements germanium detectors can be effectively windowless, in that no separate vacuum system is used, but a physical barrier is usually present to exclude visible and infrared.

5.4.2 Lithium-drifted silicon detectors

These detectors, often called Si(Li) detectors (pronounced "silly") are p-i-n devices that are formed by lithium drifting of p-type silicon. They require cooling to liquid nitrogen temperatures, in part to prevent diffusion of the lithium dopant within the diode.

---

7 While beryllium provides an excellent material for X-ray windows, thin windows under vacuum are easily broken, hence great care must be taken when using detectors equipped with beryllium windows. The compounds of beryllium are toxic, and dangerous when inhaled; should a window break, stringent safety precautions are required.

Si(Li) detectors used to be commonplace, but have now been essentially supplanted by silicon drift detectors (see Section 5.4.3). Nonetheless, they do provide advantages over other solid-state detectors in that the silicon can be made up to 5 mm thick, which means that they are more efficient than silicon drift detectors. We will discuss escape peaks in Section 5.4.11, but we note here that escape peaks are less pronounced in silicon detectors than germanium detectors. This is because the low silicon X-ray fluorescence energy means that fluorescent photons are less likely to escape the detector. But silicon drift detectors are thin devices, and hence escape peaks are present, but thicker Si(Li) detectors show much smaller escape peak amplitudes by several orders of magnitude.

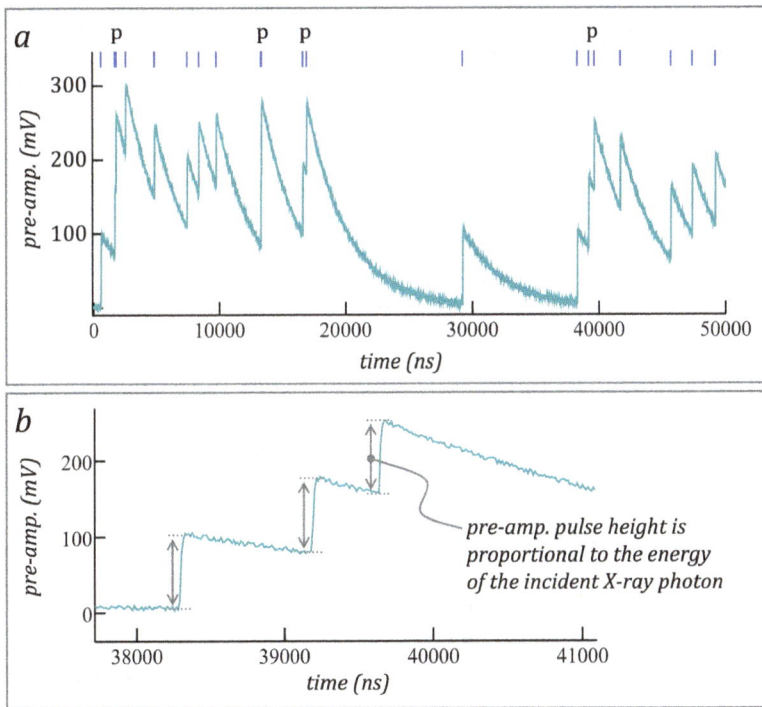

Figure 5.6: Pre-amplifier pulse train from a germanium detector. (a) The arrival of photons denoted by vertical blue lines (**p** denotes pileup events) and (b) detail for three individual pulses.

5.4.3 Silicon drift detectors

These devices, sometimes called "SDDs" or just "drift detectors", are now very well established and can be found at many beamlines. A schematic of an SDD is shown in Figure 5.7. The device consists of a core of fully depleted high-resistivity silicon, in which an electric field with a strong component parallel to the surface of the device drives electrons generated by the absorption of X-ray photons towards the small central anode. The correctly oriented electric field is generated by a number of increasingly reverse-biased field strips known as drift rings, which are arranged along the rear surface of the detector. SDDs have a small anode capacitance, which is independent of the active area of the detector. This gives higher energy resolutions at high count rates relative to conventional photodiode energy-dispersive detectors. Because the readout is centrally located in the SDD the drift time affects the signal; photons arriving at the periphery of the device must drift through a longer path than those arriving centrally. Consequentially, the peripheral signals are relatively delayed as they must travel further and moreover these show a longer rise time due to dissipation of the charge cloud as it travels to the anode. In most cases, high resolution and high count rates are possible using SDDs without the need for the liquid nitrogen cooling that is required for both Ge and Si(Li) detectors.

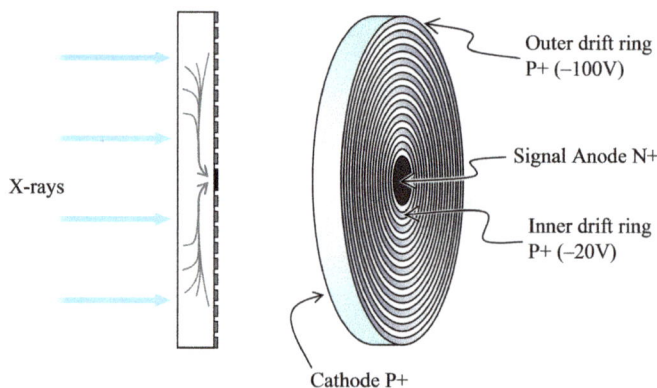

Figure 5.7: Schematic diagram of a silicon drift detector.

Another application of SDDs is for specialized studies requiring the use of a transverse magnetic field, such as X-ray magnetic circular dichroism of paramagnetic systems. With Ge detectors, for example, the Hall effect can oppose the current flow in the detector causing the signal to drop in amplitude with substantial field-dependent shifts in features. With SDD detectors, on the other hand, the electron flow is perpendicular relative to conventional photodiodes, and field-dependent effects are considerably reduced (George, S. J. personal communication).

SDDs are good general-purpose detectors providing excellent performance usually without the need for liquid nitrogen cooling and can be used for a range of applications. They have the disadvantage of having low efficiencies at high X-ray energies relative to germanium detectors, due to the lower X-ray cross-section of silicon.

5.4.4 Detector readout – analog pulse processing

Readout of solid-state detectors can use analog pulse processing or digital signal processing (described in the following section). While at one point in the past, analog readout systems of solid-state detectors were standard, digital systems have mostly replaced the older analog hardware; indeed, the user might be hard-pressed to find an analog system on an experimental floor. Nonetheless, since there is much commonality between the analog systems and the modern digital counterparts, we will describe both.

With the analog approach, the pre-amplifier pulse train is fed to a spectroscopy amplifier, typically either a Gaussian or a triangular shaping amplifier, since such shaped pulses can be more readily used. The shaping amplifier magnifies the pre-amplifier pulse, which would typically be in the mV range, to a pulse somewhere in the range of 0.1–10 V, which enables pulse height measurements with what is called a multi-channel analyser in conjunction with a peak-sensing analog to digital converter, or a single-channel analyser. A critical variable is the spectroscopy amplifier shaping time, which controls the energy resolution of the output, for which a typical value used for high-speed data acquisition might be 0.125 μs. The longer the shaping time, the better the resolution, but the lower the maximum count rate. Conversely, with shorter shaping times, count rates can be higher, but at the expense of degraded resolution. Moreover, the peak voltage is typically lower (the shaped pulses are broader), which can result in a poorer signal to noise.

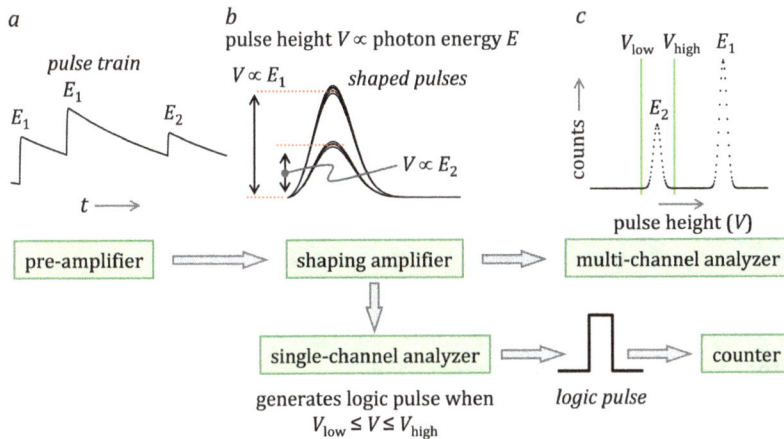

Figure 5.8: Schematic of analog pulse processing. (a) The pre-amplifier pulse train, with events from two different photon energies. (b) The shaping amplifier output as might be viewed on an oscilloscope, triggered on the rising edge of each shaped pulse; the two pulse heights correspond to the two different photon energies. (c) A multi-channel analyser output (histogramming memory), used to set the voltage discrimination levels of the single-channel analyser (b).

A block diagram of a typical analog pulse processing set-up is shown in Figure 5.8. The pre-amplifier pulse train is fed to the shaping amplifier, which generates shaped pulses, one per photon registered, having a peak voltage that is proportional to the energy of the incident photon. The hardware would typically be used in two different ways, one for set-up and the other for data acquisition. For set-up, the shaped pulses are fed to a peak-sensing analog to digital converter, the output of which is fed to a multi-channel analyser or MCA. Years ago, the MCA used to be a rack-mounted stand-alone device, but in modern systems these have been replaced by computer software. The MCA is equipped with histogramming memory, with the voltage scaled along the abscissa and accumulated (histogrammed) counts as the ordinate. This typically will give Gaussian peak shapes and can be used to set the voltage limits of a single-channel analyser or SCA. The SCA is simply a voltage discriminator that generates a logic pulse when the peak voltage of the shaped pulses falls between preset values. The MCA is used to select SCA voltages that will in turn be used to discriminate only those photons of interest by the SCA, which are registered using a counter and re-corded during data acquisition. Such SCA-discriminated counts are known as win-dowed counts, and setting the SCA voltage discrimination limits is known as setting the windows for the detector.

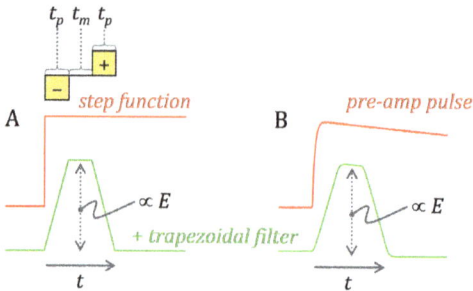

Figure 5.9: Trapezoidal filtering. (A) A trapezoidal filter applied to a perfect step function. (B) The application of a trapezoidal filter to one event in a pre-amplifier pulse train.

5.4.5 Detector readout – digital signal processing

The advent of fast ADCs makes digital signal processing a practical matter. Here, the pre-amplifier pulse train is digitized, and then can be manipulated digitally to deter-mine the step heights in the pulse train, which correspond to the energies of the inci-dent photons. A simple strategy is to use what is called a trapezoidal filter (Figure 5.9). If applied to a pure step function, a trapezoidal filter would give a perfect trapezoid. In this filter, three time intervals are important, a **peaking time**, which we can call t_p, which must be longer than the rise time of the pulses in pre-amplifier pulse train, followed by a **plateau time** t_m, followed again by another interval of t_p. For the first time interval of t_p data points from the ADC are summed, resulting in a trapezoidal rise, followed by a gap interval of t_m, which results in the plateau of the trapezoid, then followed by another interval of t_p in which the ADC data are now subtracted. For a real pulse train, this gives an approximately trapezoidal function in which the plateau height corresponds to the energy of the incident photon (Figure 5.9). Events (photon arrivals) that are separated by less than $2t_p + t_m$ may not be separable by this method; a workaround is to use two trapezoidal filters, one with a long peaking time and a second with a much shorter peaking time. Pileup events may be detectable because the response of the trapezoidal filter is linear. In some cases, peaking times are dynamically adjusted with the count rate, so that shorter peaking times would be used at higher count rates. Unfortunately, this causes a count-rate-dependent energy resolution for the detector readout, which may be problematic for some applications (see Section 12.5.2).

Contrasting this simple filter-based approach are more sophisticated approaches that use fitting of the digitized pulse train $p(t)$ (the ADC output) to an analytical expression of the form (5.4), where τ is the photon arrival time, α and β are rate constants, respectively, specifying the rise and decay of the pre-amplifier pulse and a is the amplitude, which is proportional to the incident photon energy:

$$p(t) = a \begin{cases} -e^{-\alpha(t-\tau)} + e^{-\beta(t-\tau)}, & t \geq \tau \\ 0, & t < \tau \end{cases} \tag{5.4}$$

Since to evaluate eq. (5.4) for every photon arrival would be a relatively computationally expensive procedure, in practice, the average of a number of photon arrival events (e.g. 100) is used to determine a pulse shape model on live startup of the acquisition system. This model function is then used to digitally model the pulse train in a dynamic high-speed manner. With both trapezoidal filter and model based approaches, the energy resolution typically will degrade with increasing count rate, although with the trapezoidal filter approach the resolution degrades close to exponentially with increasing count rates (at least initially), whereas with a model-based approach this degradation can be much less dramatic and close to a linear function of count rate.

The output of the digital signal processing process is essentially the same histogrammed MCA-type spectrum discussed above for analog processing, in which the photon energy is plotted on the abscissa and the counts at a particular energy plotted on the ordinate. Indeed, although many users of modern synchrotron facilities may never have actually seen a stand-alone MCA, the term MCA spectrum, referring to the spectral output of a solid state detector, is now embedded in the nomenclature of beamlines and synchrotron science.

Figure 5.10 shows a typical MCA spectrum, which in this case is of an arsenic-containing dilute aqueous sample. The MCA spectrum is essentially a low-resolution (relative to the natural linewidths, or *ca.* 150 eV FWHM) X-ray emission spectrum of

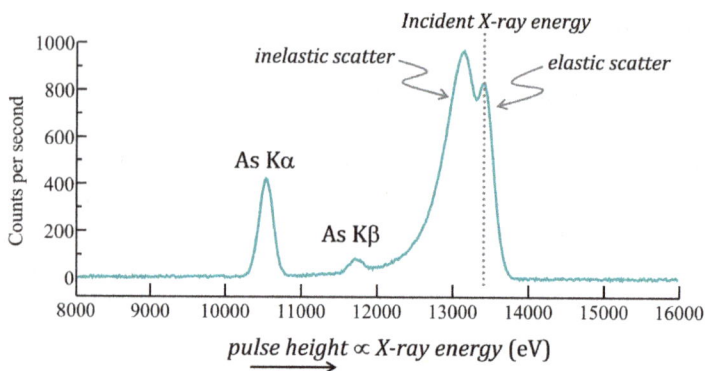

Figure 5.10: Example multi-channel analyser (MCA) spectrum. MCA spectrum of an aqueous sample containing ~5 mM arsenic, showing the As Kα and Kβ fluorescence lines, together with contributions due to the elastic and inelastic X-ray scattering by the sample. The incident energy, or the energy to which the monochromator was set to record this MCA, is indicated; the elastic scatter peak is at the same energy.

the sample, showing the spectrum of energies observed by the detector, in this case a germanium solid-state detector. Salient features of the spectrum are the elastic scatter, which falls at the energy of the incident X-ray beam, the inelastic scatter (Compton scatter), typically a broader feature with some asymmetry extending to the low energy side of the elastic peak and X-ray fluorescence lines. In the example of Figure 5.10 the As Kα and As Kβ are both clearly visible. The former is comprised of the As Kα1 ($2p_{3/2}{\rightarrow}1s$) and As Kα2 ($2p_{1/2}{\rightarrow}1s$) transitions, which are not resolved by solid-state detectors, and the latter comprised of As Kβ1 ($3p_{3/2}{\rightarrow}1s$) and As Kβ3 ($3p_{1/2}{\rightarrow}1s$), also unresolved by solid-state detectors. Although these Kα and Kβ components are not resolved in this energy region, at high X-ray energies, e.g. at the iodine K-edge, major fluorescence lines (e.g. Kα1 and Kα2) are sufficiently separated in energy (28 612.0 and 28 317.2 eV, respectively) to be scantly resolved by solid state detectors. While an entire MCA spectrum for every energy point in an XAS spectrum is in principle available with digital signal processing, a common practice is to set discrimination levels around a fluorescence line of interest, sum the counts within these limits, and subsequent to that, disregard the full MCA data, in a process analogous to the more traditional analog method.

It is informative to compare analog and digital approaches. When compared with early hardware for digital signal processing, the analog approach allowed access to higher count rates, but this is no longer the case with modern hardware solutions. Since useful readouts from digital signal processing hardware can now be obtained with count rates that run into the MHz regime, analog systems are no longer used in most cases. Nonetheless, there are many similarities between the end-result of the two approaches, in that the broadening of spectra features at short shaping times (analog) is very similar to that at short peaking times (digital).

Before moving on to discuss other factors, we will briefly revisit pileup, and rejection of pileup events. With analog signal processing systems, some kind of pileup rejection is frequently built in as part of the system. However, the situation differs with digital signal processing systems. Depending upon the choice of hardware, the pileup problem is variously ignored completely, or accounted for using quite stringent efforts.

5.4.6 Energy resolution and noise

One frequently cited metric of solid-state detector performance is the energy resolution. Typically, resolution is specified by manufacturers as the FWHM of the manganese Kα line. This metric is used because it is very conveniently measured in a factory test setting using a ^{55}Fe radioactive source. ^{55}Fe decays via electron capture to ^{55}Mn; the resulting 1s core hole is filled by the processes previously discussed in Chapter 4 to generate Mn Kα and Kβ X-ray fluorescence, and Auger electrons. The energy resolution is also dependent upon the energy of the X-rays being measured and decreases with decreasing photon energy. In general, germanium detectors have slightly better (smaller) inherent energy resolution than do silicon drift detectors at the Mn Kα energy and above, but at low X-ray energies this trend is reversed, and the energy resolution of the drift detectors becomes superior. The peak shapes in a typical detector output are usually at least approximately Gaussian, but there are complexities that distort the shape, especially at low energies, the most pronounced of which is incomplete charge collection. In this case not all of the electron-hole pairs are measured, and the apparent photon energy will be lower than the energy of the incident X-ray, which typically manifests as a tail on the low energy side of the peak. At low X-ray energies the photons penetrate less deeply into the detector and charge collection is usually poor near the front contact; consequentially the peak in the MCA spectrum for such low energy X-rays will be broad with a low-energy tail, and will give a mean energy lower than expected. Figure 5.11 shows this low-energy tail for the Al Kα line, detected with a germanium array detector. We note that modern germanium detectors, such as the so-called ultra-LEGe detectors (LE stands for low energy) have considerably better performance than this example, with much superior charge collection at low X-ray energies, and excellent performance down to a few hundred electron volts.

All types of solid-state detectors show what is known as **leakage current**, which tends to be greater when using higher bias voltages. This leakage current results in a positive slope in the pre-amplifier output, opposing the effects of sensing photons, and is one source of statistical noise in the detector output. The leakage current is proportional to both the temperature and the active area of the detector, and tends to be slightly larger for silicon drift detectors than for Si(Li) and Ge detectors. Opposing this trend is the so-called voltage noise, which is lower for silicon drift detectors and higher for Ge and Si(Li) detectors. The voltage noise depends upon the various capacitances in the sensor electronics, including that of the sensor anode. Because silicon drift detectors have relatively small anodes (e.g. Figure 5.7), they have a low capaci-

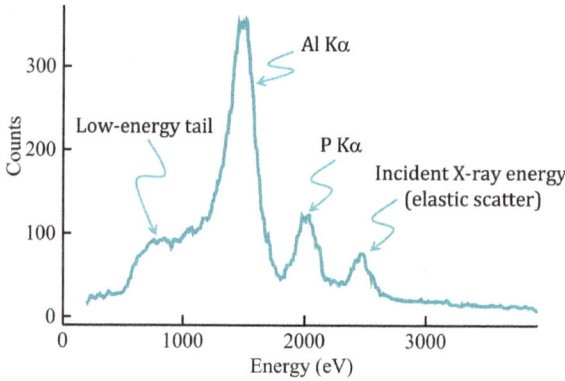

Figure 5.11: MCA spectrum from the first germanium array detector [3] showing Al and P K-edge X-ray fluorescence. The figure shows a low-energy tail arising from incomplete charge collection, predominantly at the Al Kα energy (~1,486.6 eV). We note that the Al Kβ at ~1,557.5 eV is incompletely resolved from the Kα. Data replotted from Cramer et al. [4].

tance and show a correspondingly low voltage noise, even when using the short processing times needed for high count rates. This makes silicon drift detectors especially useful for high count-rate applications.

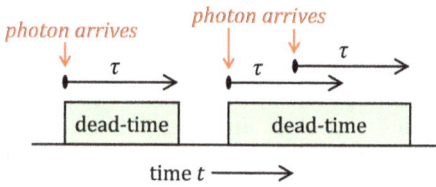

Figure 5.12: Schematic of paralyzing electronic dead time τ. The figure shows the dead time extension on arrival of a second photon.

5.4.7 Count rates and detector dead times

With all solid state detectors, high count rates can cause detector non-linearity due to the dead time that is associated with electronic signal processing. Typically, the type of dead time is that known as paralyzing dead time, and can be quantified using

$$r(E) = \rho(E)e^{-\rho(E)[\bar{\tau}-\tau(E)]} \tag{5.5}$$

where $r(E)$ is the registered count rate at an X-ray energy E, $\rho(E)$ is the true count rate incident on the detector, $\tau(E)$ is the electronic processing time for a photon of energy E, and $\bar{\tau}$ is the average value of $\tau(E)$ over the entire spectrum of energies. When a photon of energy E arrives at the detector there is a dead time $\tau(E)$ associated with its processing. If another photon arrives during $\tau(E)$ then not only is the second photon not registered, but the dead time is extended by another $\tau(E)$ from the time of arrival of the second photon (Figure 5.12). This process means that, as a function of increasing true counts, the counts registered by the detector can go through a maximum with higher rates of photon arrival actually registered as decreased counts. The resulting dead-time curves can be used to correct for dead time effects at more modest count rates. Figure 5.13A shows a comparison of two dead-time curves, both taken from a single detector element (i.e. one discrete diode), one from the first 13-element array detector operating using what was at that time a short shaping time of 1 µs (*a*) and the second from a more modern (faster) germanium array detector operating using $^1/_8$ µs shaping (*b*). The degradation in energy resolution with increasing count rates is shown in Figure 5.13B.

Ideally, the detector should function at a low enough count rate to remain in the pseudo-linear regime, although in an XAS measurement what is really important is the dynamic count-rate range that the detector will sense during the experiment. For example, a very dilute sample, for which the fluorescence signal is very small compared with the background of scatter and other element fluorescence, will have a small dynamic range over change of total counts over the absorption edge and can be run at a higher total incoming count rate without significant dead-time effects. Conversely, a fairly concentrated aqueous sample for which the fluorescence dominates should be run at a lower count rate. Even the very earliest work using a detector that was useful for XAS showed how effective dead-time corrections could be and in modern experiments corrections for detector dead times are frequently used.

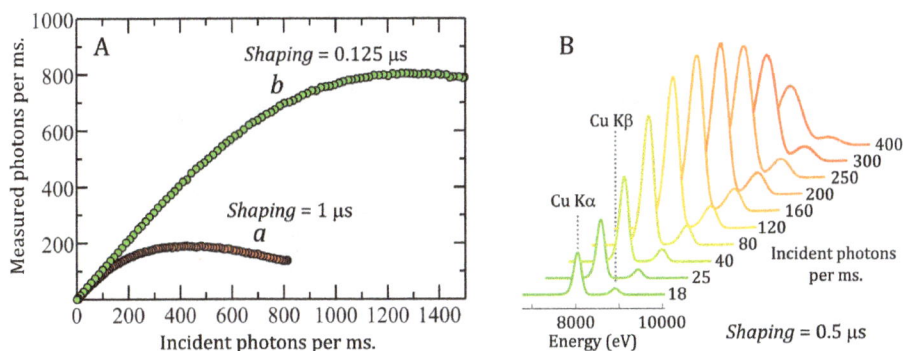

Figure 5.13: Count rates and detector dead times. **(A)** Electronic dead-time curves for different generations of Ge detectors; *a* shows the original 13 element array detector using a 1 µs shaping, and *b* a second-generation Ge array with an $^1/_8$ µs shaping. **(B)** MCA spectra showing degradation of spectroscopic resolution of Cu Kα and Cu Kβ lines at high count rates for 0.5 µs shaping.

5.4.8 Array detectors

X-ray fluorescence is isotropic over all 4π steradians, and in a fluorescence-detected XAS experiment the ideal would be to collect fluorescence over as much of this solid angle as possible. Electronic dead times are equivalent irrespective of the area of the detector used, so that for high count-rate applications, there is little advantage in having individual solid-state detectors with very large active areas, and the solution is to use array detectors. These can be either arrays of individual discrete detectors [3], for example, the 30-element array shown in Figure 5.14, or monolithic constructs with pixels comprised of individual diodes, such as a 100-pixel monolith. The disadvantage of monolithic detector systems is that events may cause partial signals to be registered by adjacent pixels. Solutions to this are to place a collimator between the sample and the detector, or to use digital signal processing hardware that examines the timing of signals from adjacent monolithic pixels, rejecting those that fit the criteria for partial simultaneous signals. Because of such issues, the pre-amplifier rise time of monolithic detectors tends to be longer than that used with discrete arrays, which places constraints upon their performance at high count-rates. Irrespective of the detector used, a modern XAS beamline provides sufficient photon flux to be able to easily saturate the detector. While the simplest way to reduce count rates is to translate the detector away from the sample and thus to reduce the solid angle, this will diminish the desired X-ray fluorescence signal to the same extent as the unwanted background signals. This means that other mechanisms must be used to reject unwanted signal components, which brings us to discuss X-ray filters and Soller slits.

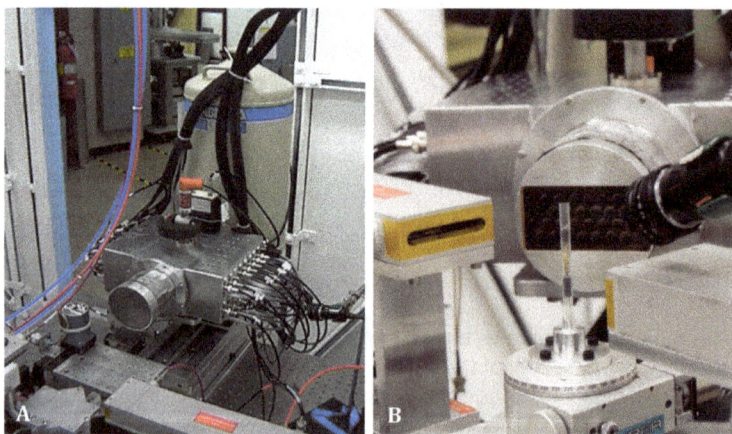

Figure 5.14: A germanium 30-element array detector. (**A**) The overall detector, with the cluster of pre-amplifiers with attached cables, and liquid nitrogen cryogenic cooling (indicated). The active end of the detector is covered by a protective beryllium window. (**B**) A close-up of the active end of the array detector, with the protective outer window removed. The dimples in the inner thin beryllium window are caused by the internal vacuum and show the positions of the individual diodes.

5.4.9 Filters and Soller slits

The use of X-ray filters and Soller slits predates the use of energy-dispersive X-ray detectors by many years, as they are also very useful with non-dispersive detector systems (see above). With a non-dispersive detector, a useful rule of thumb is that if the edge jump is less than 25% of the total signal above the absorption edge, then the use of a filter and Soller slits will give improved signal to noise. In essence the idea is to choose a filter with an absorption edge that lies above the fluorescence energy of interest, but below the energy of the X-ray scatter, resulting in the scatter being preferentially absorbed. This can be understood by reference to the schematic of Figure 5.15 and the MCA spectra shown in Figure 5.16. The X-ray filter typically is placed close to the sample, and the Soller slits between the filter and the detector. In most cases, Soller slits are often made of a relatively X-ray opaque (high Z) material such as metallic silver. Rays emanating directly from the sample that are not absorbed by the filter pass directly through the Soller slits to the detector. Rays that are absorbed by the filter can be re-emitted as filter fluorescence, which will radiate over 4π steradians, with a substantial fraction being absorbed by the Soller slits.

Figure 5.16 shows three MCA spectra. The first (Figure 5.16a) is an aqueous solution containing arsenic (the same sample as Figure 5.10) in which the most intense features are the elastic and inelastic scattered radiation. The desired signal is the arsenic Kα fluorescence line, with the goal of using filters and Soller slits being to suppress the scatter while preserving as much as possible of the arsenic Kα fluorescence. In many cases the choice of filters can be made using a $Z-1$ rule. For example, for copper ($Z = 29$), the element with $Z-1$ is nickel ($Z = 28$). The Cu Kα fluorescence is at ~8,041 eV and the Cu K-absorption edge is close to 9,000 eV; any scatter in a Cu K-edge XAS experiment will be close to this energy. The Ni K-edge is 8,350 eV, so that the nickel-containing filter will strongly absorb the scattered radiation, but only weakly absorb the Cu Kα. Ideally, none of the filter fluorescence lines should be too close to the sample fluorescence line. In our example, the position of the Ni Kβ is 8,265 eV, which can readily be discriminated from the Cu Kα. Likewise, the best filter for selenium ($Z = 34$) K-edge XAS would be an arsenic ($Z = 33$) filter, and for an arsenic ($Z = 33$) K-edge XAS experiment would be germanium ($Z = 32$). Examining the MCA spectrum of our sample in the presence of Ge filters but without Soller slits in place shows a dramatic reduction in the scattered radiation by filter absorption (Figure 5.16b). However, the unwanted scatter signal is partly replaced by filter fluorescence, much of which is eliminated when Soller silts are inserted between the filters and the detector (Figure 5.16c). There is inevitably some attenuation of the As Kα by the filters, and this can be seen as a drop in amplitudes in the lower two traces of Figure 5.16, so that if it is possible to run samples without saturating the detector then filters should not be used; using the detector alone will always give the best signal to noise. However, in the scenario presented, the overall count rate is lowered, and a greater proportion of the remaining counts are due to the desired signal.

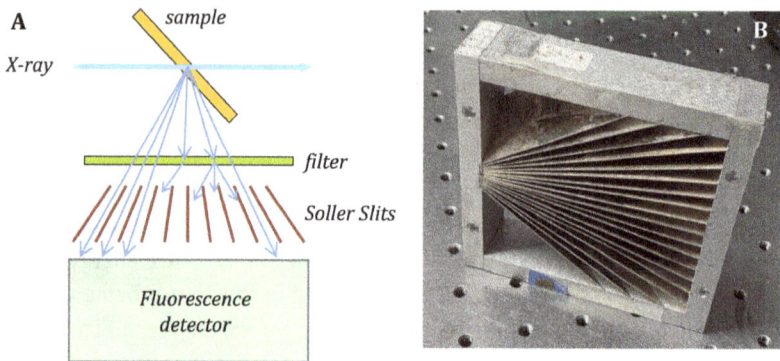

Figure 5.15: Soller slits for X-ray fluorescence measurements. (**A**) A schematic diagram of the use of X-ray fluorescence filter and Soller slits (plan view). (**B**) A photograph of a set of silver Soller slits; these slits are designed to focus on a line source of X-rays coming from a sample oriented at 45° to the incident X-ray beam. The slits are oriented to give a view from the sample side.

No matter how well-made the Soller slits are, some filter fluorescence always gets through, for example, in Figure 5.16 with Ge filter + Soller slits ~90% of the germanium fluorescence is removed but ~10% still remains. For this reason, pure metals do not always make the best filters because of their larger EXAFS at higher energies, in effect providing a modulated background signal, and in many cases a simple compound of the filter element with low-Z elements, such as an oxide, carbonate or hydroxide, and without strong EXAFS, might be a better choice. Countering these considerations is variability in the thickness. Filters made of metallic foils are typically of a precisely known and uniform thickness, so that the performance is reliable, and absorption can be carefully tuned. Conversely, filters employing metal oxides, carbonates and the like are typically made by casting slurry of finely powdered compound in cellulose nitrate/acetone-based liquid cement and allowing it to dry to form a uniform film. Even with great skill, there is an inevitable variability in thickness, so that despite more intense EXAFS, metallic filters may actually be a better choice.

In the very early days of solid-state detector arrays, the available flux on many beamlines could readily saturate the detector. Approximately 10 years after the first array [4], new detector technology increased the available count rates, which meant that solid-state detector arrays could be used mostly without filters, but beamline technology then improved, so that filters were once again needed for most measurements.

5.4.10 Neutral X-ray filters

The use of filters as described in Section 5.4.9 will not help if the solid-state detector is saturating because the sample contains significant quantities of elements with a lower absorption edge than that being studied. If the energy of the unwanted fluorescence is significantly lower in energy than the desired fluorescence signal then, a simple neutral filter placed between the sample and the detector can be used to attenuate unwanted low-energy fluorescence contributions to the overall signal. The most common material to use is aluminium foil or a piece of aluminium sheet. Rolls of aluminium foil are commonplace at synchrotron beamlines, in part because vacuum beamlines use aluminium foil to help attach electrical heating tape used for vacuum chamber bake-out. Typically, the thickness of the foil is given on the container (e.g. 25 μm) and layers of folded foil can be built up until the desired thickness is obtained. Thus, Pushie et al. [5] used Mo K-edge XAS to examine a bromide complex of a molybdenum enzyme, which required a large ~200-fold excess of Br⁻, the fluorescence from which would normally saturate the fluorescence detector. In order to record the Mo K-edge XAS, Pushie et al. employed a 0.5 mm thick metallic aluminium neutral filterer positioned between the sample and the detector. The higher energy of the Mo Kα (17,444 eV) is much more penetrating, so that about 50% of the Mo K was absorbed while more than 85% of the Br Kα (11,909 eV) was absorbed, allowing good quality Mo K-edge XAS data to be collected even in the presence of a large excess of bromide.

5.4.11 Escape peaks

All solid-state detectors generate what are known as escape peaks, which can be quite noticeable with solid-state detectors when operating with incident photon energies that are just above the above the K absorption-edge energy of the detector material (silicon or germanium). At such energies, X-ray photons arriving at the detector will generate X-ray fluorescence. X-ray fluorescence is isotropic, and while some of the fluorescence photons will be directed inwards and be absorbed within the detector, some will be lost through the front of the detector, or in other words they escape. For such escaping photons the energy registered will be that of the incident photon displaced to lower energy by the fluorescence energy.

For silicon-based detectors the Si Kα and Kβ fluorescence at approximately 1,740 and 1,836 eV, respectively, will generate escape peaks. The escape peaks will typically be small features displaced by the fluorescence energies below major features in the spectrum. At these low X-ray energies most materials are reasonably X-ray opaque, and for the Si Kα only 50 μm of silicon will absorb about 98% of X-rays. Because of this, not much fluorescence actually escapes, and escape peaks are thus typically small in silicon-based detectors. With germanium, on the other hand, the Ge X-ray fluorescence is more penetrating, and escape peaks from any major features in the MCA spectrum that fall above the Ge K-edge at 11,103 eV are easily observed, shifted to low energies by the Ge Kα and Kβ energies (9,876 and 10,982 eV, respectively). The X-ray fluorescence escape peaks for a germanium detector are illustrated by the example in Figure 5.17 and appear as two sets of low energy, low intensity copies of the major features. The presence of escape peaks can often obscure the fluorescence lines

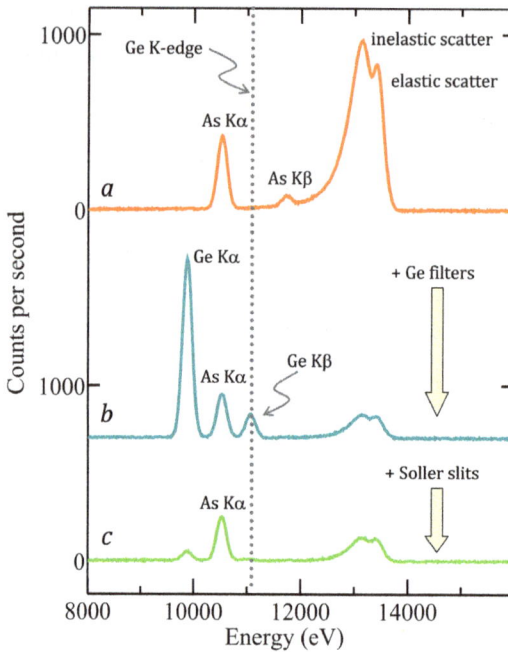

Figure 5.16: MCA spectra with and without filters and Soller slits. Curve *a* shows an MCA spectrum of the same 5-mM aqueous arsenic sample shown in Figure 5.10, with no filters or Soller slits. The effect of adding 3-absorption length germanium filters is shown in *b*, where the intense scatter peak is greatly diminished through filter absorption and is replaced by intense Ge filter fluorescence (Ge Kα and Kβ). The As Kβ peak, whose energy is also above the filter absorption edge, is mostly eliminated. The As Kα peak is also diminished due to absorption by the filter, but less so because its energy falls below the filter absorption edge energy (Ge K-edge). Curve *c* shows the effects of adding Soller slits, which largely eliminate the Ge Kα and Kβ filter fluorescence without affecting the As Kα signal. The overall count rate is substantially lower, with a greater proportion of photons reaching the detector being those desired (i.e. the As Kα).

of lighter elements in the sample; for example, in Figure 5.17 sulfur and potassium are difficult to observe due to superimposed escape peaks.

Figure 5.17 shows X-ray fluorescence escape peaks falling at lower energies separated from the main peaks by the Kα and Kβ energies, with structure from both the elastic scatter and inelastic (Compton) scattering from the sample. Another mechanism by which photons can escape the detector is through Compton X-ray scattering within the detector itself. Compton escape features are not usually observed in MCA spectra recorded within the normal operating energy range of X-ray absorption spectroscopy, and are not visible in Figure 5.17, but at higher X-ray energies this phenomenon can generate what is known as a Compton edge, a broad feature corresponding to the broad nature of the Compton spectrum, lacking the distinctive structure characteristic of X-ray fluorescence escape peaks.

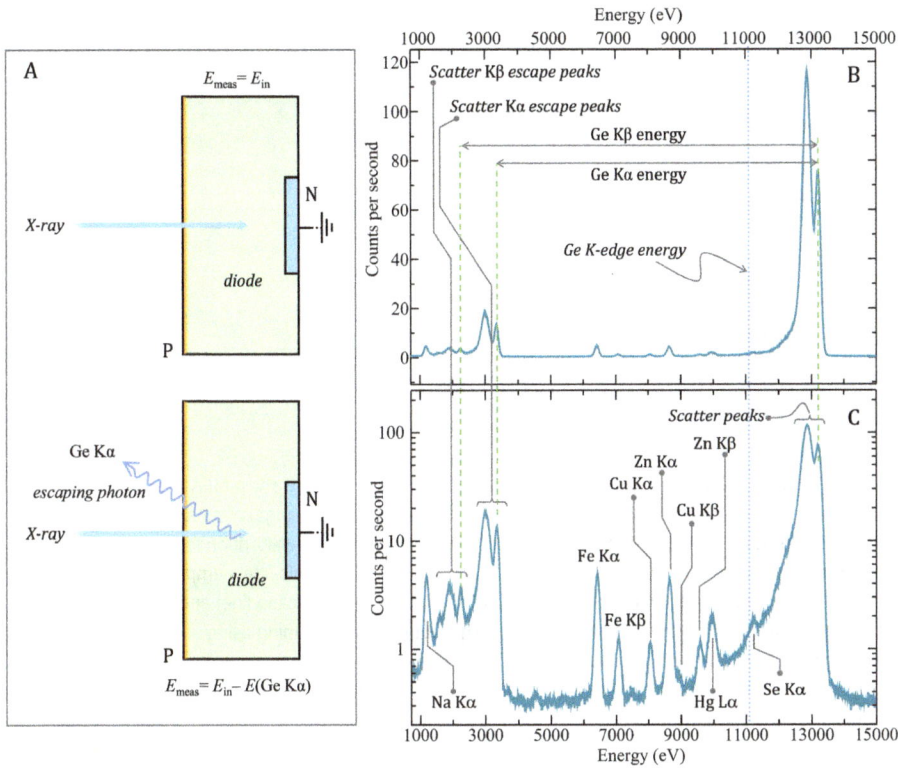

Figure 5.17: Solid-state detector escape peaks. (**A**) A schematic of normal operation (upper) where the incident X-rays are entirely absorbed and (lower) where a Ge Kα photon escapes. (**B**) The MCA spectrum of a tissue sample with the Ge Kα and Kβ escape peaks from the scatter peaks clearly visible, and (**C**) the same MCA spectrum using a logarithmic ordinate scale, which more clearly shows small features in the spectrum.

5.4.12 Solid-state detector readout – binning versus fitting

The digital signal processing technology discussed in Section 5.4.5 can give an MCA spectrum at every energy point in an XAS experiment. This gives the possibility of quantifying the fluorescence signal of interest using a parametric analysis, in which the MCA spectrum is fitted to a sum of the various fluorescence peaks arising from constituents of the sample, plus an appropriate model for elastic and inelastic (Compton) scattering, filter fluorescence, and other background signals. The conventional approach is an integrative analysis, often called binning, in which all detected photons within a specific energy range are summed. In X-ray fluorescence imaging applications [6], there are substantial advantages in such peak fitting endeavors in that this accurate quantification of overlapping features [7]. With XAS, however, the situation can be made complicated by the presence of a fluorescence filter, discussed in

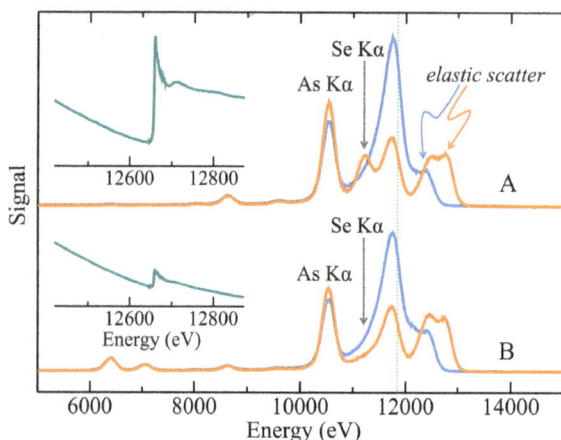

Figure 5.18: MCA spectra of dilute selenium-containing samples. A and B show MCA spectra from two samples with incident X-ray energy at the start (blue) and at the end (orange) of an XAS scan. Spectra were collected using a 100-pixel germanium monolithic detector with 6-absorption lengths arsenic filters plus Soller slits to maintain the count rates in a linear regime. The arsenic filters employed have ~45% transmittance below the arsenic K-edge (shown by the vertical green broken line) and less than 1% transmittance above. The insets show the XAS spectra obtained with binning using a window encompassing the Se Kα peak. In B the Se Kα peak is hard to identify. In the blue low-incident energy traces for both A and B the most intense feature is due to the low-energy part of the Compton scattering, abruptly cut by the arsenic filter.

Section 5.4.9. The filter can abruptly cut the Compton scattering part way through, especially at the low-energy end of the scan, giving rise to unusual looking skewed peaks in the MCA spectra appearing just below the filter edge energy, which, because they can overlap the fluorescence of interest, must be very accurately fitted if fitting is used for detector readout. This can be especially challenging for dilute samples, as is shown in Figure 5.18, and we know of no successful applications of XAS of dilute samples using peak deconvolution fitting of MCA data that have been published to date.

5.5 Scintillation detectors

At one time scintillation detectors could be found on almost every beamline, but nowadays one would be hard-pressed to find one in use on any modern beamline. We consider these here primarily for the sake of completeness, and as such our description will be brief. These detectors use a range of different scintillator materials, such as thallium-doped sodium iodide, or NaI(Tl), crystal which gives off a flash of visible light, or scintillation, in response to an X-ray photon. For NaI(Tl) this occurs in the blue-violet at a wavelength of 415 nm. A range of other scintillators can be used, such as CsI(Tl), but the most common is NaI(Tl). NaI(Tl) crystals are deliquescent and hence are hermetically sealed behind a beryllium window, and interfaced to a photomultiplier tube consisting of a photocathode and electron multiplier. The latter consists of sequentially biased dynodes (secondary electron emitters) that serve to electrostatically accelerate electrons to give a 10^4–10^7 fold signal amplification at the anode. The photomultiplier output can be fed to a spectroscopy amplifier and then to a single-channel analyser and a counter providing an interface to the data acquisition computer. The energy resolution of these detectors is poor compared to solid state detectors. Their maximum photon count rates also are limited by the decay time, which for NaI(Tl) is ~250 ns, although newer scintillation materials such as $LaBr_3(Ce)$ are much faster (~16 ns) and have a higher scintillation yield.

5.6 Advanced photon detectors

We now turn to two different advanced X-ray detector technologies that are capable of dramatically improved energy resolution. The crystal optics that are employed for advanced X-ray spectroscopy will be considered in the next section; here we will focus upon the transition edge sensor (TES) detector and the superconducting tunnel junction (STJ) detector, both of which find most use in the soft X-ray energy regime.

5.6.1 Transition edge sensor detectors

These detectors, known as TES detectors, are effectively microcalorimeters employing a superconducting thin film operating at a temperature just below the transition between its superconducting and normal states, using a constant voltage. The arrival and absorption of a photon heats the TES causing its resistance to increase; as the voltage is constant, the current flowing drops, and this can be measured using a special current preamplifier. The challenge with TES detectors is stabilizing the temperature in the narrow range required; the use of a superconducting quantum interference current amplifier (called a SQUID) provided the breakthrough that allowed the TES detectors to be adopted as practical beamline instrumentation [8]. TES devices provide outstanding energy resolution; for example, Iyomoto et al. [9] have reported a FWHM of 2.37 eV using a ^{55}Fe source, and a similar resolution has been reported by Uhlig et al. [10], easily resolving the Mn Kα1 and Kα2 lines. However, the devices are limited to low count-rate applications. Because of this, large arrays have been developed, such as the 240 pixel TES Mo/Cu bilayer array reported by Lee et al. [11].

5.6.2 Superconducting tunnel junction detectors

For applications that require a higher count rate than possible with a TES detector, the STJ detector may provide advantages. STJs are athermal devices, essentially consisting of two superconducting electrodes separated by an insulating barrier, which is thin enough to permit quantum mechanical electron tunneling. Absorption of an X-ray photon by one of the electrodes excites free charges above the superconducting energy gap, which tunnel through the insulating barrier creating a current pulse that can be registered using an FET-based pre-amplifier in a similar manner to conventional reverse-bias diode-based solid-state detectors. With soft X-rays STJs have resolutions approaching those of TES detectors, but STJ pulse decay times are typically measured in μs, which is much faster than TES detectors, so that individual detectors can operate at ~10^4 counts per second. For X-ray spectroscopy using synchrotron radiation these detectors are also easy to saturate, hence STJ arrays have been constructed, such as the 112 pixel array reported by Friedrich et al. [12]. For hard X-rays STJ detectors can be problematic to manufacture as the first electrode is typically insufficiently thick to permit much hard X-ray absorption, and an additional absorber must be incorporated [13]. Like TES detectors these systems find most use in the soft X-ray regime. Table 5.2 compares the energy resolution at the Mn Kα of the various energy-dispersive detector systems that we have discussed here.

Table 5.2: X-ray fluorescence detector energy resolution.

Detector type	Energy-resolution FWHM (eV) at Mn Kα
Scintillator NaI(Tl)	3,000
Gas proportional counter	1,200
Lithium-drifted silicon Si(Li)	160
High-purity germanium	140
Silicon drift detector (SDD)	130
Superconducting tunnel junction (STJ)	12
Transition edge sensor (TES)	3

5.7 Crystal optics-based photon detectors

Strictly speaking, the systems that we are about to discuss might not really belong in a chapter on detectors, because the detectors are the devices that actually detect the X-ray photons, and these are not what we are about to discuss. Nonetheless, just as we have discussed filters and Soller slits, we will also include detector systems based on diffractive optics here; these devices provide important alternative detector systems and, in some cases, provide genuinely transformative capabilities. Such capabilities arising from high energy resolution fluorescence detection and associated advanced spectroscopies are discussed in more detail in Chapter 14. In this section we focus on the crystal optics and choices of geometric configurations.

We first recap Bragg's law, previously discussed in Section 3.2 (eq. (3.1)):

$$n\lambda = \frac{hc}{E} = 2d\sin\theta_B \qquad (5.6)$$

Here n is the order of the diffraction, λ is the X-ray wavelength, h is Plank's constant, c is the velocity of light, E is the X-ray energy, d is the lattice spacing and θ_B ithe Bragg angle. For a cubic crystal structure, the lattice spacing can be found from the lattice parameter a and the Miller indices h, k, l using the relation $d_{hkl} = a_0 / \sqrt{h^2 + k^2 + l^2}$ (eq. (3.3)). In almost all cases with hard X-ray spectroscopy, the crystals concerned are silicon or germanium, with a_0 = 5.43102 and 5.657582 Å, respectively.

5.7.1 Types of crystal diffractive optics

The types of diffractive optics can be broadly separated into Bragg and Laue, with the basic differences between these shown in Figure 5.19. Laue optics differs from Bragg optics in that the X-ray beam is transmitted through the diffracting crystal. In both types of crystal optics, Bragg diffraction from a specific crystallographic plane is involved.

5.7.2 Log-spiral bent Laue detector systems

In these detector systems a silicon crystal wafer, usually about 100 μm thick, of a particular cut is bent into a carefully chosen logarithmic spiral shape. The log-spiral shape is configured so that X-rays with a particular energy range are diffracted towards the detector, whereas unwanted X-rays, such elastic scatter or fluorescence from other elements, are blocked by the Soller slits (Figure 5.20). The study reported by Kropf et al. [14], used a silicon wafer with an Si(100) surface normal and employed the Si(111) planes for diffraction. These detector systems are individually tuned to the fluorescence energies of particular metal ions, and only a limited range can be studied using a particular detector system; for example, the Hg Lα, As Kα and Se Kα can be examined using the same detector system, but for the Zn Kα one would need to change to a different log-spiral Laue system. Their advantages include the absence of count-rate linearity limitations because the diffracted X-rays normally can be recorded using a non-dispersive detector. They also more effectively discriminate low-energy fluorescence from other elements in the sample, but require a small beam (e.g. ~100 μm × 2–3 mm). For spectroscopy at lower X-ray energies the absorption of the X-rays by the silicon crystal itself becomes significant and these detectors may not be practical for energies that are below the Mn K-edge. The disadvantage of these systems is the need to swap detector systems to study XAS of different elements.

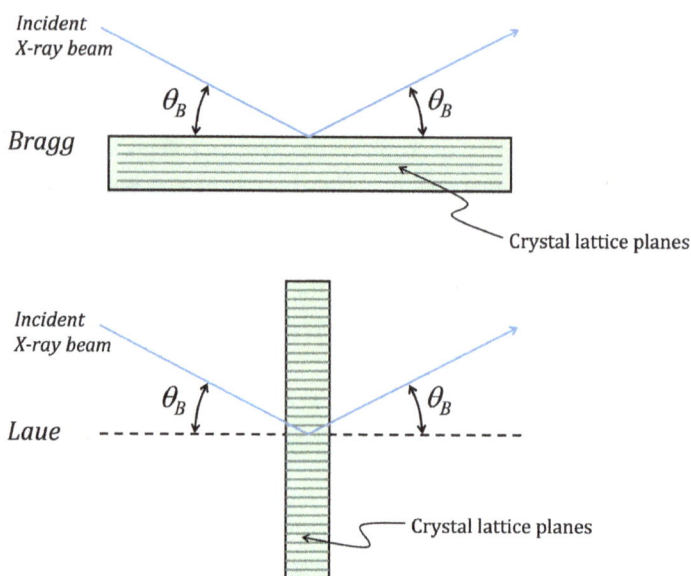

Figure 5.19: Bragg versus Laue diffraction optics.

5.7.3 Bent crystal Bragg optics

The use of Bragg diffraction using flat crystals (e.g. Figure 5.19) has been described in Chapter 3 for X-ray monochromators (Section 3.2). Such devices have excellent resolution, approaching the limits defined by the crystal cut, and are an excellent choice with X-rays of low beam divergence, but are of limited use for phenomena such as X-ray fluorescence because they can only accept a very narrow solid angle. In experiments where the X-ray fluorescence must be measured with good energy resolution, bent crystal optics provide an excellent solution.

The angular width of a Bragg reflection, called the Darwin width, ω_D, decreases with higher order reflections (i.e. decreases with values of higher n in eq. (5.6)). The energy resolution is given as follows:

$$\left|\frac{\Delta E}{E}\right| = \omega_D + \Delta\,\theta_B \cot\theta_B \tag{5.7}$$

This shows that angular divergence $\Delta\theta_B$, arising from the source size (the size of the incident X-ray spot illuminating the sample) and the acceptance of the detector, has a

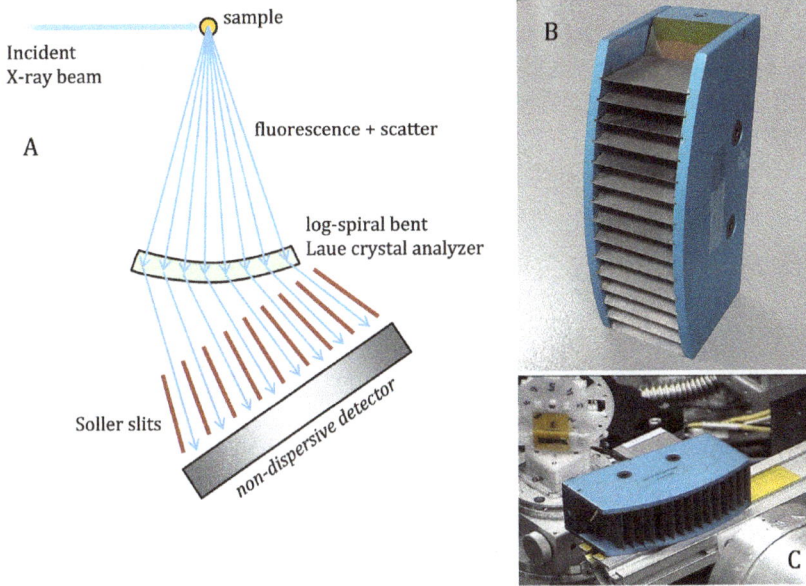

Figure 5.20: Log-spiral Laue detector systems. Panel A shows a plan view of a log-spiral Laue detector system. B and C show photographs of a real system (photos courtesy of Roman Chernikof, Canadian Light Source).

contribution to the energy resolution $\Delta E/E$ that becomes smallest as $\theta_B \rightarrow 90°$. This is important and we will return to it when we discuss advanced X-ray spectroscopy (Chapter 14).

5.7.4 Johann geometry

For a bent single-crystal analyser it is helpful to consider the Rowland circle (Figure 5.21) defined by its radius R_R that is half the bending radius of the diffracting crystal planes R_P, so that $R_P = 2R_R$, with, in the Johann geometry, the bent crystal surface having the same radius as the crystallographic planes in the bent crystal. Since we are interested in the fluorescence, the source of the X-rays is the spot on the sample that is illuminated by the incident X-ray beam; both the source of X-rays and the detector are positioned on the Rowland circle, which typically has a 1 m diameter. This is shown in Figure 5.21A, along with a comparison of experimental data from both a Johann analyser and a germanium solid-state detector (Figure 5.21B), which shows the striking improvement in energy resolution. Importantly, such Johann crystal analysers can measure the X-ray emission from a sample with much better resolution than the natural linewidth. The X-ray detector itself (Figure 5.21A) is typically a silicon drift detector, which can help reject spurious background signals, or another type of photodiode optimized for rapid data acquisition such as an avalanche photodiode.[8] For convenience of display the whole arrangement is shown in Figure 5.21 in the horizontal plane, whereas in most experimental set-ups the Rowland circle is vertically oriented, with the sample and the detector positioned above one another. In many cases an array of analyser crystals might be used to increase the solid angle accepted.

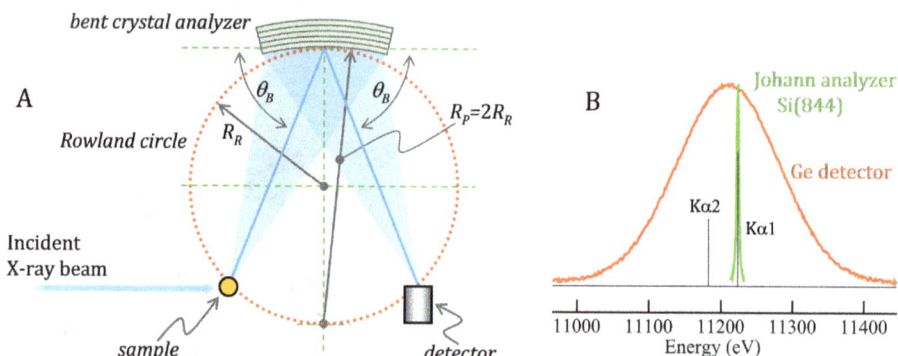

Figure 5.21: Johann geometry bent crystal analyser. A. Schematic diagram (plan view) of a Johann geometry bent crystal analyser, showing the Rowland circle (red dashed line) and Bragg diffraction. For clarity, the values of the Bragg angle θ_B in the figure are deliberately shown as much smaller than the ideal condition of $\theta_B \rightarrow 90°$. B. Experimental data comparing the energy resolutions at the Se Kα of a high-purity germanium solid-state detector (red) and a Johann geometry crystal analyser (green) using a Si(844) crystal ($\theta_B = 85.3°$). Only the Se Kα1 is shown for the Johann geometry analyser, as the Se Kα2 is just outside the practical spectrometer working range for the Si(844) cut analyser crystal.

5.7.5 The Von Hamos spectrometer

Another type of spectrometer geometry that has seen increasing use on various beamlines is Von Hamos geometry. Here the axis of bending for the crystal analyser, the source (the illuminated spot on the sample) and the detector are all aligned on a common axis. The cylindrically bent analyser crystal disperses the X-ray emission spectrum and focuses it along the axis of curvature onto a line-like feature, which is read out using a charge-coupled device (CCD) detector or similar pixel-based readout system. Figure 5.22 shows a schematic of a Von Hamos detector system in which the analyser crystal is cylindrically bent about the axis shown. The advantage of such a Von

---

8 Avalanche photodiodes are high-speed devices with a P+/I/P/N+ structure operated under reverse bias that effectively provide a photo diode with a gain. Photons enter through the P+ contact, electron-hole pairs are generated in the I layer and are effectively amplified in the P layer. They provide high-speed capabilities with response times measured in ns.

Hamos spectrometer system is that the entire emission spectrum can be read out in a single shot, which contrasts with the Johann analyser system described in 5.9.3 which typically needs to be scanned to record an emission spectrum, and hence takes much longer to acquire. The disadvantage is that the energy resolution is typically not as good as a spectrometer employing Johann geometry, so this detector system may not be suitable for some advanced spectroscopy applications. Nevertheless, the energy resolution of the Von Hamos is more than sufficient to separate fluorescence lines which, under the modest resolution of a solid-state detector, would interfere with each other, such as Co in the presence of Fe. This allows conventional XAS and EXAFS to be measured on dilute samples, which would otherwise be intractable with a solid-state detector.

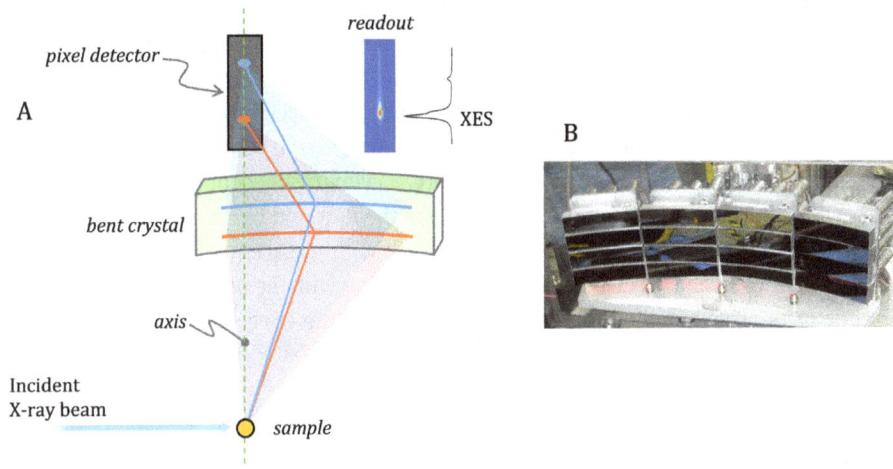

Figure 5.22: Von Hamos X-ray spectrometer. (A) Schematic diagram of a Von Hamos X-ray spectrometer. (B) A 12-crystal Von Hamos array developed for the linear coherent light source (photo courtesy of Dimosthenis Sokaras, Stanford Synchrotron Radiation Lightsource).

5.7.6 Johansson geometry

The last bent crystal analyser system considered here is the Johansson geometry. With the previously discussed Johann geometry, the bent crystal surface has the same radius R_P as the crystallographic planes, which is twice the radius of the Rowland circle R_R. With Johansson geometry, the bent crystal analyser surface is specially shaped so as to match the Rowland circle. This means that with all of sample, analyser and detector on the Rowland circle the angle of incidence equals the angle of reflection, with constant θ_B over the analyser crystal surface, reducing aberrations that are inherent in the Johann design (for example). In this case the aforementioned ideal condition of $\theta_B \rightarrow 90°$

for good energy resolution still holds. A schematic diagram comparing similar analyser crystals with Johann and Johansson geometry is shown in Figure 5.23.

Figure 5.23: Schematic diagrams comparing Johann and Johansson geometry analysers.

With the Johansson geometry, the center of the crystallographic bend radius need not coincide with the centre of the Rowland circle, and considerable variation in designs are possible. The caveat to these advantages is that fabrication of Johansson analyser crystals can be significantly more challenging than for comparatively simple Johann geometry analysers, and in particular surface imperfections in the Johansson analyser can significantly degrade performance. Of interest are tender X-ray spectrometers that have used the so-called off-Rowland or inside-Rowland geometry [15, 16], which allows some previously discussed limitations to be overcome. Here, a cylindrically bent Johansson analyser is used with the source (the sample) placed inside the Rowland circle. Because the source is displaced, θ_B changes along the analyser surface with a corresponding energy dependence, enabling an energy-dispersive mode of operation with the output recorded using a pixel detector. This mode of operation also suppresses the effects of source size on the energy resolution, even for quite low values of θ_B [15, 16].

5.8 Electron detectors

Finally, we turn to electron detectors. Electron yield XAS differs in a fundamental manner from the photon-based detection methods discussed so far, because electrons generally are much less penetrating than photons. This means that electron yield shows surface sensitivity, as the signal derives from the top 20 to 100 Å of the sample. Figure 5.24 shows an example of the surface sensitivity of electron yield. Moreover, no windows are possible as even very thin windows will completely wipe out the electron yield signal. As we have discussed in Chapter 4, electron yield decreases with increasing atomic number; thus, electron yield is more useful for light elements.

5.8.1 Gas-amplified total electron yield detector

While most hard and tender X-ray XAS measurements depend upon X-ray transmittance and X-ray fluorescence, electron yield using gas amplification detectors are sometimes used. Specific examples are to deliberately confer surface sensitivity, or to

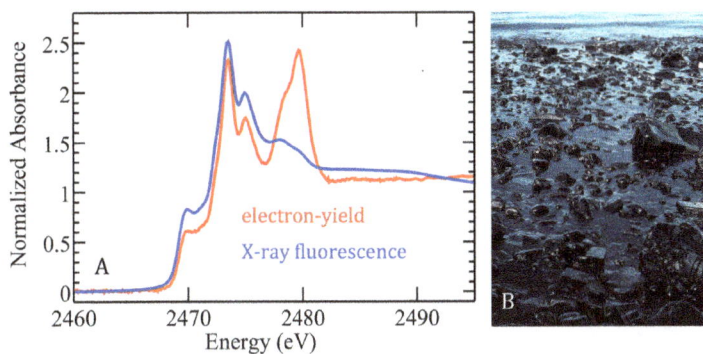

Figure 5.24: Surface sensitivity of electron yield versus X-ray fluorescence. Panel A compares the sulphur K-edge XAS of the same sample measured using electron yield and X-ray fluorescence detection. The sample is a weathered rock from a contaminated beach sampled several months after the 1989 Exxon Valdez oil spill (B). Note that B, showing a heavily oiled beach, was taken just after the oil spill. The electron yield shows surface oxidation of sulphur forms in the crude oil (at ~2,480 eV), whereas the fluorescence probes deeper into the sample, and shows less oxidized products.

measure an undistorted spectrum of a concentrated solid in the tender X-ray range, if a thin film is impractical and fluorescence signal exhibits self-absorption distortion. These detectors operate under atmospheric pressure using an appropriate low cross-sectional gas, such as hydrogen or helium.[9] High-energy photoelectrons from the sample interact with the helium gas to give a cascade of ions and electrons, with the latter picked up by a metal grid or wire placed in front of the sample and biased relative to the sample. The current flow is then fed to a current-to-voltage amplifier and readout proceeds just as with a gas ionization chamber. These detectors are easy to use and can be relatively trouble free. A schematic diagram of a simple gas-amplified total electron yield detector is shown in Figure 5.25. Typically, the detector would be operated with flowing helium gas surrounding the parts shown, or in an open configuration in a helium flight path. Samples, in the form of powdered solids, would typically be mounted on conductive carbon adhesive tape to ensure good electrical contact with the backplane. In some cases, samples might need to be ground with graphite to increase conductivity.

5.8.2 The channeltron electron multiplier

Channeltron detectors (Figure 5.26) are standard on the vacuum beamlines and are typically used for soft X-ray XAS experiments. The technology of a vacuum soft X-ray

---

9 Hydrogen, H_2, is actually a better choice than helium, but due to its combustible nature in the presence of atmospheric oxygen, it is not typically used.

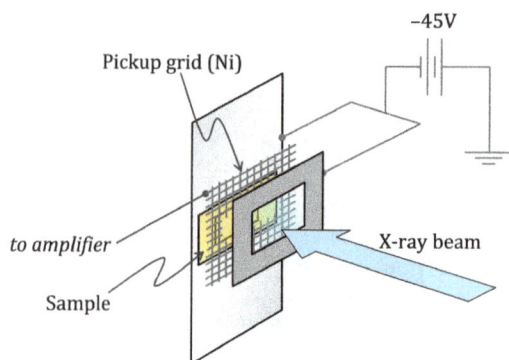

Figure 5.25: A schematic diagram of a gas-amplified total electron yield detector. The pickup grid has a hole cut in it to prevent it being struck by the incident X-ray beam. In many cases traces of the element of interest might be present on the mesh, resulting in a background signal when the mesh is illuminated – for example, with sulphur K-edge XAS the nickel mesh will typically show a trace sulphate signal, which is prevented when a hole allowing passage of the beam is present.

beamline differs somewhat from the hard X-ray environment and merits a brief description of a typical set-up. In contrast to hard X-ray beamlines there can be very few apertures and windows in most vacuum soft X-ray beamlines. Typically, instead of the ion chambers used in hard X-ray beamlines, I_0 is measured as the photocurrent from a gold grid, whose surface can be refreshed through evaporation of small quantity of gold to form a fresh surface. XAS is measured either by fluorescence using a solid state detector or by monitoring total electron yield either directly using a channeltron or indirectly by measuring the drain current. In essence, the channeltron consists of an electron multiplier, similar in some ways to the electron multiplier in the photomultiplier tube discussed in Section 5.4, but cleverly made in a single compact piece. Thus, instead of multiple discrete dynodes used in a photomultiplier tube the channeltron uses what amounts to a continuous dynode. A channeltron is fabricated by treating lead glass precursor with hot hydrogen gas to create a semiconducting layer along which a high voltage is applied to drive the electron multiplier. Channeltron devices usually have a funnel-shaped entry point, and are curved to minimize noise from desorbed gasses. The technology of channeltron devices is decades old, with the first reliable devices dating from 1958. Typically they have a useful functional lifetime in a beamline environment of 2–3 years, after which their accumulated damage may necessitate their replacement, but when functioning normally they can give excellent quality electron yield data.

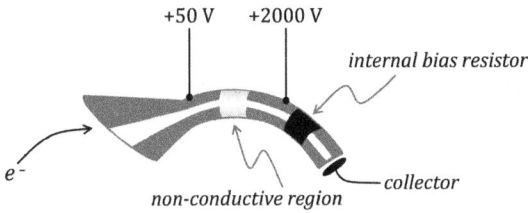

Figure 5.26: Schematic simplified diagram of a channeltron electron multiplier. The external non-conductive region is needed to internalize the electrostatic gradient within the channeltron.

5.8.3 Drain current

In a vacuum beamline environment, with a well-isolated sample connected to earth through a current-to-voltage amplifier, XAS can be detected by monitoring the current flow to earth. The drain current provides a readily reliable method of detection.

5.8.4 Energy-dispersive electron energy analysers

We briefly discuss three methods for energy-dispersive electron analysis. The first is a simple partial electron yield detector, which operates using electrically biased grids, usually two, placed between the source (here, the illuminated spot on the sample) and the electron detector (e.g. a channeltron). The first grid is operated at ground potential and the second at a sufficient potential to elicit a cut-off energy in the electron yield, resulting in the desired partial electron yield. More sophisticated devices include so-called cylindrical mirror analysers, and hemispherical electron energy analysers, which are now used on many beamlines.[10] Hemispherical electron analysers contain two concentric hemispheres to which different voltages are applied; the resulting electric field alters the trajectories of electrons, depending upon their kinetic energies. These devices are essential for techniques such as angle-resolved photoemission spectroscopy but are little used for X-ray absorption spectroscopy.

---

10 The shining metallic domes of hemispherical electron energy analysers are a familiar sight on many vacuum beamlines; their identity is usually the second question that the novice will ask about these beamlines.

Further reading

Knoll, G. F.; *Radiation Detection and Measurement*. Fourth Edition. John Wiley and Sons: Hoboken, New Jersey, USA. 2010

References

[1]   Soller, W. A new precision X-ray spectrometer. *Phys. Rev.* **1924**, *24*, 158–167.
[2]   Tse, J. J.; George, G. N.; Pickering, I. J. Use of Soller slits to remove reference foil fluorescence from transmission spectra. *J. Synchrotron Radiat.* **2011**, *18*, 527–529.
[3]   Müller, O.; Stötzel, J.; Lützenkirchen-Hecht, D.; Frahm, R. Gridded ionization chambers for time resolved X-ray absorption spectroscopy. *J. Phys.: Conf. Ser.* **2013**, *425*, 092010/1–4.
[4]   Cramer, S. P.; Tench, O.; Yocum, M.; George, G. N. A 13-element Ge detector for fluorescence EXAFS. *Nucl. Instrum. Meth.* **1988**, *A266*, 586–691.
[5]   Pushie, M. J.; Doonan, C. J.; Wilson, H. L.; Rajagopalan, K. V.; George, G. N. Nature of halide binding to the molybdenum site of sulfite oxidase. *Inorg. Chem.* **2011**, *50*, 9406–9413.
[6]   Pushie, M. J.; Pickering, I. J.; Korbas, M.; Hackett, M. J.; George, G. N. Elemental and chemically specific X-ray fluorescence imaging of biological systems. *Chem. Rev.* **2014**, *114*, 8499–8541.
[7]   Crawford, A. M.; Sylvain, N. J.; Hou, H.; Hackett, M. J.; Pushie, M. J.; Pickering, I. J.; George, G. N.; Kelly, M. E. A comparison of parametric and integrative approaches for X-ray fluorescence analysis applied to a Stroke model. *J. Synchrotron Radiat.* **2018**, *25*, 1780–1789.
[8]   Irwin, K. D.; An application of electrothermal feedback for high resolution cryogenic particle detection. *Appl. Phys. Lett.* **1995**, *66*, 1998–2000.
[9]   Iyomoto, N.; Bandler, S. R.; Brekosky, R. P.; Brown, A.-D.; Chervenak, J. A.; Finkbeiner, F. M.; Kelley, R. L.; Kilbourne, C. A.; Porter, F. S.; Sadlier, J. E.; Smith, S. J. Close-packed arrays of transition-edge X-ray microcalorimeters with high spectral resolution at 5.9 keV. *Appl. Phys. Meth.* **2008**, *92*, 013508/1–3.
[10] Uhlig, J.; Fullagar, W.; Ullom, J. N.; Doriese, W. B.; Fowler, J. W.; Swetz, D. S.; Gador, N.; Canton, S. E.; Kinnunen, K.; Maasilta, I. J.; Reintsema, C. D.; Bennet, D. A.; Vale, L. R.; Hilton, G. C.; Irwin, K. D.; Schmidt, D. R.; Sundström, V. Table-top ultrafast X-Ray microcalorimeter spectrometry for molecular structure. *Phys. Rev. Lett.* **2013**, *11*, 138302/1–5.
[11] Lee, S.-J.; Titus, C. J.; Alonso Mori, R.; Baker, M. L.; Bennett, D. A.; Cho, H. S.; Doriese, W. B.; Fowler, J. W.; Gaffney, K. J.; Gallo, A.; Gard, J. D.; Hilton, G. C.; Jang, H.; Joe, Y. I.; Kenney, C. J.; Knight, J.; Kroll, T.; Lee, J.-S.; Li, D.; Lu, D.; Marks, R.; Minitti, M. P.; Morgan, K. M.; Ogasawara, H.; O'Neil, G. C.; Reintsema, C. D.; Schmidt, D. R.; Sokaras, D.; Ullom, J. N.; Weng, T.-C.; Williams, C.; Young, B. A.; Swetz, D. S.; Irwin, K. D.; Nordlund, D. Soft X-ray spectroscopy with transition-edge sensors at Stanford Synchrotron Radiation Lightsource beamline 10-1. *Rev. Sci. Instrum.* **2019**, *90*, 113101/1–11.
[12] Friedrich, S.; Harris, J.; Warburton, W. K.; Carpenter, M. H.; Hall, J. A.; Cantor, R. 112-Pixel arrays of high-efficiency STJ X-ray detectors. *J. Low Temp. Phys.* **2014**, *176*, 553–559.
[13] Angloher, G.; Hettl, P.; Huber, M.; Jochum, J.; Feilitzsch, F. V.; Mößbauer, R. L. Energy resolution of 12 eV at 5.9 keV from Al-superconducting tunnel junction detectors. *J. Appl. Phys.* **2001**, *89*, 1425–1429.
[14] Kropf, A. J.; Finch, R. J.; Fortner, J. A.; Aase, S.; Karanfil, C.; Segre, C. U.; Terry, J.; Bunker, G.; Chapman, L. D. Bent silicon crystal in the Laue geometry to resolve X-ray fluorescence for X-ray absorption spectroscopy. *Rev. Sci. Instrum.* **2003**, *74*, 4696–4702.

[15] Kavčiča, M.; Budnar, M.; Mühleisen, A.; Gasser, F.; Žitnik, M.; Bučar, K.; Bohinc, R. Design and performance of a versatile curved-crystal spectrometer for high-resolution spectroscopy in the tender X-ray range. *Rev. Sci. Instrum.* **2012**, *83*, 033113/1–8.

[16] Nowak, S. H.; Armenta, R.; Schwartz, C. P.; Gallo, A.; Abraham, B.; Garcia-Esparza, A. T.; Biasin, E.; Prado, A.; Maciel, A.; Zhang, D.; Day, D.; Christensen, S.; Kroll, T.; Alonso-Mori, R.; Nordlund, D.; Weng, T.-C.; Sokaras, D. A versatile Johansson-type tender X-ray emission spectrometer. *Rev. Sci. Instrum.* **2020**, *91*, 033101/1–12.

6 The X-ray absorption spectroscopy experiment – I

6.1 Introduction

Previous chapters have examined the sources of synchrotron radiation (Chapter 2), X-ray beamlines (Chapter 3), some fundamentals of interactions (Chapter 4) and detector technology (Chapter 5). This chapter will examine a typical experiment from the point of view of a user arriving at a facility. We will consider the various equipment arrangements that users might expect to find at the beamline, and how they are used together in an X-ray absorption spectroscopy experiment. We will first consider a typical experimental set-up for the hard X-ray regime, followed by tender and then, soft X-ray experiments. The division between these types of experiment is quite arbitrary; the designation of the region between soft X-ray and hard X-ray has recently become known as "tender X-ray" and this is now widely accepted, and we adopt this nomenclature here. In part, the differences in experimental approach are governed by how penetrating X-rays are at the different energies. Figure 6.1 shows calculated curves for the transmittance of helium gas and air both at a pressure of 1 atmosphere.

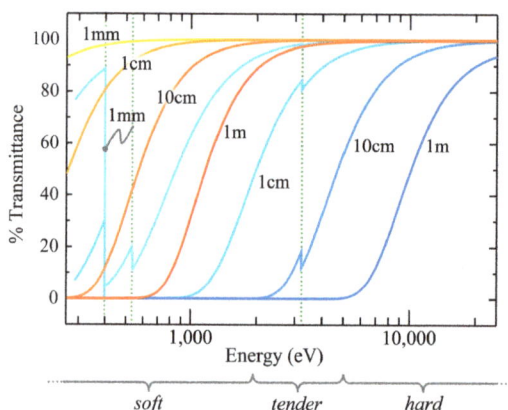

Figure 6.1: Percentage transmittance vs X-ray energy for different thicknesses of air and helium gas at 1 atmosphere. The blue curves show the transmittance for air with path lengths of 1 mm, 1 cm, 10 cm and 1 m, with the K-edges of oxygen, nitrogen and argon indicated by the green broken vertical lines, visible as the step discontinuities in the transmittance. The orange curves show the transmittance for the same path lengths of helium gas. The arbitrary experimental divides between soft, tender and hard X-rays are shown below the plot.

Figure 6.1 illustrates some of the challenges faced by experimenters using soft X-rays and to a lesser extent, tender X-rays. With incident X-ray energies below 1 keV, even

https://doi.org/10.1515/9783110570441-006

1 cm of air will severely attenuate the X-ray beam. We comment further on these challenges below.

6.2 Hard X-ray experiments

Almost all hard X-ray experiments are carried out in an experimental hutch.[1] The experimental hutch has already been briefly mentioned in Section 3.6, under radiation shielding and personnel protection. Hutches are needed with hard and tender X-ray infrastructure, but typically not with soft X-ray beamlines. In the early days of XAS experiments, users would arrive at a synchrotron facility to a bare hutch into which the experimental set-up would need to be installed in the hours that followed. Fortunately, nowadays, the arriving user typically will instead be presented with a hutch that contains a working experimental set-up, almost ready for measurements. Figure 6.2 shows a schematic of typical experimental set-up for XAS, while Figure 6.3 shows photographs of XAS set-ups inside a hard X-ray beamline hutch.

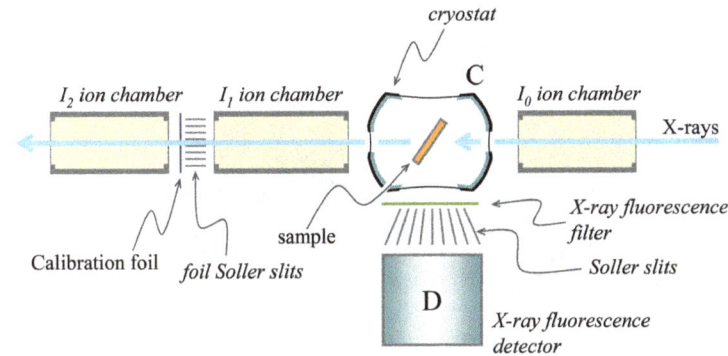

Figure 6.2: Schematic plan view of a hard X-ray XAS experimental set-up. C, cryostat; D, detector.

While a number of experimental arrangements for XAS of bulk samples are available, many use, close to what could be considered, a standard experimental configuration, shown in Figure 6.2. The X-ray beam enters the experimental set-up on the right of both figures. Its intensity is first measured by a simple detector, which might be a gas ioniza-

---

1 One of us was surprised during a 2006 visit to the Siam Photon source that the XAS beamline there (BL-8) had no experimental hutch. This beamline accessed both tender and hard X-rays, using a double crystal monochromator, with a choice of InSb(111), Si(111) or Ge(220) crystals. It is powered by a 1.44T bend magnet, and the energy of the storage ring is 1.2 GeV, giving ~10^9 photons per second at 10 keV. This is a relatively low photon flux, and consequently, at that time, in the opinion of the facility, a hutch was not needed for personnel protection.

tion chamber, or a photodiode, which we call I_0. Some experiments use a scatter target, which might be a polymer film inclined at 45° to the incident beam, or even a volume of air, with a photodiode placed at 90° to the incident beam monitoring I_0, which is proportional to the X-ray scatter. Other experimental set-ups use photodiodes that have been specially thinned in order to only slightly attenuate the incident X-ray beam, transmitting most photons to illuminate the sample. Following I_0, the beam then passes through the sample within the cryostat (in our example), with the downstream intensity of the beam then measured by another ion chamber, or perhaps a thinned photodiode, as I_1. The beam then passes through an energy calibration standard, often a metal foil, made of the element being studied, and the intensity is again measured at I_2. The calibration standard is used to precisely and simultaneously calibrate the energy of the X-ray beam throughout a series of measurements. In our example (Figure 6.2), the X-ray fluorescence is monitored by a solid-state detector, which is positioned at 90° to the incident beam to minimize the scattered radiation. Photographs of two experimental set-ups on the same beamline are shown in Figure 6.3.

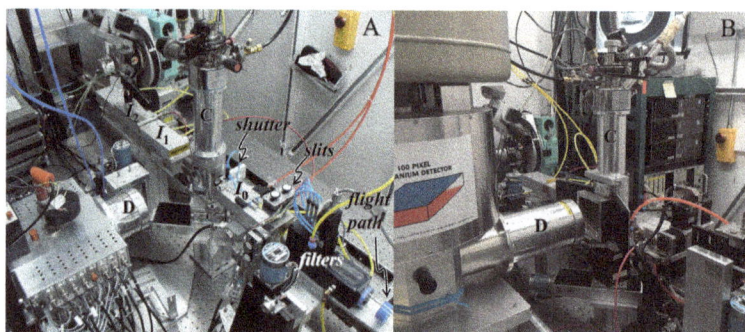

Figure 6.3: Hard X-ray experimental set-ups on beamline 9-3 at the Stanford Synchrotron Radiation Lightsource (SSRL). (A) The set-up with a 30-element germanium array detector and (B) a similar set-up but using a 100-pixel germanium monolithic detector (photograph (A) courtesy Dr Matthew Latimer, SSRL; and photograph (B) courtesy Dr Simon George, Simon Scientific). In (A), ion chambers (I_0, I_1 and I_2) are labelled, and in both (A) and (B), cryostat C and solid-state fluorescence detector D are shown. During measurements, the fluorescence detector is positioned close to the cryostat; in (A), the detector D has been moved back to allow visualization of the large cryostat window, and the Soller slit/X-ray fluorescence filter assembly has been removed for the same reason, while in (B), the Soller slit/X-ray fluorescence filter assembly is in place. The set-up on this beamline includes several motorized stages to facilitate alignment, incorporates various cameras to monitor the experiment and is enclosed within a lead-lined radiation-proof hutch.

Even in the hard X-ray regime, long passages through air can attenuate the X-ray intensity (Figure 6.1) and helium-filled or evacuated flight paths are usually employed for any long passages of the X-ray beam, and in Figure 6.3, such a flight path can be seen in the photograph (the transparent tube entering bottom right of the picture). Other hardware that is typically present include precision slits to define the X-ray beam, upstream of I_0, neutral X-ray filters to decrease the photon flux and a fast shutter downstream of I_0 that can be closed to protect the sample from the beam when data is not being collected.

The advantages of low temperatures with XAS are now widely realized. With some applications, such as operando experiments on catalyst systems, the experimenter may not have a choice. But in general, for static samples, cooling samples to cryogenic temperatures gives substantial advantages. With EXAFS, freezing out of vibrational modes gives lower Debye-Waller factors, and hence large amplitudes and more pronounced features in the Fourier transform. Figure 6.4 shows an example illustrating the effects of sample temperature on the Bi L_{III} EXAFS Fourier transforms of a mixed-metal bismuth oxide. Low temperatures can also offer some protection against radiation damage. In both cases, the use of liquid helium cryogenics provides improvements, compared with liquid nitrogen.

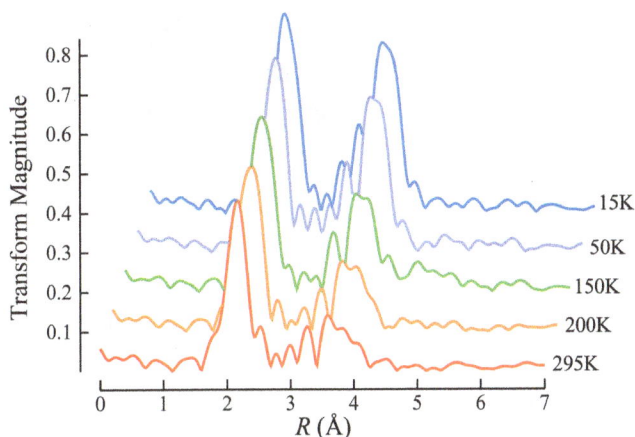

Figure 6.4: Effects of sample temperature on the EXAFS Fourier transforms of a mixed-metal bismuth oxide. Transforms were computed over the k-range 0–16 Å^{-1}, using Bi–O phase correction. The outer shell EXAFS at $R{\sim}4$ Å become increasingly prominent at low temperatures.

Our example set-up is designed to measure dilute samples by measuring the X-ray fluorescence (Figure 6.2, corresponding to the photographs shown in Figure 6.3), which means that the cryostat must be equipped with suitably thin windows and large windows to transmit the X-ray fluorescence with an acceptable solid angle to the fluorescence. Liquid helium flow cryostats for XAS were first used by one of us in 1984, and apart from some minor differences, the basic design has changed very little since those early days. Various other options are available for achieving cryogenic temperatures, each with its own advantages and disadvantages for the XAS experiment. These alternatives include pulse-tube refrigerators, which are in use at Diamond (UK) and the Australian Synchrotron, and both helium and nitrogen cryojet or cryostream coolers, which might be used for small samples. The helium flow cryostat has the advantages of low

mechanical vibration and rapid sample change. The major disadvantage is the requirement for liquid helium, which is expensive and is becoming increasingly precious as
global supplies are depleted. Figure 6.5 shows one model of helium flow cryostat, with
collection of good solid angle X-ray fluorescence and which allows rapid sample change
(e.g. 10 min.). This particular cryostat can be used with a compact liquid helium generation system (Figure 6.5C), which takes the helium gas exhaust from the cryostat and re-
generates liquid helium in a closed cycle, making the use of the cryostat both cost-
effective and environmentally appropriate by conserving helium.

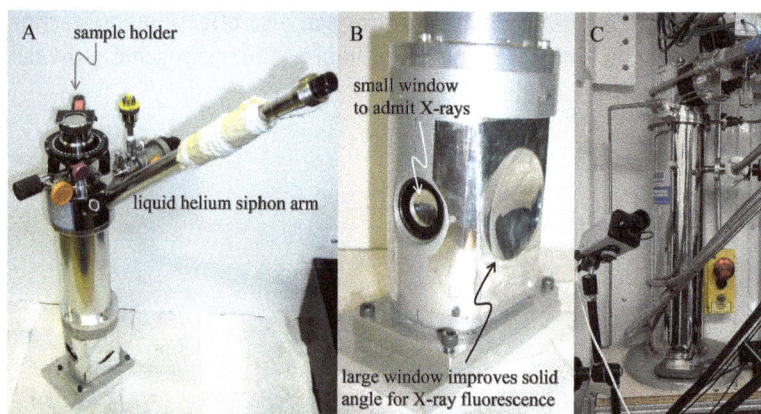

Figure 6.5: Liquid helium flow cryostat. (A) The cryostat prior to installation in the experimental hutch.
(B) A close-up of the lower part of the cryostat, including both a small window to admit/transmit the
incident X-ray beam and a large window to allow measurement of the emitted X-ray fluorescence. Both
are typically fabricated from aluminized Kapton (polyimide) film. The internal vacuum pulls the thin
windows inwards. (C) A closed cycle liquid helium generator installed in the beamline 7-3 experimental
hutch at SSRL. Photographs (A) and (B) are provided with courtesy of Dr Matthew Latimer, SSRL.

6.3 Tender X-ray experiments

The experimental set-up on a tender X-ray beamline is similar in many respects to the
hard X-ray example. At tender X-ray energies, helium is reasonably transparent,
while air absorbs the beam significantly (Figure 6.1). While short air-filled path-
lengths of perhaps one or two mm are sufficiently transmitting to be tolerated, a helium-filled flightpath is typically needed. Thin polymer windows must be used in
place of the rather thicker windows that are found on hard X-ray experiments. For
example, 6.3 μm-thick polypropylene windows are often used, which transmit ~89%
of an X-ray beam at the phosphorus K-edge (2,145 eV) and ~92% at the sulfur K-edge
(2,470 eV). In practice, the sample is usually maintained in a chamber containing an
atmosphere of helium, with many experiments being carried out at room temperature. With tender X-ray experiments, it can be challenging to make samples suffi-

ciently thin to transmit the X-ray beam, thus few measurements are done using transmittance in these energy regimes. In addition, since most samples will effectively
block the X-ray beam with no transmittance, the simultaneous transmission measurement of the calibration foil placed behind the sample is not feasible. Figure 6.6 shows
a plan-view schematic of a tender X-ray experimental set-up and Figure 6.7 shows a
photograph.

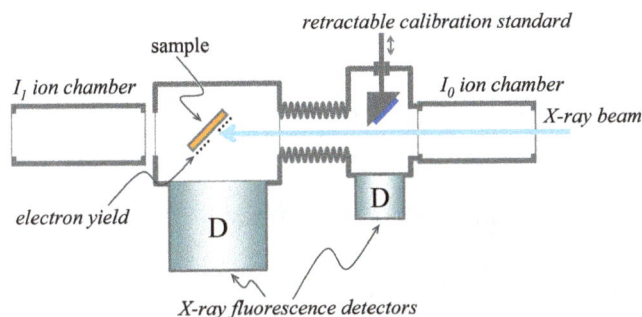

Figure 6.6: Schematic plan view of a tender X-ray experimental set-up. The X-ray beam enters the set-up
from the right side of the figure, passes through the I_0 gas ionization chamber, then through a thin
polymer X-ray window and enters a separate chamber where a retractable calibration standard is located
(retracted to pass the X-ray beam in the figure). In Figure 6.7, this calibration standard is located vertically
and approaches the X-ray beam from above, but is shown in a horizontal location in this schematic to
simplify the figure. The beam then passes through a flexible set of bellows, which function to allow the
sample chamber to be moved, and then intersects the sample in the sample chamber. The X-ray
fluorescence detectors D are shown for both the calibration standard (smaller, upstream) and the sample
(larger, downstream). Also in the sample chamber is a pickup grid to detect helium gas-amplified electron
yield (Section 5.8.1). All of these compartments are maintained in an atmosphere of helium. Downstream
of the sample chamber, additional gas ionization chamber detectors can be placed for use with any
samples that do transmit X-rays. In our schematic, just the I_1 detector is shown.

The use of photodiodes, which are sensitive to visible light, normally means that the
experimental hutch must be darkened during measurements. We note that the set-up
shown in Figure 6.7 is relatively impervious to stray light, with possible entry locations
effectively limited to the exit window of the sample chamber. Nonetheless, if the hutch
is not darkened and any windows are not covered, then the sensitive photodiode detectors will pick up stray signal from the small quantities of visible light present.

We have noted that transmittance samples for tender X-rays are difficult to prepare
because they need to be physically thin. For this reason, and as discussed by George
et al. [1], the use of transmittance detection with tender XAS measurements can be
fraught with problems and potential artefacts (see Section 12.2). Notwithstanding these
difficulties, with suitable care in preparation, good transmittance samples can be prepared. Figure 6.8 shows an example of a measurement of dibenzyl disulfide, cast from
an acetone solution with nitrocellulose, in a 5 μm thick film. The figure shows simulta-

neous detection of fluorescence, electron yield and transmittance using an experimental set-up very similar to those shown in Figures 6.6 and 6.7.

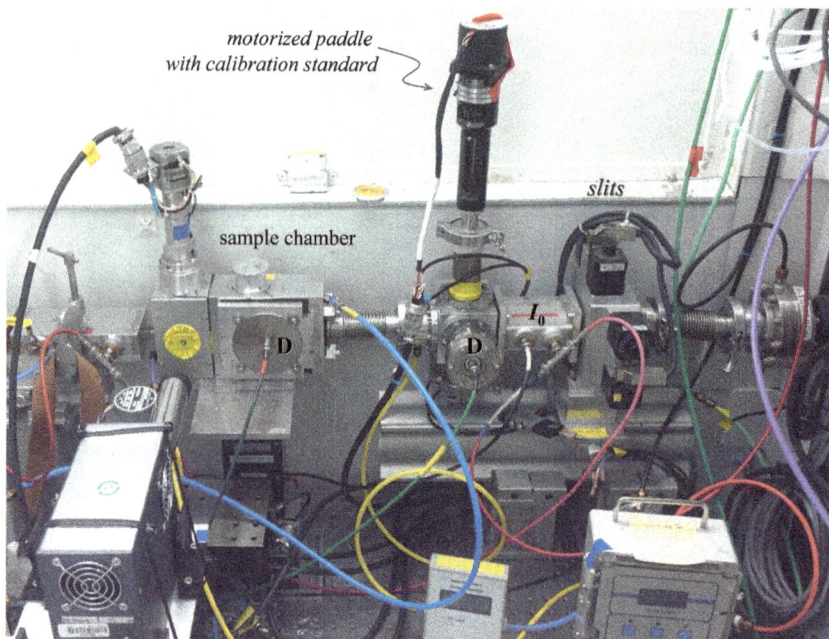

Figure 6.7: Photograph of a tender X-ray experiment on SSRL's beamline 4–3. The experiment involves room temperature measurements at the sulfur K-edge. Beamline 4-3 is a wiggler side station and accepts the outboard portion of the radiation fan from the wiggler source. Because of this, the hutch wall is close to the experimental set-up and can be seen directly behind it. The monochromatic X-ray beam enters the set-up through the beam pipe on the right side of the photograph, where its size is defined by motorized precision slits (indicated). I_0 is then measured using a helium-filled gas ionization chamber, and downstream of this, the beam enters the chamber containing the vertically retractable calibration standard with its own detector (D), which in this case is a photodiode. The beam then passes through a flexible set of bellows to the sample chamber, where it impinges on the sample, with the resulting fluorescence measured by the large photodiode (PIPS) detector (also D). In this experimental set-up, a second chamber is present with a silicon drift detector array, which is retracted and closed off. No equipment for electron yield detection is installed in the set-up shown. Photograph, courtesy of Ms. Linda Vogt, University of Saskatchewan.

Cryogenic temperatures in the tender X-ray regime can be achieved by various means; various cold finger designs using circulating liquid nitrogen have been used. Notwithstanding these efforts, probably the most effective is a commercial helium cryostream cooler, which can usually approach 30 K. In this case, the environment around the sample must be enclosed but vented to maintain a helium atmosphere (from the exhaust of the cryostream) and to prevent X-ray attenuation by air; this is usually accomplished using a simple plastic bag.

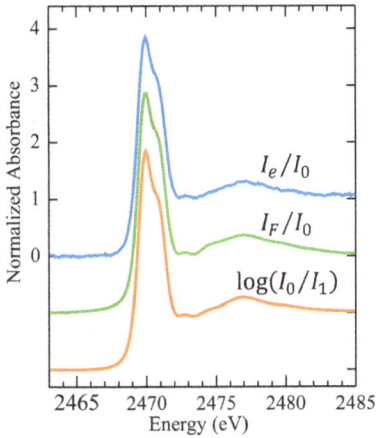

Figure 6.8: Comparison of electron yield (blue line), fluorescence (green line) and transmittance (orange line) for a thin film sample of dibenzyl disulfide, showing near-equivalence of the experimental data for the different detection types. The experimental data reported by George et al. [1] have been re-plotted in this figure.

With some experiments, it may be useful to maintain samples in an atmosphere of air, and for this purpose, what is known as a cut-off box may be employed, as illustrated in Figure 6.9. Here, what would normally be a more rectangular sample chamber (see Figure 6.6) is cut through at 45° (the cut-off box). The beam exits through a thin polymer window, impinging upon the sample, which can be in air with short air pathlength of ca. 1 mm (Figure 6.9). If a different atmosphere is desired, this can be accomplished by arranging an additional chamber outside the cut-off box with a different gas flowing through it. Such an arrangement has been used for *in situ* sulfur K-edge XAS

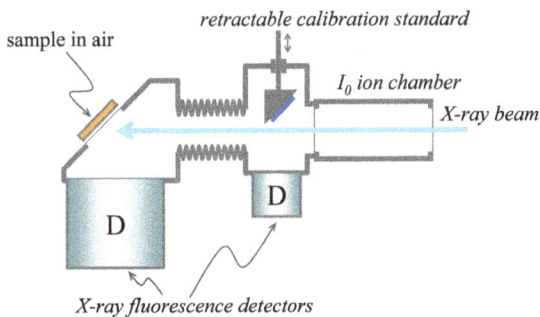

Figure 6.9: Schematic plan view of the cut-off box for tender X-ray XAS measurements with sample in air or alternative atmosphere. The thin window at 45° to the incident beam serves to separate the helium-filled flightpath for both incident and fluorescence beams from the atmosphere of the sample. The flightpath thus both transmits the beam to the sample and accepts the X-ray fluorescence from the sample where it can be registered by the detector (D).

measurements of living cell mammalian cultures [2], which allowed visualization of the full complement of sulfur-containing metabolite species present, which is called sulfur metabolomics. This experimental arrangement has other advantages: for example, positioning a reflectance spectrophotometer behind the sample allows *in situ* monitoring of any spectroscopic changes.

6.4 Soft X-ray experiments

With soft X-rays, there can be much greater variation in the experimental set-ups found at different beamlines. Typically, there is no hutch, since the entire experiment is enclosed by a vacuum chamber, whose walls effectively ensure no beam escapes. Irrespective of this, even if it were possible for X-rays to somehow escape the chamber, the air would effectively wipe out the beam in a fraction of a millimetre. In almost all cases, soft X-ray experiments are conducted in a vacuum, and samples must be loaded into a vacuum chamber, and hence must be vacuum compatible. The measurement of I_0 is typically done by allowing the monochromatic X-ray beam to pass through a gold grid of 85–90% transmittance, which undergoes X-ray-stimulated electron yield, which is relatively unstructured because the incident energies are far from the gold absorption edges, and hence is a good measure of I_0. The electron yield is usually measured as the current passing to ground, also called the drain current (Section 5.8.3). Whereas transmission XAS is unusual for tender X-ray measurements (Section 6.3), it is even more unusual in soft X-ray measurements. In most cases, the XAS is detected by electron yield or X-ray fluorescence. Electron yield is typically monitored using a channeltron electron multiplier (Section 5.8.2), or by measuring the sample drain current (Section 5.8.3). X-ray fluorescence can be measured using a photodiode, a solid-state detector or an advanced photon detector of some type. Typically, the vacuum chamber will have several detectors present, with more than one stream of data being used, depending on the nature of the sample. Arrangements are typically present to accommodate loading of several samples at once in the vacuum chamber, since breaking vacuum and pumping down can be a time-consuming process, although with some beamlines, this has been streamlined. Figure 6.10 shows a schematic of a soft X-ray experimental set-up, together with a photograph of a modern experiment.

Measurement of vacuum-compatible solid materials is relatively simple, for which samples can be prepared as finely ground powders spread on conducting carbon tape, or perhaps pressed into disks and attached to the sample paddle. The design and fabrication of vacuum-compatible sample holders for liquids can be complicated, but even for elements such as oxygen, for which contamination is a major problem, tractable solutions are possible [3]. The significant background absorption by soft X-rays precludes all but the thinnest windows, but Vogt et al. have succeeded in developing a sample containment system based on unsupported 100 nm thick silicon nitride (Si_3N_4) windows

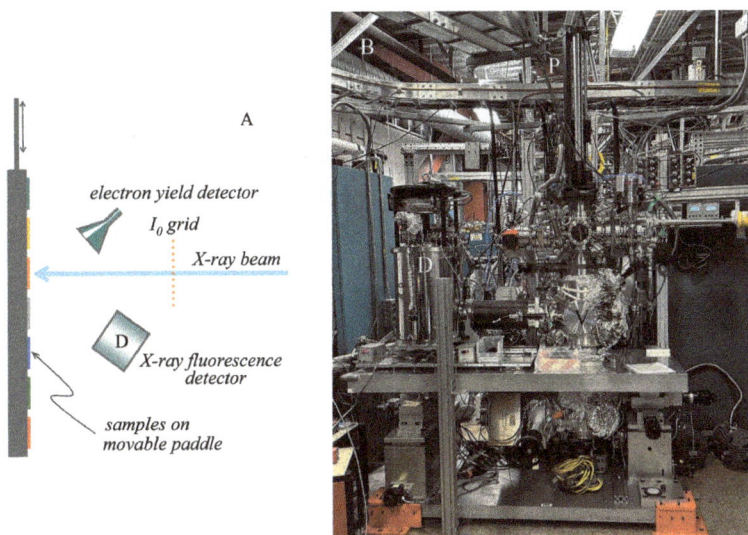

Figure 6.10: Soft X-ray XAS experimental set-up. A shows a schematic of the components inside the vacuum chamber in a typical soft X-ray XAS experiment. Samples, including an X-ray calibration standard of some type, are mounted on a movable paddle (shown side-on) that can be used to select which sample is being illuminated. B shows a photograph of an experimental set-up (beamline 13.2 at SSRL, photograph courtesy of Dr Simon George, Simon Scientific). In B, the viewpoint is along the incident X-ray beam and the samples are moved within the vacuum chamber V using the manipulator P. The set-up is inevitably complex, with the design of the vacuum chamber and sample handling capabilities being a large part of developing a successful beamline. The detector labelled D is a 240-pixel transition edge sensor array X-ray fluorescence detector (Section 5.6.1).

for oxygen K-edge XAS, removing surface oxygen contamination by etching with hydrogen fluoride (HF) [3].

Cryogenic cooling of samples has been achieved by means of a cold finger arrangement, which includes liquid helium temperatures [4]. With frozen aqueous samples, using only liquid nitrogen temperatures will typically cause windowless samples to lyophilize, resulting in loss of water from at least the surface layers, which with biological and other samples might lead to chemical changes. Conversely, use of liquid helium gives temperatures sufficiently cold that significant loss of solvent does not occur.

6.5 *In situ* experiments

We have already discussed the use of XAS with mammalian cell cultures for sulfur metabolomics, which required careful control of atmosphere and temperature, and limiting beam exposure with microscopic monitoring of the cells [2]. This is an example of an *in situ* experiment. The majority of *in situ* XAS work has been done on catalyst samples [5], with careful control of reactants, temperature, electrochemical conditions, etc.; considerable ingenuity has gone into constructing these *in situ* or *operando* experimental set-ups. Of significant interest may be applications of XAS to electrocatalyst development [6] and a technique known as modulation excitation [7]. This last is poorly named because there is no inherent excitation related to the modulation, but instead refers to cyclic changes in one or more of the catalytically relevant sample conditions, such as temperature or electrochemical conditions, and examining the resulting changes in the XAS as a difference. Because driving chemical changes by altering sample conditions is too broad a topic for a chapter on XAS experimental set-ups, in general, recent reviews can provide more information on this topic [5–7].

References

[1]   George, G. N.; Gnida, M.; Bazylinski, D. A.; Prince, R. C.; Pickering, I. J. X-ray absorption spectroscopy as a probe of microbial sulfur biochemistry: The nature of bacterial sulfur globules revisited. *J. Bacteriol.* **2008**, *190*, 6379–6383.

[2]   Gnida, M.; Sneeden, E. Y.; Whitin, J. C.; Prince, R. C.; Pickering, I. J.; Korbas, M.; George, G. N. Sulfur X-ray absorption spectroscopy of living mammalian cells: An enabling tool for sulfur metabolomics. In situ observation of taurine uptake into MDCK cells. *Biochemistry.* **2007**, *46*, 14735–14741.

[3]   Vogt, L. I.; Cotelesage, J. J. H.; Titus, C. J.; Sharifi, S.; Butterfield, A. E.; Hillman, P.; Pickering, I. J.; George, G. N.; George, S. J. Oxygen K-edge X-ray absorption spectra of liquids with minimization of window contamination. *J. Synchrotron Radiat.* **2021**, *28*, 1845–1849.

[4]   George, S. J.; van Elp, J.; Chen, J.; Peng, G.; Mitra-Kirtley, S.; Mullins, O. C.; Cramer, S. P. Soft X-ray absorption and X-ray magnetic circular dichroism in biology. In *Synchrotron Radiation in Biosciences* (Proceedings of the 4th International Conference on Biophysics and Synchrotron Radiation) (Eds. Chance, B.; Diesenhofer, J.; Ebashi, S.; Goodhead, D. T.; Helliwell, J. R.; Huxley, H. E.; Iizuka, T.; Kirz, J.; Mitsui, T.; Rubenstein, E.; Sakabe, N.; Sasaki, T.; Schmahl, G.; Stuhrmann, H. B.; Wutrich, H.; Zaccai, G.) Clarendon Press: Oxford, **1994**, pp. 313–332.

[5]   Bordiga, S.; Groppo, E.; Agostini, G.; van Bokhoven, J. A.; Lamberti, C. Reactivity of surface species in heterogeneous catalysts probed by *in situ* X-ray absorption techniques. *Chem. Rev.* **2013**, *113*, 1736–1850.

[6]   Timoshenko, J.; Roldan Cuenya, B. In situ/operando electrocatalyst characterization by X-ray absorption spectroscopy. *Chem. Rev.* **2021**, *121*, 882–961.

[7]   Müller, P.; Hermans, I. Applications of modulation excitation spectroscopy in heterogeneous catalysis. *Ind. Eng. Chem. Res.* **2017**, *56*, 1123–1136.

7 The X-ray absorption spectroscopy experiment – II

7.1 Introduction

Following on from the descriptions of configurations in Chapter 6, this chapter addresses some practical aspects of the XAS experiment. We will discuss details of sample preparation, calculation of the X-ray absorbance, incident X-ray energy calibration and some aspects of data acquisition. We will consider two common types of samples – powdered solids and liquid solutions.

7.2 Solid sample preparation for hard X-ray transmittance measurements

Unfortunately, this is an area in which bad practices are widespread. We will discuss common methods, both good and bad, and provide examples and anecdotes to illustrate the "dos and don'ts" of sample preparation. As we will discuss in Chapter 12 of this book, poor sample preparation will result in distorted data. We wish to emphasize that collection of good, distortion-free data is the single most important aspect in performing an XAS experiment. No matter how sophisticated the analysis techniques applied are, one can never do perfect analysis of inadequate data. In the hard X-ray regime, the most useful method for data acquisition on solid concentrated samples is to use transmittance. In many cases the solid sample might be sufficiently concentrated to require dilution in some way in order to achieve a sample absorbance between 1 and 2. Three different methods are in common use: (i) samples as pressed discs (Section 7.2.2.1); (ii) powder on tape (Section 7.2.2.2); and (iii) powders packed into plates, called powder plates (Section 7.2.2.3). These are discussed in detail below.

7.2.1 Powder diluents

For methods discussed in Sections 7.2.2.1 and 7.2.2.3, a suitable diluent must often be found. The diluent should be high-purity and have a low X-ray cross-section to minimize contribution to the absorbance. The diluent is typically comprised of low Z elements, with suitable materials including sucrose, boron nitride, silica, lithium carbonate, and polymers with suitable mechanical properties. Very often, air-sensitive or moisture-sensitive samples may need to be prepared in an anaerobic glove box; we will discuss some of the issues related to this below. Since preparation typically involves intimate grinding of the sample with the diluent, it is very important to employ a diluent that does not react chemically with the sample, and to avoid mixing chemically incompatible

https://doi.org/10.1515/9783110570441-007

substances. One of us vividly recalls witnessing a user from an adjacent beamline in shared laboratory space at NSLS-I attempting to prepare a transmittance sample of sodium chromate (a strong oxidizer) by grinding it with sucrose in a mortar and pestle; fortunately, nobody was injured in the resulting firework display. Each diluent brings both advantages and potential problems. As already mentioned, organic materials such as sucrose and cellulose will ignite, hence mixing these materials with strong oxidizers should definitely be avoided. Boron nitride (BN) is perhaps the most widely used diluent for XAS; comprised of light elements it has a low X-ray cross-section of only 2.6 cm^2·g^{-1} at 10 keV. Structurally, BN is commonly found in cubic and hexagonal polymorphs; the cubic form resembles diamond with each B bound to four N and vice versa, while the more common hexagonal polymorph (thermodynamically favoured at high temperatures) has a structure that is similar to graphite, with layers containing adjoined six-membered rings comprised of alternating B and N atoms, with each B bound to three N and vice versa. Most commercial BN is amorphous, which probably contains hexagonal-like domains, giving BN its slippery feel, and providing motivation for its incorporation into various cosmetic products.[1] While generally BN can be regarded as a safe material, there are some health concerns as inhalation of some forms has been linked to adverse effects. Notwithstanding such issues, BN is chemically and thermally stable, not flammable in air and can withstand temperatures above 1,000 °C without degradation. It will react with strong alkali (e.g. KOH) at elevated temperatures, and also with some strong oxidizers, again at high temperatures and/or pressures. In most cases the only major issue with BN as a diluent may be its tendency to align crystallites of sample with the aforementioned hexagonal planes giving a preferred orientation in the X-ray beam. We now consider the different methods of sample preparation.

7.2.2 Adhesive tape

For methods discussed in Sections 7.2.2.2 and 7.2.2.3, an appropriate tape must be used, which minimizes contaminant elements in both plastic and adhesive since the beam will penetrate both. The film is typically made of Mylar (polyester) or Kapton (polyimide) film and is available as a specialist tape for X-ray applications.

7.2.2.1 Samples as pressed discs

Here, a hydraulic press is used with a pressing die loaded with pre-weighed sample intimately mixed with diluent, to produce a pressed-disc sample of cardboard-like consistency. Diluent materials that have suitable mechanical properties for pressed discs include cellulose or polyvinyl pyrrolidone. Unfortunately, BN, discussed above,

---

1 Boron nitride is used in a range of cosmetics and personal care products, including face powder and lipsticks mostly because it imparts a smooth and silky feel to the products.

lacks mechanical suitability when pressed into discs, which tend to be of a fragile consistency, flaking into layers, and challenging to remove intact from the die. With stable matrix compounds (diluents), discs prepared in this way can persist for years, or even decades. The caution here is that in the press samples and diluents are exposed to considerably elevated pressures and consequently elevated temperatures, which has the potential to modify the sample. Moreover, the caution expressed above about not combining chemically incompatible substances is therefore especially important in the case of pressed-disc samples.

7.2.2.2 The powder-on-tape method of solid sample preparation

This method uses layers of adhesive tape with powder spread between each layer, often using the forefinger of a latex- or neoprene-gloved hand, or perhaps a brush or a rubber policeman to spread the powder on the tape. Often several layers of tape with powder are sandwiched together to achieve the desired sample thickness. The advantage of this method is that it effectively screens out grains of powder that are too large, as these tend not to stick to the tape. This is important because the sample should not have individual particles that are greater than one absorption length, or the data can be distorted. The powder plus tape combination can be weighed on the application of each tape layer, to estimate how much sample is actually present, but in practice users tend to use a combination of guesswork and experience in preparing such samples. In our opinion the disadvantages of this method considerably outweigh the advantages, and hence this is our least favourite method of solid sample preparation. There are a number of problems associated with powder-on-tape. First and foremost, powder-on-tape tends to be a messy procedure, and any method where fingers (gloved or otherwise) and samples are in close proximity has potential safety issues. The use of brushes or a rubber policeman to spread the powder on the tape is also problematic because these can be difficult to clean (especially brushes) and in our experience powder will typically end up contaminating the locality where the sample is prepared. Secondly, adhesive tape can contain metals, which in some cases may be the element that is being studied by XAS. Multiple layers of tape can add undesirably to the background X-ray absorbance. For example, zinc and manganese compounds are added by manufacturers to some formulations; and while any sample preparation method using adhesive tape must guard against this, powder-on-tape uses multiple layers of tape compounding the potential for contaminating signals. Thirdly, the adhesive can react with the compound. Common adhesives found on tapes are acrylics and silicones. The former frequently contain methacrylic acid esters or methacrylic amides, while the latter are often polysiloxanes, frequently polydimethylsiloxanes. Both categories of adhesive have potential reactivity with some samples. This is much more of an issue with powder-on-tape than with powder plates (which we will consider next) because of the intimate contact between adhesive and finely powdered sample. Fourthly, if the samples are not carefully made, the possibility of heterogeneity may cause distortions of the data, and the amount of sample in a powder-on-tape preparation is hard to quantify. In short, the powder-on-tape method for preparation of transmittance samples is a poorly quantitative, messy, and unsafe method that we strongly advise the reader not to use.

7.2.2.3 Powder plates

The powder plate is a metal or plastic plate, usually 0.5–1 mm thick, with a slot milled in the middle to contain the sample (Figure 7.1). The sample, which in most cases has been suitably diluted, is sealed and enclosed by a layer of adhesive tape on either side of the powder plate. There can be no holes in tape, as the sample may leak out, no gaps or holes in the packed powder, no wrinkles in the tape windows, no bubbles between the tape and the powder plate, and tape must cleanly adhere to the powder plate with no sample in between. One method for filling powder plates is as follows. An adhesive tape window is applied to one side of the powder plate, and the plate is positioned with the exterior side of the window down onto a 35 mm glass microscope slide. An adhesive tape guard with a hole matching the slot of the powder plate is then placed on top of the powder plate, which serves to keep the powder plate clean of powdered sample during filling. The sample is then packed into the slot using a flexible spatula that is wider than the slot like a trowel to give a smooth surface flush with the top of the guard tape. The adhesive tape guard can then be carefully removed, revealing the clean upper surface of the powder plate. Finally, a fresh adhesive tape window can then be applied to the powder plate to enclose the sample. While static electricity can be a problem, causing the finely powdered sample to jump out of the slot just as the final tape window is applied, judicious use of anti-static guns[2] can effectively mitigate these frustrations. Filled powder plates should be clearly labelled indicating their contents and should not be stored for long periods under liquid nitrogen in case liquid seeps into the powder, which would cause the adhesive tape windows to burst off upon warming.

Powder plates can be prepared in a quantitative manner because slots of a particular volume generally pack approximately the same quantity of (typically) BN diluted compound. Powder plates can also be used inside anaerobic or dry glove boxes, using the plate mounted on the 35 mm glass microscope slide as a convenient and portable work surface, allowing the user to work with oxygen or moisture sensitive compounds. We have already complained that the powder-on-tape is our least favourite method of sample preparation, but even worse can be the poorly prepared powder plate. Figure 7.2 shows a comparison of a correctly loaded powder plate and one that has been improperly prepared.

---

2 These use piezoelectric crystals compressed by a hand trigger to create a spray of ions and help to eliminate static charge.

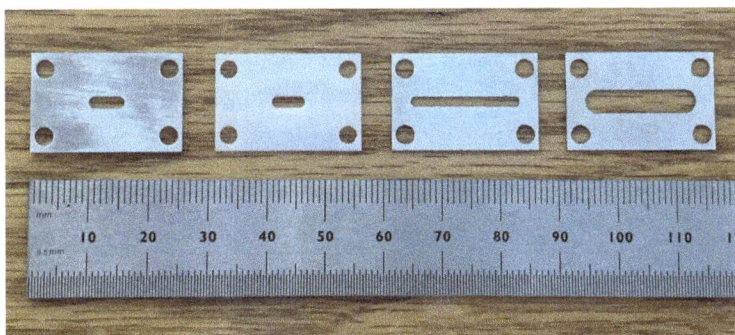

Figure 7.1: Powder plates for XAS of solid samples with slots of different sizes. The smaller slot sizes are more suitable for beamlines with a focused X-ray beam, when samples are precious. The white-coloured powder plate second from the left is made of nylon while the other three are made of aluminium. The typical thickness is between 0.5 and 1 mm.

Figure 7.2: Examples of correctly and incorrectly made powder plate samples. The sample on the left was prepared by one of the authors and contains 1,2,3,4,5,6-hexakis-phenylhexaarsinane (Ph_6As_6). The sample is clearly labelled, so that its contents are known, and it is adequately sealed within the powder plate using metal-free Mylar adhesive tape. The sample on the right was found by the authors at the beamline, left there by a previous user group. It is unlabelled and thus its contents are completely unknown, it needed to be disposed of as hazardous material. It has also been carelessly made and its contents have clearly leaked from the powder plate, potentially contaminating both the cryostat and the beamline area. Moreover, the tape window is wrinkled giving a variable sample thickness. Holding the sample up to the light showed that it was half-empty, either due to leakage or careless preparation or probably both.

7.3 Calculation of total X-ray absorbance

As we have previously discussed in Chapter 4, the absorbance A of a sample of thickness x is given as follows (eq. 4.14):

$$A(E) = \mu(E)x = \log\frac{I_0}{I} \tag{7.1}$$

where I_0 and I are the incident and transmitted X-ray intensities, and $\mu(E)$ is known as the X-ray absorption coefficient. We briefly mentioned X-ray cross-sections in Chapter 4, and $\mu(E)$ is equal to the product of the absorption cross-section $\sigma(E)$ and the sample density ρ by

$$\mu(E) = \sigma(E)\rho \tag{7.2}$$

$$A(E) = \sigma(E)\rho x \tag{7.3}$$

The absorption cross-sections for most of the chemical elements are available in tables such as the McMaster tables, and, as we have previously mentioned, are provided in odd-seeming units of $cm^2 \cdot g^{-1}$. However, for ρ in $g \cdot cm^{-3}$ and x in cm, these make some sense, because the dimensionless $A(E)$ can be calculated from the tabulated $\sigma(E)$ using eq. (7.3). For a compound, one can calculate the total cross-section from the chemical composition by summing all the elemental $\sigma(E)$ weighted by the mass fractions f_i:

$$\sigma(E) = \sum_{i}^{n} f_i \sigma_i(E) \tag{7.4}$$

where n is the total number of elements in the compound. The mass fraction can be calculated from a chemical formula by multiplying the atomic weight of each element, M_i by its stoichiometry N_i and dividing by the sum of these:

$$f_i = \frac{N_i M_i}{\sum_i^n N_i M_i} \tag{7.5}$$

Thus, using Mn_2O_3 as an example, with respective molecular weights for manganese and oxygen of 54.94 and 16.00, yields f_{Mn}= $2 \times 54.94/(2 \times 54.94 + 3 \times 16.00) = 0.696$. For physical mixtures of compounds, we can calculate the total absorption cross-section of the mixture σ_{tot}, from the cross-sections of each component $\sigma_j(E)$ and their mass fractions f_j, summing over j with total components N:

$$\sigma_{tot}(E) = \sum_{j}^{N} f_j \sigma_j(E) \tag{7.6}$$

As previously mentioned, an "ideal" transmittance sample has a maximum absorbance of about 2, which generally occurs just after the absorption edge, with the minimum absorbance typically occurring just below the absorption edge. For a sample absorbance that is very much larger than 2, the transmitted intensity cannot be measured sufficiently accurately, and the signal will be poor and prone to artifacts and distortions. For a sample absorbance that is very much smaller than 1, the edge jump may be too weak. As an example, we will consider a pure metal foil made of copper. One can look up the maximum absorption cross-section just above the K-edge from the McMaster tables as approximately 300 $cm^2 \cdot g^{-1}$. The absorbance just below the absorption edge is ~30 $cm^2 \cdot g^{-1}$. The density ρ of copper is 8.96 $g \cdot cm^{-3}$. From eq. (7.3) for

$A = 2.0$, we can find the thickness in cm to be $x = A/(\sigma\rho) = 2.0/(300 \times 8.96) = 7.4 \times 10^{-4}$ cm = 7.4 μm. Fortunately, copper foils very close to this thickness are commercially available. Similarly, tabulated absorption cross-sections can be used to calculate an ideal sample thickness for powder samples, based on composition of the sample, and neglecting any spectroscopic structure. In the case of powder plate samples, the actual density of the packed powder must be used, which differs from the densities of the sample and the diluent complicated by inclusion of air in the packed sample. In cases where the sample is dilute, and the diluent used is BN, a density ρ of approximately 1 g·cm^{-3} works reasonably well for the diluted sample. In other cases the effective density can be estimated from the actual mass packed, divided by the volume of the sample slot. We note that for pressed discs, the situation is simpler; the density and the thickness of the final sample are unimportant, and one only needs the composition and masses of sample and diluent, together with the area of disc that will emerge from the die.

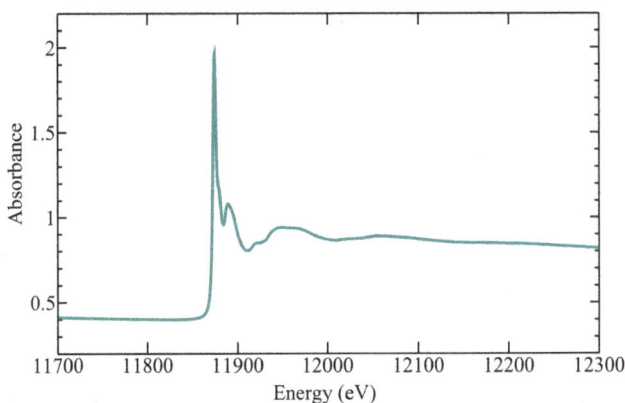

Figure 7.3: Raw transmittance X-ray absorption spectrum of solid $Na_2HAsO_4·7H_2O$ diluted 10 wt% in BN and packed into a 1 mm thick sample plate similar to that shown in Figure 7.2. The intense peak at 11,874.7 eV would cause the sample to be too absorbing at this energy if it were diluted to give a post-edge absorbance of 2.0 (ca. 32 wt% in BN).

As we have already noted, the calculations above neglect any structure in the spectrum aside from the absorption edge. In many cases, near-edge spectra can contain intense features which, for a sample formulated to have a post-edge absorbance of 2, would have a peak absorbance outside acceptable bounds. An example is shown in Figure 7.3; $Na_2HAsO_4·7H_2O$ diluted in BN, which was deliberately made more dilute than our discussion so far would suggest, because we correctly expected that the As K-edge near-edge spectrum of this compound would contain an intensely absorbing feature.

7.4 Solid samples for tender and soft X-ray fluorescence/electron yield

Having just critiqued the use of powder on tape for hard X-ray transmittance measurements, we need to point out that the use of adhesive tape is entirely acceptable for tender X-ray spectroscopic measurements when using X-ray fluorescence detection of samples that are sufficiently dilute. For XAS fluorescence measurements the requirements are that solids must be both sufficiently dilute and finely powdered that no significant self-absorption is present (see Section 12.3), and hence that meaningful measurements can be made. For electron yield measured either directly by using a channeltron electron amplifier or sample drain current (soft X-ray, vacuum) or by using gas-amplified electron yield, self-absorption is not a problem. But there is a different requirement that the sample must conduct sufficiently, so that it will not accumulate sufficient electrical charge to distort the signal (see Section 12.4). The required current flow is typically not much, usually fractions of a nano-ampere. Typically, mounting the sample on conducting carbon tape or by mixing the sample with graphite or even overlaying a metal grid will allow sufficient conductivity. In these cases, powder on tape is useful for both fluorescence and electron yield measurements. For soft X-ray XAS of solids the finely ground sample might alternatively be prepared by embedding in a film of indium metal, which is a reasonable electrical conductor and sufficiently malleable to easily hold powdered samples.

7.5 Solution sample preparation

In general, sample preparation for solutions is less problematic than for solids, especially if the solutions are dilute. In the hard X-ray regime, solutions having element of interest concentrations in the vicinity of 100 mM have been used for transmittance measurements; these would be poorly suited for X-ray fluorescence measurements because of self-absorption artifacts (Section 12.3). More often, however, solution samples are dilute, meaning in the vicinity of 0.1–1 mM. Solutions are usually held in small plastic sample cuvettes, with thin Mylar or Kapton adhesive tape windows. Figure 7.4 shows a range of different cuvettes that can be used for holding solution samples.

For low Z-samples that need low temperatures the same sample cuvettes can be used but with a much thinner window, such as a 2 μm thick Mylar window, glued to the plastic cuvette. We have stated that solutions are simpler, but this is not to say that there are no bad practices. One bad practice is to construct a small envelope out of adhesive tape and introduce a pocket of solution by poking a hole with a hypodermic syringe needle and injecting solution. There are a number of problems with this practice, not the least of which is that the sample will be of uneven thickness so that the data will be prone to distortions. A more substantial issue is once again one of safety, related to the potential for sample spillage, and for this reason we strongly discourage such practices. Indeed, in the past we have needed to clean up such samples

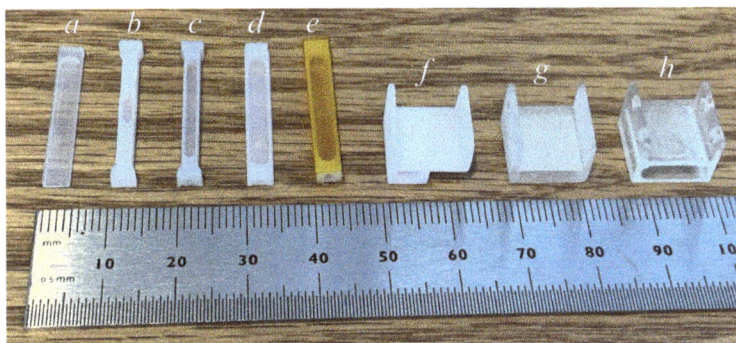

Figure 7.4: Liquid sample cuvettes. Cuvettes *a* through *d* show different mid-range hard X-ray energy designs: *a*, 1 mm path-length cuvette; *b*, 2-mm path-length small volume cuvette; *c*, 60 μL, 2 mm path length (or "dogbone") cuvette; *d*, 200 μL, 2 mm path-length cuvette; *e*, same type as *d* with a Kapton adhesive tape window; *f*, *g* and *h*, show cuvettes suitable for higher X-ray energies (e.g. Mo K-edge), all with 10 mm path lengths.

left by an earlier user group at our beamline, with the tape-pocket liquid samples sitting in a puddle of leaked contents and without any label indicating what they were.

7.5.1 Filling the sample cuvette

In many cases solution samples are run frozen in a cryostat. There are a number of reasons for this, such as freezing out of vibrational modes to decrease the Debye-Waller factor, making the EXAFS more visible at high *k*-values. We will consider filling a cuvette with a window applied as in Figure 7.4e. Typically, solution is introduced into the cuvette with a hypodermic syringe through the fill hole in the plastic cuvette end, which is visible in the figure. For aqueous solutions the best type of syringe to use is a glass precision syringe, with a Teflon plunger, such as those made by the Hamilton Company (Reno, NV, USA). These are easy to clean and to use, and if the sample is precious, there is little waste. For aqueous samples that are plentiful and thus not particularly precious, a disposable medical-type syringe may also be used. For non-aqueous solutions in organic solvents the plastic of the cuvette and the adhesive tape for the window must be carefully chosen, and a glass filling syringe with a stainless steel plunger may be a better choice. A typical cuvette filling procedure would be as follows. The syringe needle should be dry on the outside (clean it with a disposable wipe, if needed). The cuvette should be held so that the fill hole is uppermost, and the syringe needle should be introduced into the cuvette without touching the adhesive tape window, until the tip is very close to the bottom of the cuvette. The cuvette should then be filled by depressing the plunger on the syringe while withdrawing the needle at such a rate so as to keep pace with rising level of liquid within the cuvette. If a meniscus of liquid is allowed to form between the syringe needle and the cuvette fill hole then this will block the escape of atmosphere from the cuvette as it is filled, and a clean fill with no bubbles may not be possible. Bubbles in the sample are undesirable since they can cause problems with the data; they should be avoided at all costs. For anaerobic samples, this procedure can be carried out in an anaerobic glove box.

7.5.2 Freezing the sample

As we have already mentioned, frozen solutions are often used. For aqueous samples these must be prepared in such a way as to prevent formation of sizable ice crystals. As we discuss in Section 12.5.1, crystalline ice in the sample causes X-ray diffraction which, for mid X-ray energy experiments, e.g. the Cu K-edge at ~9 keV, can be problematic if X-ray fluorescence detection is used. As the incident X-ray energy is changed during an XAS experiment the angle of the ice-diffracted X-rays will sweep across the fluorescence detector, causing saturation and eliciting a non-linear response. At high X-ray energies, such as at the Mo K-edge at ~20 keV, the diffraction angles are smaller, and ice diffraction is typically not a substantial problem. Two common methods are used to prevent ice crystal formation. The first would be to add a glassing agent to disrupt the water network, preventing ice crystal formation. Traditional glassing agents include sucrose, glycerol, ethylene glycol and polyethylene glycols. All of these have been shown to exacerbate photo-reduction by the X-ray beam (see Section 12.7). Moreover, in some cases it is not possible to add a glassing agent for chemical or biochemical reasons. For both these reasons, a second and better method might be to use flash freezing. In any case, sample freezing should be as fast as possible, even when a glassing agent is being used.

Immersion of a sample in liquid nitrogen gives relatively slow freezing due to what is known as the Leidenfrost effect. Here, the sample surface is at substantially higher temperature than the cryogen's boiling point, causing the formation of a layer of gas or vapour between the sample and cryogen and the familiar sizzling effects. The vapour layer has low thermal conductivity and consequentially the rate of heat exchange between the sample and the cryogen is low. The Leidenfrost effect can be observed in the kitchen when drops of water are introduced into a heated cooking pan; the droplets persist, rolling around on the hot surface on a cushion of steam. The temperature at which the Leidenfrost effect begins is cryogen-specific and is called the Leidenfrost point; at the Leidenfrost point a stable vapour film will persist. In order to freeze a room temperature sample rapidly a cryogen with a Leidenfrost point that is above room temperature must be used. Convenient cryogens include a partly frozen isopentane slurry, and liquid propane obtained from an inverted camping gas cylinder. Unfortunately, propane camping gas typically has a small quantity of ethane thiol present as an odorant, which makes it less amenable for such laboratory use.[3] The partly frozen isopentane slurry can be made by gradually adding liquid nitrogen to isopentane with stirring. Isopentane freezes at −160 °C but boils at around +28 °C; dropping an aqueous sample directly into partly frozen slurry causes freezing very rapidly with essentially no boiling of the cryogen and high thermal conductivity, so that large ice crystals typically do not form.

7.6 Safety considerations

We have already mentioned the importance of cleanliness and prevention of contamination of both work areas and beamline equipment. Here we briefly stress the importance of some safety precautions. Many of the compounds that researchers are called to study are hazardous in some way; for example, some of the compounds of mercury are highly toxic, and particularly so if they are organometallic. It is imperative for researchers to understand the possible risks associated with their experiments; in our opinion there is no excuse for ignorance of the materials that you are working with – if experimenters have failed to equip themselves with a knowledge of the risks of particular materials then they have no place in a laboratory setting. Lack of training in a particular scientific discipline is no excuse.[4] Thus, an essential prerequisite to any experiment must be for experimenters to read about the chemistry and the possible hazards of all materials to be used. For toxic materials the amount of material used should be kept to a minimum, and the quantity of waste should also be kept within reasonable bounds. Appropriate personal protective equipment (PPE) must also be used.

We have already discussed the use of cryogens such as liquid nitrogen and cold isopentane, and these deserve a special mention. Samples requiring cryogenic protection are typically shipped in so-called dry-shipping dewars. These are insulated containers lined with high heat capacity materials, which are pre-cooled by filling with liquid nitrogen. Before shipping, the liquid is removed and the samples transported dry, at temperatures below −190 °C. It is common practice to refill the dewar with liquid nitrogen upon receipt, and for this reason we caution about the use of samples in sealed vials. While in many cases screw-top "cryotubes" or crimped-seal vials will remain liquid free, on rare occasions liquid nitrogen will gain admittance to the vial

---

3 Both isopentane and liquid propane are highly flammable with respective flash points of −51 and −104 °C, and we advise using these materials in a laboratory fume hood (the flash point is defined as temperature above which there is sufficient vapor pressure to allow formation of an ignitable air/vapor mixture).

4 One of us recalls observing a graduate student from another research group using a solution of hydrofluoric acid in an open beaker on a regular bench in a shared laboratory space at a synchrotron light source. Upon being informed that these procedures were extremely unsafe, the student offered ignorance as a defence, proudly proclaiming to be a physicist and not a chemist, the implication being that the knowledge required for chemical safety was somehow beneath this student. Safety is everyone's business, and ignorance can never excuse unsafe practices.

and upon warming will cause an explosion.[5] We therefore recommend that in general vent holes be included in all vials that might be stored under liquid nitrogen, as even the most oxygen-sensitive of metalloenzyme samples will retain their activity aerobically at liquid nitrogen temperatures. Transport of these into an anaerobic glove box can be achieved easily using a metallic block containing holes of the correct size to contain a screw-top vial. The block can be cooled in liquid nitrogen before inserting the cryotubes into the holes, and will remain cold for sufficient time to transport frozen air-sensitive proteins into an anaerobic glove box for sample preparation.

7.7 Signal contamination

Contamination of the signal can be a major problem for fluorescence measurements of dilute samples. Our goal in this section is to point out some of the possible sources. As we have already discussed, many adhesive tapes are deliberately formulated to contain small quantities of metal oxides, such as manganese and zinc oxides, to give them a translucent rather than clear appearance. Many types of polymers contain sulfur (e.g. polysulfones) and some types of acrylic adhesive used on tapes contain chlorine. Zinc occurs at readily detectible levels in fingerprints, hence care should be taken when handling solution cuvettes intended for dilute samples, including in absence of any samples during window application. Disposable laboratory gloves are generally dusted internally with a mixture of corn-starch and talc, $Mg_3Si_4O_{10}(OH)_2$, often with some $CaCO_3$ and $MgCO_3$, and users carrying out Ca or Mg XAS might be concerned about potential contamination. Many plastics contain bromine compounds, added as a free radical polymerization initiator as part of their manufacture, which if present in sample cuvettes can effectively truncate the k-range of (for example) Se K-edge EXAFS experiments. Unfortunately copper and iron are found just about everywhere; in beam pipes, beryllium windows, hutch walls, cryostat components, etc. Zinc is also often present at low levels in the aluminized coatings of Mylar and Kapton windows used in X-ray cryostats. One of the worst types of contamination is leaking of a poorly prepared solid sample into the cryostat (e.g. Figure 7.2). In short, there are many types of possible signal contamination, which can be especially important in the case of very dilute samples. Some of these can be guarded against, but for others, such as a cryostat that has been contaminated by a previous user, there may be little

---

5 One of us witnessed the aftermath moments after such an explosion. In this case a heavy-duty glass crimped-sealed vial had been used to store a highly air-sensitive metalloprotein as a frozen solution. The user, who wore no PPE, was peppered with glass shards causing multiple lacerations, requiring treatment in the emergency room of a local hospital. The research group had observed that thin-walled crimped-seal glass vials occasionally "popped" after being removed from liquid nitrogen storage. Unfortunately, their attempt to resolve this was to use vials made from much thicker glass, which resulted in a considerably more energized "pop".

that the experimenter can do. Our advice is to always record a sample blank – a solution containing none of the element of interest, but otherwise identical to the samples to be investigated – and to record this near to the beginning of the beamtime. This way, the experimenter should be aware of the problem at the outset and may perhaps be able to do something about it.

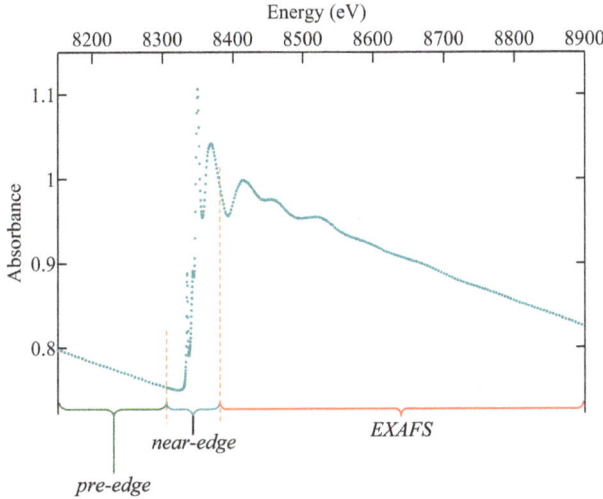

Figure 7.5: XAS data collection strategy. Different energy steps are used in different parts of the spectrum. The pre-edge segment uses relatively large energy steps, the near-edge segment uses small energy steps to capture the fine details, while the EXAFS steps in k-space, with a count time that is usually weighted by k^2.

7.8 Data acquisition strategies for XAS

With standard XAS experiments the time per point and the energy step between points is often varied. The exceptions to this are measurements using rapid scanning of some kind, such as so-called qXAS (quick-XAS) or QEXAFS (quick-EXAFS) experiments. Here, the monochromator is moved continuously during the measurement, often as fast as possible, with data being collected on the fly, as the monochromator slews across the energy range required for the scan. In this case the point spacing is more likely to be close to equal energies, and the count time will be some sample time, with a degree of spectroscopic blurring, depending on the timing. In many cases and particularly for signal-to-noise limited experiments, however, what we call a standard XAS scan is used. Here, the monochromator is moved to some desired energy, then halted and a short time is allowed for any vibrations to settle (often called the monochromator rest-time or settle-time). The various detectors are then made live, and signal is accumulated for a pre-defined time (often called the count time). The spacing between energy points is varied depending upon the part of the spectrum being recorded. An example is shown in Figure 7.5. The low-energy part of the spectrum before the absorption edge is called the **pre-edge**. Here, the absorbance should vary smoothly with energy, and consequentially large point spacing is used, such as 5 or 10 eV for hard X-ray experiments. Following this, the near-edge region has structure that can vary rapidly with energy, and a much smaller incremental energy point spacing is used, which might be 0.1–0.3 eV depending upon the X-ray energy and the resolution of the experiment, so that the details in the spectrum are properly captured. The start of the fine point spacing must be sufficiently low in energy that details are not missed, but to start too low wastes precious beam time as too much time is spent counting on the relatively featureless pre-edge region. At energies greater than E_0, steps typically are no longer at fixed energy increments but instead with fixed k increments, remembering that $k = \sqrt{(2m_e/\hbar^2)(E-E_0)}$, or $k \approx \sqrt{0.262(E-E_0)}$. Because the EXAFS data will be k-weighted, the high-k end of the spectrum can require better statistics than the low k end, and it is common practice to increase the count time, usually as a function of k^2. Because sudden changes in count time are considered undesirable, the same count time, t_{min}, is normally used for pre-edge and near-edge, after which the EXAFS count time t ramps from this minimum value, t_{min}, up to some maximum, t_{max} at the maximum k (k_{max}), according to eq. (7.7), in which n is the power by which the count time is weighted (usually 2):

$$t = t_{min} + (t_{max} - t_{min})\left(\frac{k - k_{min}}{k_{max} - k_{min}}\right)^n \tag{7.7}$$

Typical count times might be t_{min}= 0.5 to 2 s (for pre-edge and near-edge) with t_{max} ranging from 5 to 12 s.

It is considered good practice to collect a minimum of two scans to ensure reproducibility, even when the signal to noise is excellent. With transmittance of concentrated solids there is no serious limitation in signal to noise; the biggest sources of (apparent) noise tend to be non-statistical in nature, and typically arise from such things as mechanical vibrations in beamline hardware (e.g. monochromator crystals). Consequently, techniques such as qXAS/QEXAFS are possible and a modern qXAS setup can scan a 1,000 eV data set in only a few seconds. For fluorescence XAS of dilute samples, the signal to noise can be limited, since the signal can be quite small and it is in this case that signal averaging can become very important. It is also good practice to collect the EXAFS to as high a k-value as possible as this offers several advantages. These include better accuracy of determination of structural parameters, such as interatomic distances, improved ability to distinguish between different types of backscatterer atoms, improved sensitivity to heavier backscatterers and improved bond-length resolution for similar backscatterer atom types at similar distances. Having made the case for long k-ranges, we feel compelled to point out that a common novice mistake is to collect too long a k-range on dilute samples, so that the high k-end of the EXAFS is just noise with no discernable structure; obviously an appropriate balance must be struck.

Finally, we add a word about so-called dark currents, which are sometimes called offsets. There is a modern tendency to neglect these, but this is ill-advised and doing so can give rise to data artifacts, which we will briefly discuss in Chapter 12. Many detectors give small non-zero background counts when no X-rays are present; dark currents typically are measured with the incident X-ray beam turned off and collected for longer than the longest count-time in the scan, t_{max}. A long count time will give a good statistical measure of the dark currents, which are typically subtracted by the data acquisition software before it is saved. Typically, dark currents do not change appreciably during the course of a series of measurements and would be collected only following a change in the experimental configuration.

7.9 Incident X-ray energy calibration

Good energy calibration is an essential part of a competent XAS experiment. Typically, we calibrate with reference to the spectrum of a standard, whose edge energy is tabulated, such as can be found in the so-called orange book (https://xdb.lbl.gov/) or otherwise conforms to some accepted standard value. It is important to note that the calibration energy is not absolute and may be slightly offset from the true value, and that even the orange book values may be offset from the real energies. For example, the orange book value for copper is 8,979 eV, but the actual energy is probably closer to 8,980.5 eV. In our experience the novice experimenter may ask such questions as "why do we need to do the energy calibration – didn't the beamline scientist do that already?". Even if the beamline is perfectly calibrated when the user goes online, the energy calibration may change with time, as minute changes in the angle between the incident X-ray beam and the monochromator crystal may cause subtle energy offsets, and other factors such as thermal equilibration of the optics may play a role. Hence, frequent checks of energy calibration are required, even at the most advanced facilities. Typically, the energy calibration will be done twice; the first time will be at the beamline when the experimental run is started, or on moving the beamline to a new energy range, with the energy set close to the table value being used. The second calibration is a recalibration done during the data analysis phase. As we have discussed in Chapter 6, a reference standard foil is typically measured with the experiment, so that each scan of a sample will have an accompanying reference spectrum that is measured simultaneously with the data.

The energy recalibration should ideally be done using a monochromator Bragg angle offset, assuming that a double crystal X-ray monochromator is being used, rather than simply shifting energy for the whole spectrum. We can compute an angle offset $\delta\theta = \theta_{obs} - \theta_{true}$, where θ_{obs} and θ_{true} are the respective Bragg angles corresponding to the observed and tabulated calibration energies. The issue is that miscalibration due to a Bragg angle offset will translate to different energy offsets at the absorption edge compared with at the end of the scan. Hence, the application of an energy shift can directly translate to errors in bond lengths. In most cases the difference will be small and should not make much of a difference, but the cautious experimenter should be aware that this is not always true. We will illustrate this point with an example, for which we choose a titanium K-edge EXAFS measurement using an Si(111) double crystal monochromator. Here, a table value for Ti metal is 4,966.0 eV, which we assume corresponds to the measured first inflection point of the metal foil, by which we mean the lowest energy K-edge first derivative peak. Let us assume that the EXAFS data extends to 16.2 Å^{-1}, and that the data happens to be miscalibrated by +10 eV at the foil position, so that the observed inflection point is 4,976.0 eV, rather than 4,966.0 eV. We can further assume that there are no issues with point spacing, etc. and that the data will be useful upon energy recalibration. Here, the correct method of using $\delta\theta$ corresponds to an energy offset of +10 eV at the foil edge and +14.8 eV at the high-k end of the scan. This means that if an energy shift recalibration of +10 eV was used throughout the whole data set, then this would give an artificially widened k-range for the EXAFS (by about +0.12 Å^{-1}) with correspondingly and erroneously shorter derived bond lengths (by about −0.01 Å). While these errors may seem small, this example is certainly much larger than a typical precision. Moreover, this particular source of uncertainty is one that is easily disposed of, and we encourage the reader to work through a few examples by hand to underline the importance of correct energy calibration.

In some cases, a standard compound for which the energy of a prominent transition is well-established may be more convenient. An example is provided in Figure 7.6, sulfur K-edge XAS experiments frequently use sodium thiosulfate ($Na_2S_2O_3 \cdot 5H_2O$), which has a prominent low-energy peak arising from the terminally bound sulfidic sulfur. With some XAS experiments, for example, those in the tender or soft X-ray regime, the beam may not penetrate the sample. In this case, a quick scan of the standard might be measured just before and just after data collection on the sample.

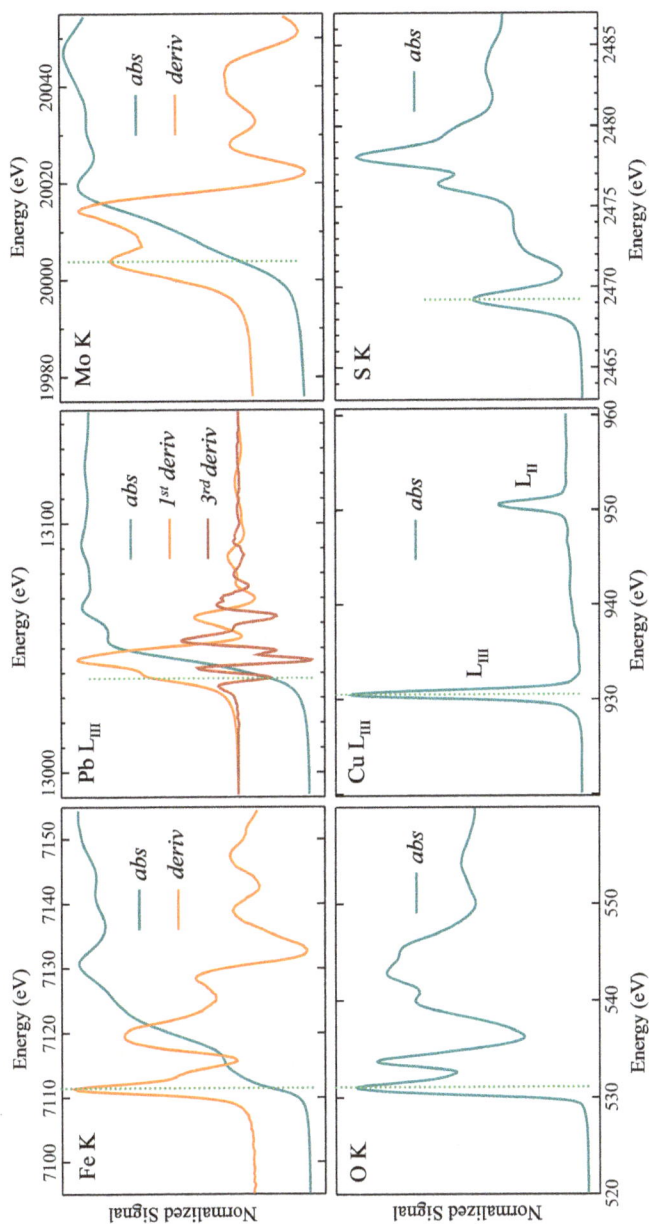

Figure 7.6: Examples of incident X-ray energy calibration with reference to different standards for hard, tender and soft X-ray experiments. Top row: For Fe K-edge XAS (left panel) using the first derivative to determine the lowest energy inflection point of an iron metal foil. For Pb L_{III}-edge XAS (middle panel) with Pb metal foil, the first derivative is insufficient, and the third derivative is used instead. For Mo K-edge XAS (right panel) using the first derivative to determine the lowest energy inflection point of a molybdenum metal foil, which in the first derivative spectrum is the lowest energy peak but, unlike Fe K, not the highest intensity peak. Bottom row: For O K-edge XAS (left panel) with reference to the lowest energy absorption peak of a TiO_2 standard. For Cu L_{III}-edge XAS (middle panel) with reference to the lowest energy Cu L_{III} energy absorption peak of a CuF_2 standard. For S K-edge XAS (right panel) with reference to the lowest energy absorption peak using a sodium thiosulfate standard. The vertical green broken lines show the calibration points.

8 The EXAFS

8.1 Introduction

In Chapter 4, we made a promise to return to both the near-edge and the EXAFS with additional rigor. In this chapter, we meet this promise for the EXAFS and provide some additional background for near-edge spectra. This book is intended for an audience with diverse backgrounds. Some material in this chapter will be more straightforward for readers with some background in physics or chemistry than for those coming from some other fields. While we do not encourage this, Sections 8.2 and 8.3 could be skipped without too much loss in terms of phenomenology.

8.2 X-ray photoabsorption

We consider the transition probability w_{if} between two quantum states, which we refer to as i and f (initial and final). Transitions happen rapidly in the presence of strong coupling between the initial and final states; the coupling is known as the transition **matrix element**. The term matrix element comes from the quantum mechanical formulation that uses matrices instead of the differential equations of Schrodinger. We note that when compared to electronic transitions in the ultraviolet or visible range, X-ray transitions tend to be very weak, since the overlap between the wavefunctions of the core-level initial state (relatively, very compact) and the final state (relatively much larger) is small. Dirac and, subsequently, Fermi described what is now known as Fermi's golden rule, which describes the transition probability w_{if} between the two quantum states, i and f, with the transition probability being proportional to the square of the transition matrix element M_{if}:

$$w_{if} = \frac{2\pi}{\hbar}\left|M_{if}\right|^2 \rho_f \qquad (8.1)$$

Here, symbols have their usual meanings, with $\hbar$ being the reduced Planck constant, also known as the Dirac constant ($\hbar = h/2\pi$), and ρ_f is the density of final states. The final state may be composed of several states with identical energies, called **degenerate states**, and in cases where there is a continuum of final states, the density of final states will be expressed as a function of energy and for a photoelectron, will be proportional to the photoelectron wave vector k. The matrix element M_{if} can be expressed in the form of an integral over all space τ in which the interaction that stimulates the transition is expressed through the operator H' that couples the initial and final states. In the case of photoabsorption, H' is often called the light-matter interaction operator:

https://doi.org/10.1515/9783110570441-008

$$M_{if} = \int \psi_f^* H' \psi_i d\tau = \left\langle \psi_f | H' | \psi_i \right\rangle \tag{8.2}$$

Here, ψ_i and ψ_f are the wavefunctions for the initial and final states, using Dirac notation (Appendix C) on the right of eq. (8.2). We can write eq. (8.1) as follows:

$$w_{if} = \frac{2\pi}{\hbar} \left| \left\langle \psi_f | H' | \psi_i \right\rangle \right|^2 \rho_f \tag{8.3}$$

For X-ray absorption, we can therefore write

$$\mu(E) \propto \left| \left\langle \psi_f | H' | \psi_i \right\rangle \right|^2 \tag{8.4}$$

Here the operator H' can be written

$$H' = (\mathbf{e} \cdot \mathbf{p}) \exp i(\mathbf{k} \mathbf{r}) \tag{8.5}$$

in which $\mathbf{e}$ is the X-ray electric field vector, $\mathbf{p}$ the electron momentum vector and $\mathbf{k}$ the forward propagation vector. The first term in the series expansion of the exponential term gives $H' \approx (\mathbf{e} \cdot \mathbf{p})$, which is the so-called dipole approximation.

8.3 A simple expression for the EXAFS

With the EXAFS, we can consider the final state wavefunction as being composed of two parts consisting of a final state wavefunction without any photoelectron backscattering, which we call ψ_0, and the change in the final state wavefunction due to photoelectron backscattering, which we call ψ_s:

$$|\psi_f\rangle = |\psi_0 + \psi_s\rangle \tag{8.6}$$

We remember from Chapter 4 (eq. (4.14)) that $\mu(E) = \mu_0(E)(1 + \chi(E))$ and we can write

$$\mu_0(E) \propto |\langle \psi_0 | H' | \psi_i \rangle|^2 \tag{8.7}$$

It can be shown that

$$\chi(E) \propto |\langle \psi_s | H' | \psi_i \rangle|^2 \tag{8.8}$$

We also now swap from considering our system in terms of the energy E to considering our system in terms of the photoelectron wave vector $k = \left[(2m_e/\hbar^2)(E - E_0) \right]^{1/2}$:

$$\chi(k) \propto |\langle \psi_s | H' | \psi_i \rangle|^2 \tag{8.9}$$

We now make a simplifying approximation. The core electron of the initial state is very tightly bound and is very compressed in space to a first approximation, mostly within a_0/Z, where a_0 is the Bohr radius,[1] and Z is atomic number. Compared to the rest of the picture, such as the distances between atoms, we will approximate this as a delta-function[2] $\delta(r)$ centred at the absorber atom. This allows us to simplify our expression for $\chi(k)$ greatly, and we can now write:

$$\chi(k) \approx \int \delta(r) e^{ikr} \psi_s(r) dr \tag{8.10}$$

$$\chi(k) \approx \psi_s(0) \tag{8.11}$$

The EXAFS $\chi(k)$ is thus essentially due to oscillations in the photoelectron wavefunction **centred at the absorber atom** arising from scattering by the neighbour atom or backscatterer atom. We can now build a simple model of the EXAFS from an absorber and backscatterer, separated by a distance R, and to keep things simple, we will assume the following:

- That the propagating photoelectron can be represented by a spherical wave
- That the backscatterer atom gives a scattering amplitude, specified by $|f_s(k)|$ with a phase shift[3] specified by $\varphi'(k) = \varphi(k) + \pi$

After a little work, it can be shown that

$$\chi(k) = \left(\frac{e^{ikR}}{kR}\right) 2k f_s(k) + \text{c.c.} \tag{8.12}$$

where c.c. is the complex conjugate and the function $f_s(k)$ is given by

$$f_s(k) = |f_s(k)| \exp i\varphi'(k) \tag{8.13}$$

We will return later to the important point that the scattering amplitude $|f_s(k)|$ and phase shift $\varphi'(k)$ are respectively the modulus and argument of the complex expression for $f_s(k)$. Combining terms in eq. (8.12), including the complex conjugate (c.c.), we can obtain

$$\chi(k) = \frac{|f_s(k)|}{kR^2} \sin(2kR + \varphi(k)) \tag{8.14}$$

This is a simple version of the EXAFS equation. The first term in eq. (8.14) $|f_s(k)|/kR^2$ relates to the EXAFS amplitude, while the $\sin(2kR + \varphi(k))$ part of eq. (8.14) relates to the EXAFS phase.

We next build upon this simple EXAFS expression by adding various terms. The first of these is the electron mean free path term. In addition to elastic photoelectron

---

1 The value of a_0 is 0.529177 Å.

2 A delta function $\delta(r)$ has a value of zero at all r except when $r = 0$.

3 The factor of π is included for the sole purpose of making our expression conform to the standard expression for the EXAFS.

backscattering, the photoelectron can also be scattered inelastically, and hence may not be able to get back to the absorber atom. Moreover, if the photoelectron returns to the absorber after the core hole that is created by the primary photoexcitation event has decayed, then the backscattering will not contribute to the EXAFS. To account for both these effects, inelastic scattering and core-hole lifetime losses, we include a damping function, $\exp(-2R/\lambda(k))$, in which $\lambda(k)$ is the photoelectron mean free path, not to be confused with the de Broglie wavelength, which (unfortunately) uses the same symbol. This gives the following equation:

$$\chi(k) = \frac{|f_s(k)|}{kR^2} e^{-2R/\lambda(k)} \sin(2kR + \varphi(k)) \tag{8.15}$$

Next, we approximate thermal and static disorder as a Gaussian probability distribution function P in R, where σ is the root-mean-square deviation in R:

$$P(R) = \exp\frac{-\left(R - \bar{R}\right)^2/2\sigma^2}{\sqrt{2\pi\sigma^2}} \tag{8.16}$$

And incorporating this into our EXAFS expression gives

$$\chi(k) = \frac{|f_s(k)|}{kR^2} e^{-2k^2\sigma^2} e^{-2R/\lambda(k)} \sin(2kR + \varphi(k)) \tag{8.17}$$

For N equivalent backscatterers, we can simply scale eq. (8.17) by N:

$$\chi(k) = N\frac{|f_s(k)|}{kR^2} e^{-2k^2\sigma^2} e^{-2R/\lambda(k)} \sin(2kR + \varphi(k)) \tag{8.18}$$

For systems with more than one atom type, we can now sum the individual EXAFS over atom types i to model the total EXAFS using backscatterer specific amplitude and phase functions, $f_{s,i}(k)$ and $\varphi_i(k)$:

$$\chi(k) = \sum \frac{N_i|f_{s,i}(k)|}{kR_i^2} e^{-2k^2\sigma_i^2} e^{-2R_i/\lambda(k)} \sin(2kR_i + \varphi_i(k)) \tag{8.19}$$

This is a useful expression for the EXAFS, which, given the functions $|f_{s,i}(k)|$ and $\varphi_i(k)$, could be used for analysis of real experimental data. In the early days of EXAFS analysis, the phase and amplitude functions were extracted from experimental EXAFS data collected on standard compounds with known structures, but, as we will discuss below, this requirement has been mostly lifted from modern analysis by the availability of easy-to-use and highly accurate theoretical codes.

8.4 The EXAFS equation

In our discussion of the EXAFS below, we will frequently refer to the program FEFF [1]. FEFF was originally the brainchild of Professor John Rehr of the University of Washington, with contributions from many individuals, most of whom were part of John's research group. FEFF provides *ab initio* theoretical access to what are known as theoretical standards, which we discuss below, and has become close to a *de facto* standard in the analysis of EXAFS data. While FEFF is probably the most widely used, we note that there are other competing codes that are also in popular use, such as the EXCURV and GNXAS programs, which are integrated and complete data analysis program suites. FEFF differs in being the theory engine that can be employed by a very large number of different analysis programs.

The standard EXAFS expression is given as follows:

$$\chi(k) = S_0^2 s(k) \sum_i \frac{N_i |f_{\text{eff},i}(k)|}{kR_i^2} e^{-2k^2\sigma_i^2} e^{-2R_i/\lambda(k)} \sin(2kR_i + \varphi_i(k) + \phi_c(k)) \qquad (8.20)$$

where S_0^2 is a many-body amplitude reduction factor, $s(k)$ is the total central atom loss factor, N_i is the number of backscatterer atoms for each shell of backscatterer atoms i, $|f_{\text{eff},i}(k)|$ is the effective curved-wave backscattering amplitude, R_i is the mean absorber-backscatterer distance for shell i, $\varphi_i(k)$ is the phase-shift function for shell i, $\phi_c(k)$ is the central-atom phase shift function, $\lambda(k)$ is the photoelectron mean free path and the summation is over all backscatterer atom types i. This equation contains all the components that are needed to describe the EXAFS and quantifies all the physical sensitivities that we empirically described in Section 4.9. While the original expression was formulated for single-scattering EXAFS in which only two atoms are involved – the absorber and the backscatterer, with a scattering path-length of $2R_i$ (from the absorber out to atom i, and back again), in Section 8.7 below, we will consider multiple-scattering EXAFS, where the photoelectron interacts with a number of backscatterer atoms. As we will see, eq. (8.20) can also be used for multiple scattering by replacing the number of backscatterer atoms N_i with a different quantity, the multiple scattering path degeneracy Γ_i. We will now consider the various parts of eq. (8.20), and the contribution that they make to the overall EXAFS.

8.4.1 Central atom effects

These manifest through both the central atom phase-shift function, $\phi_c(k)$, which is given by eq. (8.21), the total central atom loss factor $s(k)$, given in eq. (8.22) and S_0^2, the many-body amplitude reduction factor:

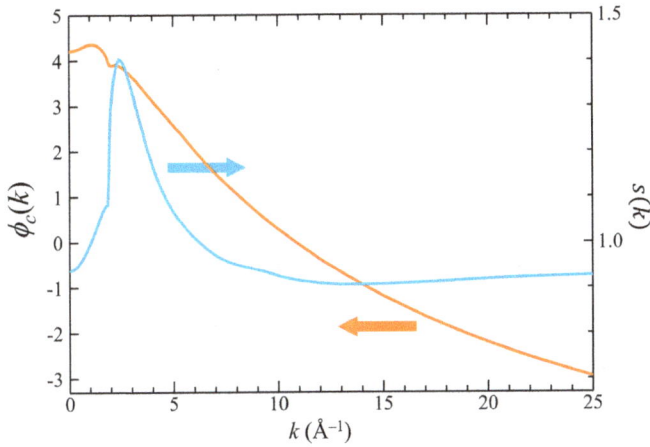

Figure 8.1: Central atom effects, showing the central atom phase-shift function $\phi_c(k)$ (orange line) and the total central atom loss factor $s(k)$ (blue line) calculated for Cu K-edge EXAFS.

$$\phi_c(k) = 2\text{Re}\left(\delta_c^l(k)\right) - l\pi \qquad (8.21)$$

Here, $\delta_c^l(k)$ is the central atom partial-wave phase shift of the final state, with l being the azimuthal quantum number; $\text{Re}()$ is the real part of this complex function (Appendix A). The factor of two in eq. (8.21) comes from the fact that the photoelectron experiences the electronic structure of the absorber atom twice, once on the way out of the atom, and once on the return after being backscattered. Note that the factor of $l\pi$ will rotate the phase of the EXAFS for different edges, for example if one were to compare the L_{III} EXAFS with the K-edge EXAFS. The central atom phase-shift typically provides the largest contribution to the overall phase shift.

The two other parts of eq. (8.20) relevant to the central atom both alter the amplitude of the EXAFS. The first of these that we will consider is S_0^2, the many-body amplitude reduction factor; which is always close to 1.0; the FEFF code can calculate an estimate for S_0^2 and it does this by computing the square of the determinant of overlap integrals for core orbitals, with and without the core hole. A typical value for S_0^2, in this example calculated for Cu K-edge EXAFS, would be 0.950, with many analyses assuming a value of 1.0 for this term. Discrete from S_0^2, $s(k)$ is the total central atom loss factor, which is related to the central atom partial-wave phase shift by

$$s(k) = \exp\left(-2\text{Im}\left[\delta_c^l(k)\right]\right) \qquad (8.22)$$

Like S_0^2, the value of $s(k)$ tends to be close to unity, especially at higher values of k; Figure 8.1 shows this function for Cu K-edge EXAFS.

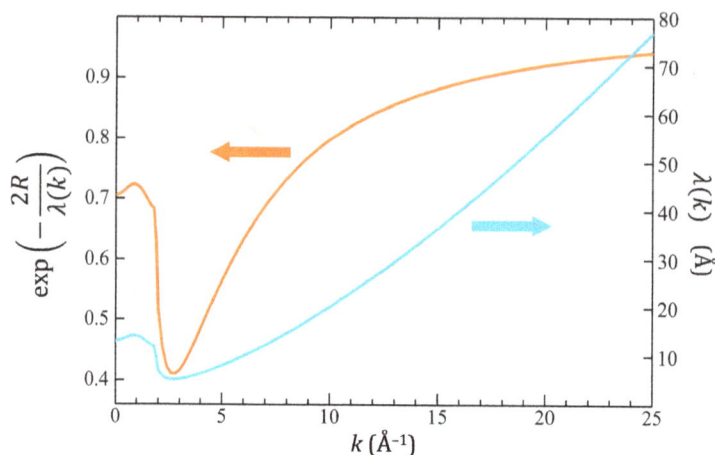

Figure 8.2: Photoelectron mean free path $\lambda(k)$ (blue line) and the corresponding amplitude adjustment (orange line) calculated for Cu–S EXAFS with $R = 2.275$ Å.

8.4.2 The photoelectron mean free path

As noted in Section 8.3, the EXAFS is due to elastic scattering. Any inelastic scattering will serve to reduce the EXAFS amplitude, as will the effects of core-hole decay; the core hole created by the primary photoexcitation event has a finite lifetime, typically measured in fractions of a femtosecond, and if the photoelectron returns to the absorber after the core hole has decayed, the backscattering will not contribute to the EXAFS. Both effects are captured within the exponential term $\exp(-2R_i/\lambda(k))$, where $\lambda(k)$ is the photoelectron mean free path. FEFF calculates $\lambda(k)$ automatically using the computed local momentum $p(k)$ with tabulated values for the core-hole lifetime built into the code using $\lambda(k) = 1/\mathrm{Im}(p(k))$. Prior to the availability of the FEFF program and others like it, values for $\lambda(k)$ were usually derived from what was known as the universal curve, which was derived from a large number of experimental measurements, and empirically given as follows:

$$\lambda(E - E_0) \approx \frac{A_0}{(E - E_0)^2} + B_0\sqrt{(E - E_0)} \tag{8.23}$$

where, with E in eV, $A_0 = 1{,}430$ and $B_0 = 0.54$, giving λ in Å. In modern EXAFS analysis, however, the universal curve is not typically used, as much better approximations of $\lambda(k)$ are available. Figure 8.2 shows FEFF calculations of $\lambda(k)$, together with the evaluated exponential term. The mean free path term serves to increase the EXAFS amplitude at high k because photoelectrons with more kinetic energy tend to have a higher mean free path. Together with the $1/R^2$ term and the Debye-Waller term, the mean free path term also contributes to the limited distance range of EXAFS, which, depending on the system, is only about 5–10 Å.

8.4.3 The photoelectron backscattering phase and amplitude

As in the simple model that we developed in Section 8.3, the backscattering is defined by a complex expression, where the backscattering amplitude is given by the modulus and the backscatterer phase-shift by the argument, as in eq. (8.22). The effective complex backscattering term $f_{eff}(k)$ gives the FEFF software its name:

$$f_{eff}(k) = |f_{eff}(k)| \exp(i\varphi(k)) \tag{8.24}$$

Figure 8.3A shows the computed phase and amplitude functions for Cu–S backscattering ($R = 2.275$ Å). The amplitude function shows a cusp at close to k of 1.5 Å^{-1}, along with a discontinuity in the phase function, which is not present in the plot of $f_{eff}(k)$ in Figure 8.3B, which shows $f_{eff}(k)$ to be a smoothly varying function, with no cusps or discontinuities. The same goes for apparent anomalies (discontinuities), which have been observed for the phase function $\varphi(k)$ for some of the heavier elements; for example, there is an abrupt shift in $\varphi(k)$ just above $k = 6$ Å^{-1} between $Z = 83$ (Bi) and 84 (Po) (Figure 8.4), and this is accompanied by a drop in the amplitude function $|f_{eff}(k)|$ to near-zero (not illustrated) at the same k-value. Such an apparent anomaly crops up for various elements in the periodic table and has been called a generalized Ramsauer-Townsend effect. [2]. However, as has been pointed out by McKale et al. for the elements of Figure 8.4 [3], the complex function $f_{eff}(k)$ shows a smooth variation with Z, with the apparent phase change being due to $f_{eff}(k)$ rotating through zero in the complex plane (Figure 8.4).

Returning now to a practical viewpoint, it is the backscatterer phase and amplitude functions that confer sensitivity to backscatterer type (Z), and which are vitally important in the viability of EXAFS as a tool for structural analysis. Figure 8.5 shows $|f_{eff}(k)|$ and $\varphi(k)$ for the complete series of backscatterers, with $5 \leq Z \leq 96$. In general, the larger the atom, the more structure is present in $|f_{eff}(k)|$ and , and in the main, these functions confer sufficient sensitivity that EXAFS can distinguish between backscatterer atoms with different atomic numbers, for example between sulfur ($Z = 16$) and oxygen ($Z = 8$). However, distinguishing between similar-sized atoms, such as between sulfur and chlorine ($Z = 16$ and 17) or between nitrogen and oxygen ($Z = 7$ and 8), while not impossible, poses considerable challenges.

8.4.4 The Debye-Waller factor

The Gaussian expression in eq. (8.20), $\exp(-2k^2\sigma_i^2)$ is known as the Debye-Waller factor. Here, σ_i^2 is the mean square deviation of the absorber backscatterer distance from the mean R_i. Although it is common practice to refer to the value of σ_i^2 as the

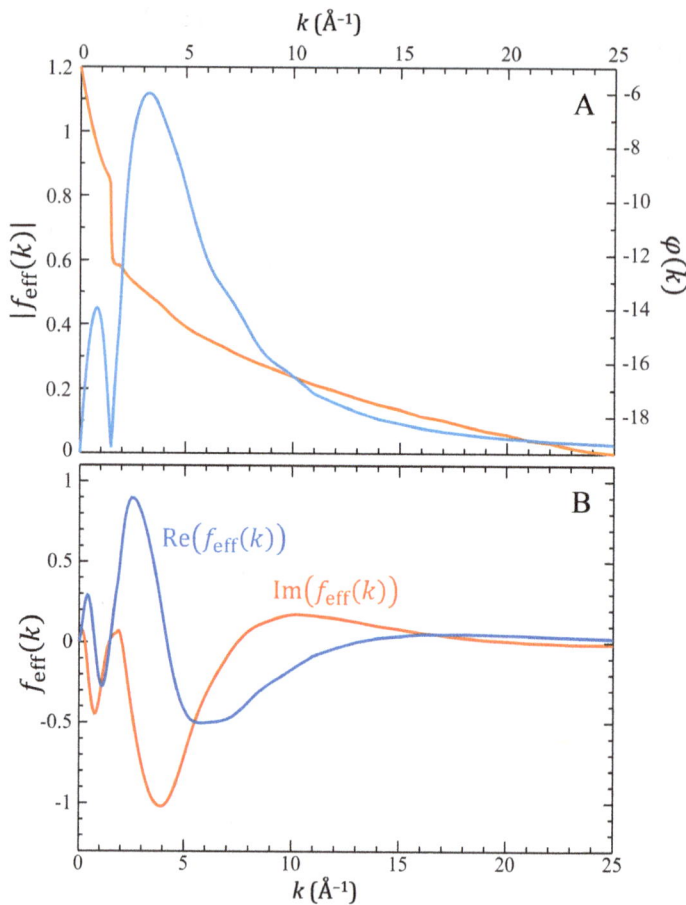

Figure 8.3: Panel A shows a comparison of the computed phase $\varphi(k)$ (orange line) and amplitude $|f_{\text{eff}}(k)|$ (blue line) functions for Cu–S backscattering. Panel B shows the real and imaginary components of $f_{\text{eff}}(k)$.

Debye-Waller term, this is strictly speaking incorrect as the whole of $\exp\left(-2k^2\sigma_i^2\right)$ is actually the Debye-Waller factor. This Gaussian model of the pair distribution function, eq. (8.16), between the absorber and backscatterer is valid at low temperatures up to somewhat above room temperature. However, at higher temperatures it becomes invalid, as we discuss below, since the function becomes anharmonic, and a cumulant expansion must be used. With our Gaussian model, which is valid for most measurements, each σ^2 is comprised of static and vibrational components, such that

$$\sigma^2 = \sigma_{\text{vib}}^2 + \sigma_{\text{stat}}^2 \tag{8.25}$$

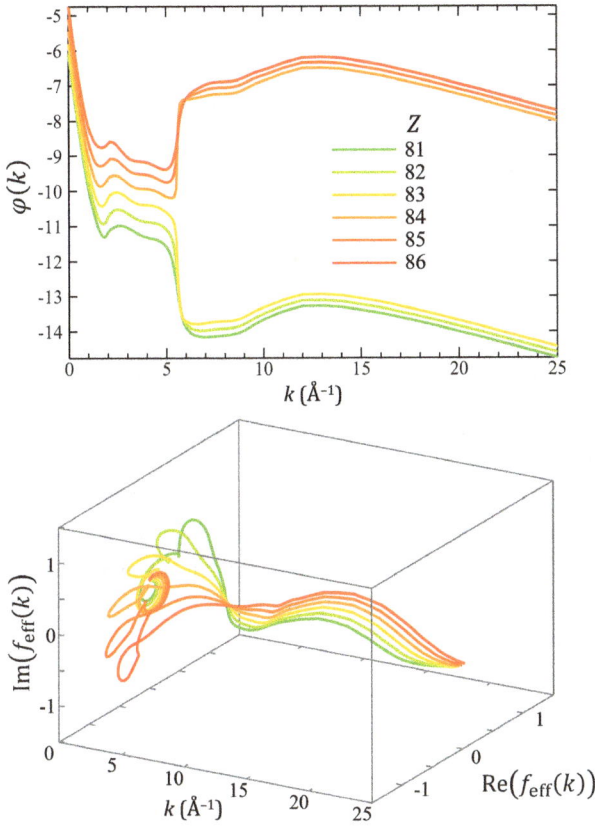

Figure 8.4: The upper panel shows a comparison of the theoretical backscatterer phase functions $\varphi(k)$ for the elements Tl, Pb, Bi, Po, At and Rn ($Z = 81$–86), showing an ~2π jump in $\varphi(k)$ above $k = 6$ between $Z = 83$ and 84. The lower panel shows real and imaginary parts of the complex function $f_{\mathrm{eff}}(k)$, showing that in fact $f_{\mathrm{eff}}(k)$ changes smoothly with Z, and that the apparent discontinuity in $\varphi(k)$ is due to $f_{\mathrm{eff}}(k)$ rotating through zero in the complex plane.

Considering the vibrational components, σ_{vib}^2, eq. (8.26) provides a general expression for these:

$$\sigma_{\mathrm{vib}}^2 = \frac{\hbar}{2\mu} \int\limits_0^{\omega_{\mathrm{max}}} \frac{1}{\omega} \rho_R(\omega) \coth\left(\frac{\hbar\omega}{2 k_B T}\right) d\omega \tag{8.26}$$

in which, ω is the angular vibrational frequency, $\rho_R(\omega)$ is the vibrational density of states projected along the absorber backscatterer pair, μ is the reduced mass, k_B Boltzmann's constant and T temperature. As a reminder, the reduced mass is the effective inertial mass, which for two atoms, is given by $\mu = m_1 m_2 / (m_1 + m_2)$ or $1/\mu = 1/m_1 + 1/m_2$, where m_1 and m_2 are the respective masses of atoms 1 and 2. The angular frequency ω, usually measured in radians per second, is related to the so-called ordinary frequency υ in cycles

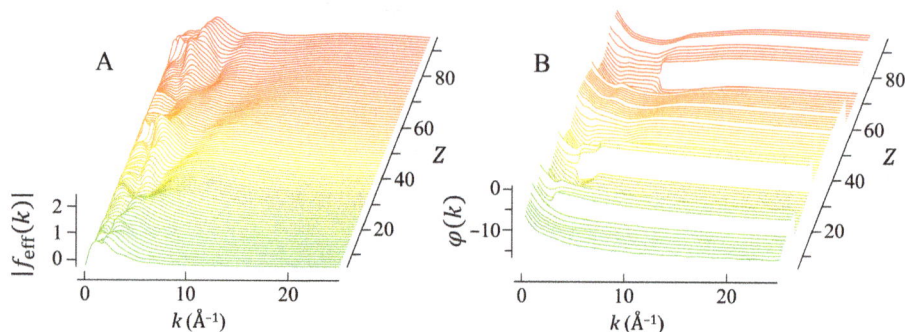

Figure 8.5: Comparison of computed EXAFS amplitude functions $|f_{eff}(k)|$ (A) and phase functions $\varphi(k)$ (B) for different backscatters as a function of atomic number Z, from boron ($Z = 5$) to curium ($Z = 96$). Several occurrences of the so-called generalized Ramsauer-Townsend effect (apparent discontinuity discussed above) can be seen in panel B with the progression of Z.

per second, by $\omega = 2\pi\upsilon$. Thus, systems with fewer vibrational modes will show lower σ_{vib}^2. This is illustrated in Figure 8.6, which shows Mo K-edge measurements of two different compounds with the same overall local Mo coordination, but with one exhibiting fewer degrees of freedom due to restricted Mo–S coordination.

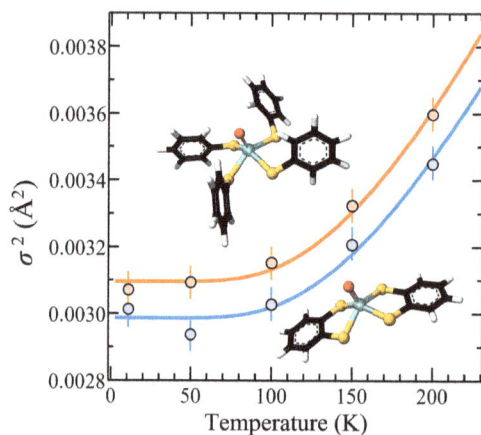

Figure 8.6: Mo–S σ^2 versus temperature for two different Mo complexes, in which the Mo has 4 Mo–S ligands and one terminal Mo = O. While coordination of the Mo site is similar, the complex with greater Mo–S freedom (upper) shows consistently larger σ^2. The insets show the structures of the compounds corresponding to the measurements of the adjacent curves.

For a diatomic harmonic oscillator, eq. (8.26) reduces to

$$\sigma_{vib}^2 = \frac{\hbar}{2\mu\omega}\coth\left(\frac{\hbar\omega}{2k_BT}\right) \tag{8.27}$$

For solids, we can use the Einstein model as follows:

$$\sigma_{vib}^2 = \frac{\hbar}{2\mu\omega_E}\coth\left(\frac{\hbar\omega_E}{2k_BT}\right) \tag{8.28}$$

where ω_E is the Einstein frequency, which is related to the force constant of the bond k_0 by $\omega_E = \sqrt{k_0/\mu}$. In all of these, the general form used for describing σ_{vib}^2 is $\sigma_{vib}^2 = a\coth(b/T)$, in which a and b are constants independent of temperature T, with σ_{vib}^2 settling to a plateau at low temperatures and increasing linearly with temperature at higher T.

Isotope effects are often not considered to be particularly important in EXAFS, but they do manifest at low temperatures in the value of σ_{vib}^2. From eq. (8.26), we can predict that for low temperatures, isotope effects might be observed, as $\lim_{T\to 0}\sigma_{vib}^2 = \hbar\sqrt{\mu/k_0}/2$, whereas for high temperatures, isotope effects are not expected, as $\lim_{T\to\infty}\sigma_{vib}^2 = k_BT/k_0$, so σ_{vib}^2 will increase linearly with T. There are very few studies showing these effects; at the time of writing, the most notable is an experimental study of the temperature-dependent EXAFS of two isotopes of elemental germanium, ^{70}Ge and ^{76}Ge, by Purans et al. [4] between 20 and 300 K. This investigation shows the expected low-temperature plateau values σ_{vib}^2 for ^{70}Ge and ^{76}Ge differing by 3.6% (as expected, ^{70}Ge being slightly higher), with this difference being quantitatively consistent with the Einstein model (8.26), although the experimental absolute values of σ_{vib}^2 were slightly lower than the calculated ones. Recently, a more detailed theoretical study, including cumulant expansions valid at higher temperatures, has been reported [5].

The static contribution to σ^2, σ_{stat}^2 is due to structural distributions of slightly different interatomic distances. σ_{stat}^2 can be calculated using the following equation:

$$\sigma_{stat}^2 = \frac{1}{n}\sum_{j=1}^{n}(R_j - \bar{R})^2 \tag{8.29}$$

Here, n is the total number of inequivalent absorber-backscatterer pairs with individual distance R_j, and $\bar{R}$ is the mean of all R_j. However, the differences between these distances must be smaller than the EXAFS resolution (to be discussed in Section 10.8.1), or they will be resolved as discrete backscatterers. To satisfy the requirement that these are unresolved, we can specify that $|R_j - \bar{R}| \leq \pi/2k_{max}$, where k_{max} specifies the extent of the experimental data. As pointed out by Cotelesage et al., [6] for systems that are not complex mixtures of many species, it is therefore possible to place physical bounds upon the value of σ^2. For example, for two bonds with lengths R_1 and R_2, we can write:

$$\sigma_{stat}^2 = \left(\frac{R_1 - R_2}{2}\right)^2 \qquad (8.30)$$

so that

$$\sigma_{stat}^2 \leq \left(\frac{\pi}{4k_{max}}\right)^2 \qquad (8.31)$$

Thus, for data with $k_{max} = 14$ Å^{-1}, the upper bound for σ_{stat}^2 for this simple case will be 0.0031 Å^2. Table 8.1 compares the maximum values for σ_{stat}^2 and the bond-length resolution δR for different values of k_{max} for the two bond-length example.

Table 8.1: EXAFS resolution and maximum σ_{stat}^2 for different extents of data.

k_{max} (Å^{-1})	δR_{min} (Å)	Maximum σ_{stat}^2 (Å^2)
10	0.157	0.0062
12	0.131	0.0043
14	0.112	0.0031
16	0.098	0.0024
18	0.087	0.0019

At high temperatures, the Gaussian model for the pair distribution function becomes invalid, and an anharmonic or asymmetric pair distribution must be considered. Figure 8.7 shows pair distribution functions for the simple Gaussian (symmetric) case and the anharmonic case. To describe the analytical approaches to this, we first consider the EXAFS of a single backscatterer, which would be conventionally described as follows:

$$\chi(k) = S_0^2 s(k) \frac{N|f_{eff}(k)|}{kR^2} e^{-2k^2\sigma^2} e^{-2R/\lambda(k)} \sin(2kR + \varphi(k) + \phi_c(k)) \qquad (8.32)$$

A cumulant expansion approach has been successful for specifying the shape of an anharmonic pair distribution function in EXAFS analysis. This employs a series of cumulants, ς_n, which are described as the mean value or position (ς_1), the variance or width (ς_2), the asymmetry (ς_3) and the sharpness (ς_4), which expresses the extent of the tails relative to a Gaussian. Thus, ς_1 will be equivalent to R in our standard expression, eq. (8.32) and ς_2 will be equivalent to σ^2.

Using the cumulant formalism, our expression for the EXAFS of a single backscatterer would be

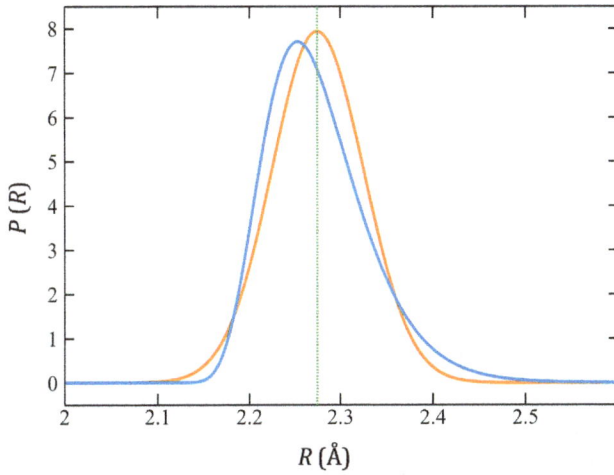

Figure 8.7: Pair distribution functions $P(R)$ for Gaussian (orange line) and anharmonic (blue line) models, appropriate for low and high temperatures, respectively. The green broken line shows the position or centroid, corresponding to $\varsigma_1 (R)$.

$$\chi(k) = S_0^2 s(k) \frac{N|f_{\text{eff}}(k)|}{k\varsigma_1^2} \exp\left(-\frac{2\varsigma_1}{\lambda(k)}\right)$$

$$\times \text{Im}\left\{ \exp i(\varphi(k) + \phi_c(k)) \exp\left(\sum_{n=1}^{n_{\max}} \frac{(2ik)^n}{n!} \varsigma_n\right)\right\} \tag{8.33}$$

in which ς_n are the cumulants. Expanding eq. (8.33) to $n_{\max} = 5$ gives the following equation:

$$\chi(k) = S_0^2 s(k) \frac{N|f_{\text{eff}}(k)|}{k\varsigma_1^2} \exp\left(-\frac{2\varsigma_1}{\lambda(k)}\right) \exp\left(-\frac{(2k)^2}{2!}\varsigma_2 + \frac{(2k)^4}{4!}\varsigma_4\right)$$

$$\times \sin\left\{ 2k\varsigma_1 - \frac{(2k)^3}{3!}\varsigma_3 + \frac{(2k)^5}{5!}\varsigma_5 + \varphi(k) + \phi_c(k)\right\} \tag{8.34}$$

Here, we can see that odd cumulants affect the phase of the EXAFS, while even cumulants affect the amplitude term. Most analyses only include cumulants up to the fourth order, and hence recalling that ς_1 and ς_2 are equivalent to R and σ^2 in our standard expression, eq. (8.32), analyses typically require two additional refinable parameters, ς_3 and ς_4, making the refinement process more cumbersome than normal and increasing uncertainties. The presence of terms involving ς_3 in the phase part of the EXAFS expression means that Fourier transform peaks will tend to shift at high temperatures unless the cumulant is included in the phase correction.

Returning now to the standard EXAFS equation, we can combine S_0^2, $s(k)$ and $|f_{\text{eff}}(k)|$ to give an overall amplitude function $A_i(k)$. The EXAFS equation in this form, eq. (8.35), is given in many publications:

$$\chi(k) = \sum \frac{N_i A_i(k)}{kR_i^2} e^{-2k^2\sigma_i^2} e^{-2R_i/\lambda(k)} \sin(2kR_i + \varphi_i(k) + \phi_c(k)) \qquad (8.35)$$

8.5 Polarized EXAFS

Here, we consider polarized EXAFS, remembering that synchrotron radiation from bend magnet or plane wiggler or undulator sources is plane-polarized (Section 2.7), with the X-ray **e**-vector in the plane of the synchrotron ring (and parallel to the direction of centripetal acceleration). With a K-edge and 1s photoexcitation, the resulting photoelectron p-wave would give rise to an additional term in the amplitude function of $3\cos^2\theta_i$, where θ_i is the angle between the X-ray **e**-vector and the absorber-backscatterer vector, as shown in Figure 8.8:

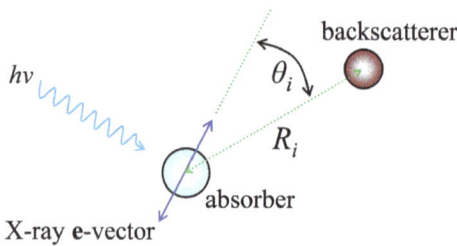

Figure 8.8: Coordinate system for polarized EXAFS of an absorber-backscatterer pair showing the angle θ_i between the direction of the X-ray electric vector **e** and the vector between the absorber and backscatterer atoms, which are separated by the interatomic distance R_i.

$$\chi(k) = S_0^2 s(k) \sum_i 3\cos^2\theta_i \frac{N_i |f_{\text{eff},i}(k)|}{kR_i^2} e^{-2k^2\sigma_i^2} e^{-2R_i/\lambda(k)} \sin(2kR_i + \varphi_i(k) + \phi_c(k)) \qquad (8.36)$$

If the polarized expression is integrated over a sphere, as would be the case for randomly oriented molecules in a powder sample or a frozen solution, then $\cos^2\theta_i$ reduces to ⅓ and eq. (8.36) becomes our original unpolarized EXAFS expression eq. (8.20).

The polarization sensitivity of the EXAFS, described in eq. (8.36), can be used to probe inter-atomic orientations within crystals or other samples with a high degree of spatial order. A schematic diagram of a polarized EXAFS experiment is shown in Figure 8.9, while Figure 8.10 shows an example of a real experiment, the polarized EXAFS of a single crystal of $CuCl_2 \cdot 2H_2O$ [7].

For the L_I edge with 2s excitation, the treatment of polarized EXAFS is the same as for a K-edge, but for L_{II} and L_{III} edges, our $\Delta l = \pm 1$ selection rule indicates that the $l = 1$ 2p initial state can go to final states that have s ($\Delta l = -1$) or d ($\Delta l = +1$) symmetry. The complete expression, eq. (8.37), contains the terms M_{10} and M_{12}, which are, respectively, the radial dipole matrix elements between the 2p ($l = 1$) and the $l = 0$ and $l = 2$ final states:

$$\chi(k) = S_0^2 s(k) \sum_i \frac{N_i |f_{\text{eff},i}(k)|}{kR_i^2} e^{-2k^2\sigma_i^2} e^{-2R_i/\lambda(k)}$$

$$\times \left\{ (1 + 3\cos^2\theta_i) \frac{1}{2} |M_{12}|^2 \sin(2kR_i + \varphi'_{2i}(k)) \right.$$

$$+ \frac{1}{2} |M_{10}|^2 \sin(2kR_i + \varphi'_{0i}(k)) \tag{8.37}$$

$$+ \left. (1 - 3\cos^2\theta_i) M_{10}M_{12} \sin(2kR_i + \varphi'_{02i}(k)) \right\}$$

$$\times \frac{1}{|M_{12}|^2 + \frac{1}{2}|M_{10}|^2}$$

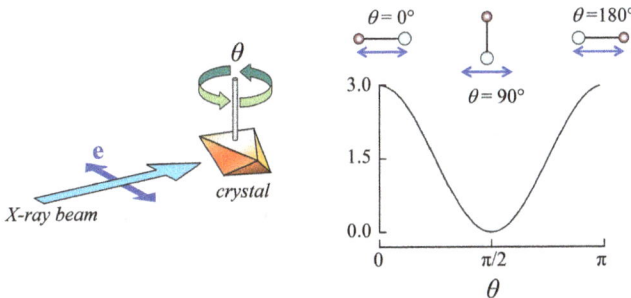

Figure 8.9: Schematic of a polarized single crystal EXAFS experiment (left) showing the polarization dependence of one absorber backscatterer (a–b) pair (right). Within the crystal, the a–b pair is aligned in the horizontal plane and is parallel to the **e**-vector (blue) at $\theta = 0$. The a–b pairs on the right are viewed from above.

Equation (8.37) contains three terms, two of which show polarization dependence. The first angular term corresponds to a de Broglie wave of d-symmetry, with an angular dependence $(1 + 3\cos^2\theta)/2$ and a phase shift $\varphi'_2(k) = \phi_c^l(k) + \varphi(k)$, where $\phi_c^l(k)$ is the central atom phase shift of the $l = 2$ outgoing photoelectron. The second term in eq. (8.37) has no polarization dependence because of the spherical symmetry of the $l = 0$ outgoing photoelectron wave and contains a phase-shift term with a central atom phase-shift $\phi'_0(k)$, with $l = 0$. The third term in eq. (8.37) is a cross-term, involving the product $M_{10}M_{12}$, again with an angular dependence $(1 - 3\cos^2\theta)/2$. For polycrystalline or other samples comprised of random orientations, the third term becomes zero as

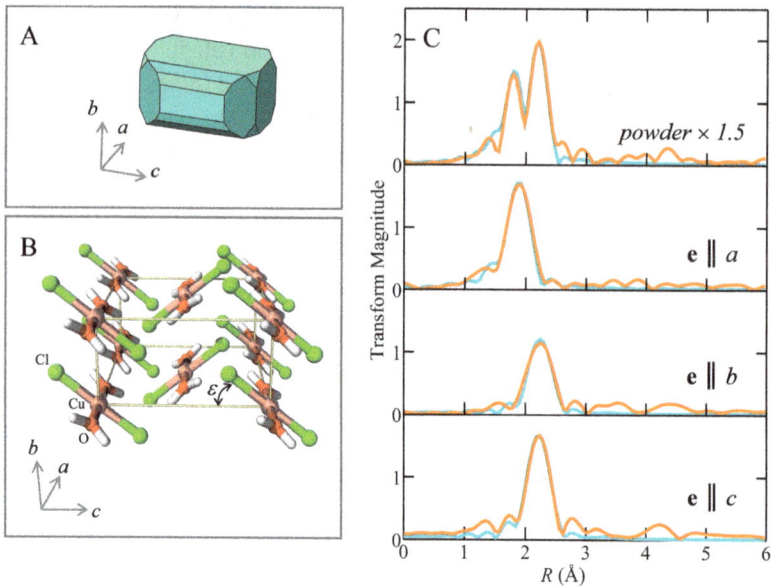

Figure 8.10: Polarized single crystal EXAFS of CuCl$_2$·2H$_2$O. (A) The crystal habit, with the relationship between the crystallographic a, b and c axes indicated. (B) The crystal structure, which consists of inclined arrangements of planar CuCl$_2$·2H$_2$O molecules, with the Cu–Cl bond tilted at an angle $\varepsilon = 38.7°$ from the crystallographic c-axis. (C) Comparison of the experimental EXAFS Fourier transforms (orange lines, Cu–Cl phase-corrected) for a powder sample (all orientations) and polarized single crystal measurements oriented so that the X-ray **e**-vector is parallel to the crystallographic a, b and c axes. The powder measurement clearly shows transform peaks due to Cu–O and Cu–Cl backscattering at 1.92 Å and 2.27 Å, respectively. With **e** ‖ a only Cu–O EXAFS is observed, while with **e** ‖ b and **e** ‖ c, no Cu–O EXAFS is observed, and the Cu–Cl EXAFS amplitude is $\propto 3\cos^2(90-\varepsilon)$, and $\propto 3\cos^2(\varepsilon)$, respectively. The blue lines in (C) show theoretical curves for the directly bonded atoms [7].

$\langle 3\cos^2\theta \rangle = 1$, and hence can be neglected. The M_{12} matrix element is expected to be somewhat larger than M_{10} because the s-symmetry final state must be orthogonal to the 1s core and is therefore more rapidly oscillatory than the $l=2$ final state in the region of the 2p wavefunction. The ratio M_{12}/M_{10} turns out to be nearly invariant with k, at least for mid-sized to heavy atoms, and has a value that is close to 5. From eq. (8.37), this means that transitions to d final states are preferred over those to s final states by a factor of about 50, and the second term in eq. (8.37) can thus be safely neglected. For randomly oriented samples, then, the treatment of L$_{III}$ and L$_{II}$ EXAFS becomes the same as for a K-edge, noting the difference in central atom phase (eq. (8.21)). But for polarized measurements, the $M_{10}M_{12}$ cross-term can be about 40% of the $|M_{12}|^2$ term and hence must be included. Figure 8.11 shows polar plots of the polarization dependence for K and L$_I$ edges, and L$_{II}$ and L$_{III}$ edges for a final state d-symmetry and for the cross-term.

We will now revert from considering phenomena associated with molecular species or extended solids to discuss a phenomenon that has its roots in the electronic structure of the atom, known as the atomic X-ray absorption fine structure, atomic-XAFS or AXAFS.

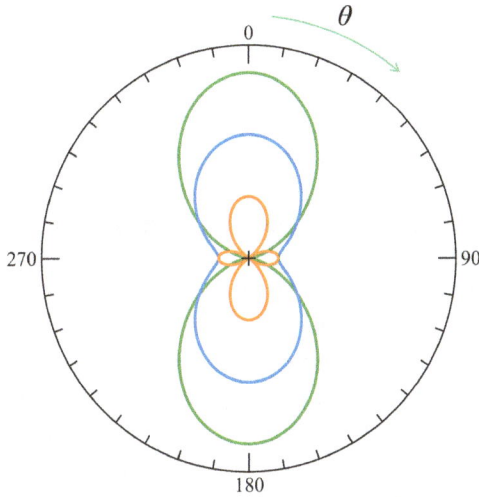

Figure 8.11: Polar plots of angular dependence of EXAFS amplitudes for K- and L_I-edges (green line), L_{II} and L_{III} edges final state d-symmetry (blue line), and the cross-term discussed in the text (orange line).

8.6 Atomic X-ray absorption fine structure

All XAS data show very short-range oscillations that are **not** due to photoelectron backscattering by neighbour atoms. This is the atomic X-ray absorption fine structure, or AXAFS, which arises from photoelectron backscattering by the interstitial charge; shells of electrons bound to the absorber atom. In order to account for the AXAFS [8], we need to adjust our equation from Chapter 4, which we stated as $\mu(E) = \mu_0(E)(1+\chi(E))$. The presence of AXAFS means that $\mu_0(E)$, as previously defined, is not a smooth function, and we must define a new smooth function $\mu_{00}(E)$ that lacks both the EXAFS $\chi(E)$ and the AXAFS $\chi_a(E)$ so that

$$\mu_0(E) = \mu_{00}(E)(1+\chi_a(E)) \tag{8.38}$$

Our expression for the absorption coefficient then becomes

$$\mu(E) = \mu_{00}(E)(1+\chi_a(E))(1+\chi(E)) \tag{8.39}$$

Modern theoretical codes, such as FEFF, can compute the AXAFS. Despite this, and in almost all cases with some notable exceptions, the AXAFS has been ignored by experimenters; it is often viewed as a nuisance or the result of a "bad spline", and as such, is often removed by background subtraction procedures (i.e. the spline). The AXAFS frequently manifests as a peak in the Fourier transform at approximately half the first-shell bond length and tends to be more noticeable in systems that have inherently small EXAFS (e.g. with light atom backscatterers). Unlike the EXAFS, AXAFS Fourier transform peaks tend to be minimized by phase correction using phases for EXAFS backscatterers.

8.7 Multiple scattering EXAFS

Up to this point, we have considered the EXAFS as originating from pairs of atoms, an absorber and one backscatterer. However, beyond this simple scattering case, multiple scattering occurs, in which the photoelectron interacts with two or more backscatterers. In many molecular cases, the single-scattering paths dominate, and considering them alone is sufficient to conduct a meaningful analysis of the data. However, in many other molecular cases, and notably in inorganic materials with small unit cells, the multiple scattering is significant and must be accounted by including different multiple scattering paths in the EXAFS analysis. Multiple scattering analyses can also allow the experimenter to do something, which is frequently said that EXAFS cannot do, which is to determine aspects of geometry, such as bond angles, albeit in a limited way.

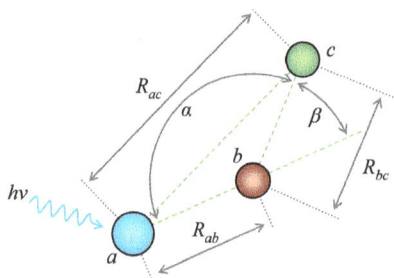

Figure 8.12: Three-atom arrangement for EXAFS multiple scattering.

As we have already mentioned, eq. (8.20) can also be used for multiple-scattering EXAFS, in the slightly modified form of eq. (8.40), which contains just a small change; the number of equivalent backscatterers N_i in eq. (8.20) is replaced by Γ_i, the scattering path degeneracy:

$$\chi(k) = S_0^2 s(k) \sum_i \frac{\Gamma_i |f_{\text{eff},i}(k)|}{kR_i^2} e^{-2k^2\sigma_i^2} e^{-2R_i/\lambda(k)} \sin(2kR_i + \varphi_i(k) + \phi_c(k)) \qquad (8.40)$$

Instead of summing over backscatterer types, as in eq. (8.20), the summation in eq. (8.40) is now over all scattering paths. Multiple-scattering EXAFS considers the different scattering paths that contribute to the overall EXAFS signal, and these paths are distinguished by the number of individual legs. Figure 8.12 shows a three-atom arrangement and defines some variables that we will use in our discussion. Figure 8.13 shows some of the different multiple-scattering paths for the same three-atom arrangement. The simplest scattering path is what we have discussed thus far and is called a **two-leg scattering path** (from absorber to backscatterer and back to absorber), otherwise known as **single scattering** and involves just the absorber atom and one backscatterer atom (Figure 8.13A and B). If we introduce a third backscatterer atom, then both **three-leg** (Figure 8.13C) and **four-leg** scattering paths (Figure 8.13D) become probable. The photoelectron phase and amplitude are affected by each backscatterer on each multiple-scattering path and is sensitive to the degree of linearity of the arrangement. For a single set of backscatterers, asymmetric paths, such as that of Figure 8.13C, have double degeneracy, since the path $a{\to}c{\to}b{\to}a$ shown in Figure 8.13C has an equivalent (reverse) path $a{\to}b{\to}c{\to}a$, yielding a path multiplicity, $\Gamma = 2$. Conversely, $\Gamma = 1$, both for the single-scattering paths of Figure 8.13A and B and the four-leg path of Figure 8.13D.

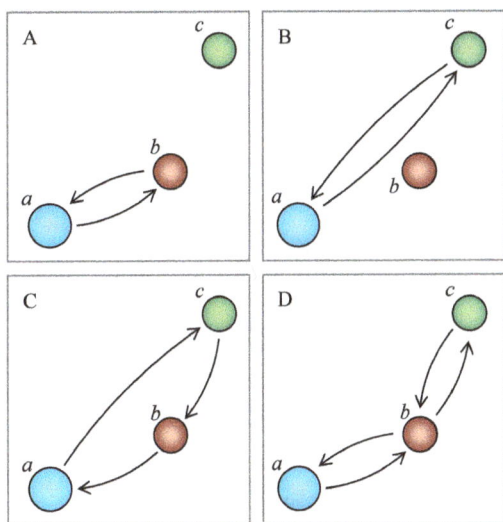

Figure 8.13: EXAFS multiple-scattering paths for a three-atom arrangement.

Neglecting the mean free path and Debye-Waller terms, the effective amplitudes A' of each scattering path A, B, C and D are given as follows:

$$A'_A(k) = \frac{|f_{\text{eff},b}(k)|}{kR_{ab}^2} \tag{8.41}$$

$$A'_B(k) = \frac{|f_{\text{eff},c}(k)|}{kR_{ac}^2} \tag{8.42}$$

$$A'_C(k,\beta) = \frac{|f_{\text{eff},c}(k)||f_{\text{eff},b}(k,\beta)|}{kR_{ac}R_{bc}R_{ab}} = \frac{|f_{\text{eff},b}(k,\beta)||f_{\text{eff},c}(k)|}{kR_{ab}R_{bc}R_{ac}} \tag{8.43}$$

$$A'_D(k,\beta) = \frac{|f_{\text{eff},b}(k,\beta)||f_{\text{eff},c}(k)||f_{\text{eff},b}(k,\beta)|}{kR_{ab}R_{bc}R_{bc}R_{ab}} = \frac{|f_{\text{eff},b}(k,\beta)|^2|f_{\text{eff},c}(k)|}{kR_{ab}^2R_{bc}^2} \tag{8.44}$$

Likewise, the phase functions of the different paths $\Phi' = \sin(2kR_i + \varphi_i(k) + \phi_c(k))$ are also modified, noting that Φ'_A and Φ'_B, like A'_A and A'_B, are simply the single-scattering relationships, and we will therefore only give phase functions for paths C and D:

$$\Phi'_C(k,\beta) = \sin[2k(R_{ac} + R_{bc} + R_{ab}) + \varphi_c(k) + \phi_b(k,\beta) + \phi_c(k)] \tag{8.45}$$

$$\Phi'_D(k,\beta) = \sin[2k(2R_{ab} + 2R_{bc}) + 2\varphi_b(k,\beta) + 2\varphi_c(k,\beta) + \phi_c(k)] \tag{8.46}$$

Both amplitude and phase are influenced by each atom in the multiple-scattering path, as a product and a sum, respectively, with sensitivity to the scattering angle β. Figure 8.14 shows calculated EXAFS Fourier transforms for a hypothetical three-atom Fe–C $\equiv$ N arrangement, for a range of bond angles, α. The EXAFS corresponding to the major paths for $\alpha = 180°$ is shown in Figure 8.15.

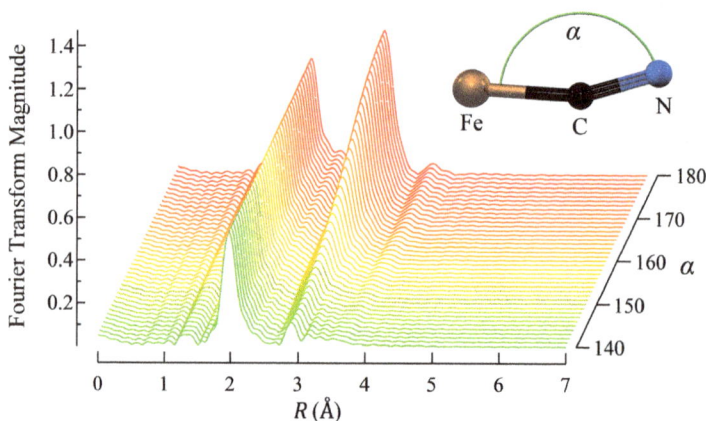

Figure 8.14: Calculated EXAFS Fourier transforms for a hypothetical three-atom Fe–C $\equiv$ N arrangement with varying bond angle α using physically reasonable Debye-Waller factors, and assuming Fe–C and C $\equiv$ N bond lengths of 1.963 and 1.158 Å, respectively. The Fourier transforms are phase-corrected using phase functions for Fe–C first-shell backscattering. The remarkable increase in the ~3.12 Å transform peak is due to the angularly dependent three- and four-leg paths discussed in the text.

The angular range in Figure 8.14 has been chosen because bridging cyanide ligands, in which the carbon links two metals, typically have a bond angle close to 140°, while terminal cyanide ligands tend to be closer to 180°. This figure clearly illustrates the capability to determine the bond angle a.

The required number of EXAFS scattering legs to adequately model the multiple scattering increases as twice the number of atoms involved in the multiple scattering. Thus, for two backscatterers, a minimum of four legs must be included, while for three backscatterers, a minimum of six legs must be included. This simplistic view excludes lower amplitude multiple scattering paths, which can become quite complex – for example, one such low probability path would be the five-leg path $a{\rightarrow}b{\rightarrow}c{\rightarrow}b{\rightarrow}c{\rightarrow}a$ – but in most cases of EXAFS analysis, such paths are sufficiently weak that they can be neglected. For low Z backscatterers of about the same atomic number and a linear geometry, a three-leg multiple scattering path will be about 90° out of phase with the two-leg path. A four-leg multiple scattering path will be 180° out of phase with the two-leg path, and so-on in increments of 90° per leg, depending on the number of backscatterers involved, as shown in Figure 8.15.

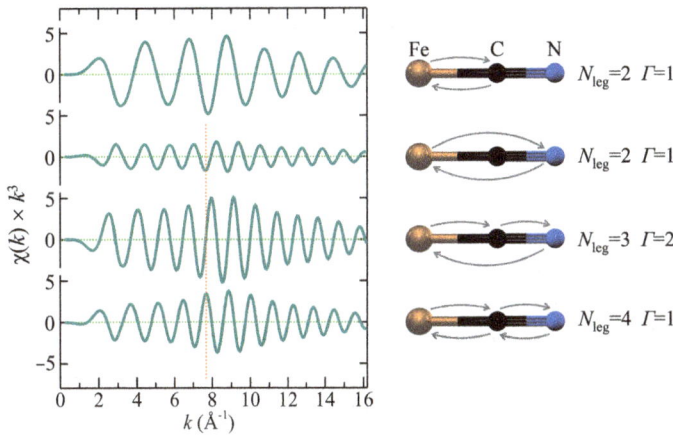

Figure 8.15: EXAFS oscillations from the different major scattering paths, calculated using the three-atom Fe–C ≡ N arrangement of Figure 8.14 with $a = 180°$. The broken orange line is shown to guide the eye to changes in the phase of the EXAFS for the different paths.

Other multiple scattering paths can also be significant; a common example is a linear arrangement of three atoms, in which the absorber is the central atom. Such arrangements occur with linear digonally coordinated Cu(I) or Hg(II) complexes, such as $HgCl_2$, or trans-**dioxo** systems such the uranyl cation [9] $[O \equiv U \equiv O]^{2+}$ or neutral the Mo(IV) species $[O \equiv Mo \equiv O]^0$, found in complexes such as $[MoO_2(CN)_4]^{4-}$. Here, three- and four-leg paths contribute significantly to the EXAFS, resulting in a small but significant

Fourier transform peak at around twice the first shell interatomic distance. An example is provided in Figure 8.16, the linear $HgCl_2$ molecule.

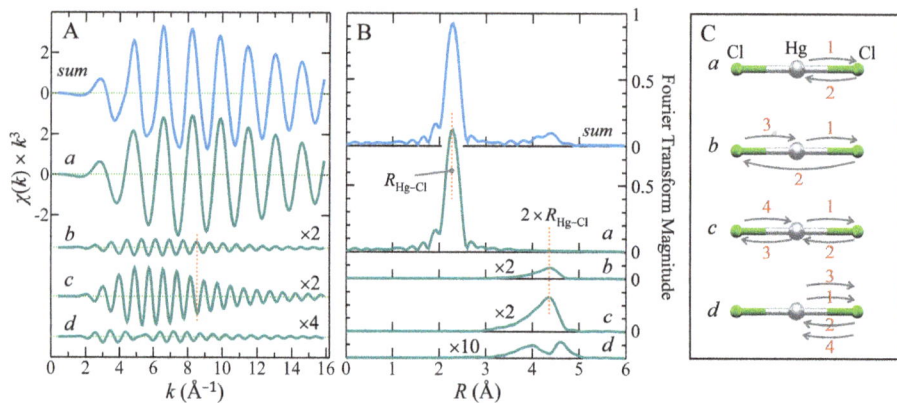

Figure 8.16: EXAFS multiple scattering in the linear $HgCl_2$ molecule. Panel A shows the Hg L$_{III}$ EXAFS oscillations, B, the Hg–Cl phase-corrected EXAFS Fourier transforms, and C shows the various scattering paths corresponding to the component curves *a–d* in A and B. The vertical broken line in A shows the relative phases of the EXAFS for paths *b* and *c*.

In Figure 8.16, the asymmetric three-leg path *b* has $\Gamma = 2$, while all others shown have $\Gamma = 1$. It is clear that path *b* and path *c* are close to 180° out of phase with each other, reducing the overall amplitude at twice the first shell Hg–Cl distance. The extent to which this cancellation occurs in related complexes will depend upon the absorber atom, acting as it does here as a backscatterer. Thus, while Hg(II) dithiolate complexes show similar multiple scattering to that of Figure 8.16, the structurally similar linear cuprous dithiolate complexes show less cancellation of the EXAFS from paths analogous to *b* and *c*. We note in passing that some multiple scattering paths have been given whimsical names, so that paths like Figure 8.16d might be called "rattle" and others have been called by more descriptive names such as "shadow", "dogleg" and "triangle" [10]. Figure 8.16 also shows an example of a low-intensity multiple scattering path that may be observed (the so-called rattle path). Such back-and-forth-type multiple scattering is less intense than the end-to-end multiple scattering. The reason for this is that true to the old name **forward focusing**, the angular profile of $|f_{eff}(k,\beta)|$ is folded sharply forward for $\beta \rightarrow 0°$, with less intensity in the reverse direction, and particularly so at high k. A polar plot of $|f_{eff}(k,\beta)|$ for chlorine at different values of k is shown in Figure 8.17. The k-dependence of $|f_{eff}(k,\beta)|$ for $\beta = 180°$ is also apparent in Figure 8.16, for example in the four-leg path *c*, where the EXAFS amplitude is maximal at close to $k = 6$ Å^{-1} (within the green of Figure 8.17) and is substantially diminished at $k = 14$ Å^{-1}.

At still lower k values, $|f_{eff}(k,\beta)|$ becomes increasingly isotropic, such that close to the absorption edge complex multiple scattering paths involving many atoms near to

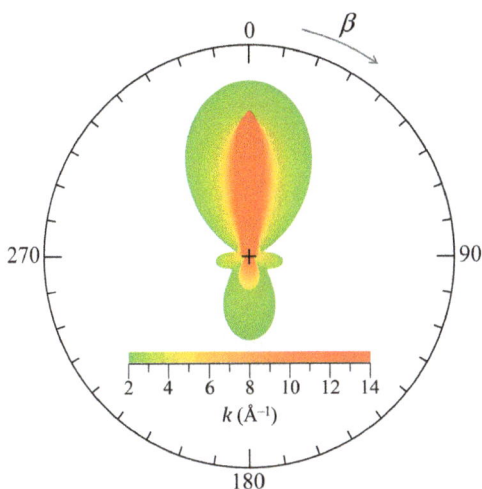

Figure 8.17: Polar contour plot of $|f_{eff}(k,\beta)|$ versus β for a chlorine backscatterer, calculated from $k = 2$–14 Å^{-1}, with the graded contour colouring indicating the k-value ($k = 2$ green, $k = 14$ red).

the absorber become important. This relates to the use of multiple scattering to calculate near-edge (XANES) spectra, which will be discussed in a later chapter. Multiple scattering analyses often necessarily require some degree of foreknowledge of the structures being investigated; hence multiple scattering contributions for a complete unknown might be missed, or perhaps miss-assigned.

All of EXAFS analysis is through model building. In the simple cases, this would consist of a postulated radial structure, testing distances and numbers of ligands and shells. When complex multiple scattering is prominent, then analysis may also need to rely on 3-dimensional models, often using other tools such as density functional theory [6], which can be used to compute reasonable structures, followed by refinement of EXAFS parameters. This is a work in progress but is perhaps the way forward. And indeed, the true power of EXAFS may only be manifested when combining methods. Such a holistic approach might include EXAFS, crystallography and computational chemistry, and would allow the development of the most accurate structural picture possible.

References

[1]   Ankudinov, A. L.; Ravel, B.; Rehr, J. J.; Conradson, S. D. Real space multiple scattering calculation and interpretation of X-ray absorption near edge structure. *Phys. Rev. B.* **1998**, *58*, 7565–7576.
[2]   Barton, J. J.; Hussain, Z.; Shirley, D. A. Generalized Ramsauer-Townsend resonance in angle-resolved photoemssion extended fine structure oscillations. *Phys. Rev. B.* **1987**, *35*, 933–938.
[3]   McKale, A. G.; Veal, B. W.; Paulikas, A. P.; Chan, S.-K. Generalized Ramsauer-Townsend effect in extended X-ray absorption fine structure. *Phys. Rev. B.* **1988**, *38*, 10919–10921.

[4]   Purans, J.; Afify, N. D.; Dalba, G.; Grisenti, R.; De Panfllis, S.; Kuzmin, A.; Ozhogin, V. I.; Rocca, F.; Sanson, A.; Tiutinnikov, S. I.; Fornasini, P. Isotoic effect in extended X-ray absorption fine structure of germanium. *Phys. Rev. Lett.* **2008**, *100*, 055901/1–4.

[5]   Huang, N. V.; Duc, N. B.; Vuong, D. Q.; Tien, T. S.; Toan, N. C. Isotopic effects in Debye-Waller factor and in EXAFS studied based on anharmonic correlated Einstein model. *Radiat. Phys. Chem.* **2021**, *180*, 109263.

[6]   Cotelesage, J. J. H.; Pushie, M. J.; Grochulski, P.; Pickering, I. J.; George, G. N. Metalloprotein active site structure determination: Synergy between X-ray absorption spectroscopy and X-ray crystallography. *J. Inorg. Biochem.* **2012**, *115*, 127–137.

[7]   Pickering, I. J.; George, G. N. Polarized X-ray absorption spectroscopy of cupric chloride dihydrate. *Inorg. Chem.* **1995**, *34*, 3142–3152.

[8]   Wende, H.; Baberschke, K. Atomic EXAFS: Evidence for photoelectron backscattering by interstitial charge densities. *J. Electron. Spectrosc. Relat. Phenom.* **1999**, *101–103*, 821–826.

[9]   Catalano, J. G.; Brown, G. E., Jr. Analysis of uranyl-bearing phases by EXAFS spectroscopy: Influences, multiple scattering, accuracy of structural parameters, and spectral differences. *Am. Mineral.* **2004**, *89*, 1004–1021.

[10] Rehr, J. J.; Albers, R. C. Theoretical approaches to X-ray absorption fine structure. *Rev. Modern Phys.* **2000**, *72*, 621–654.

9 The near-edge structure

9.1 Introduction

We have already discussed in Section 4.8 that the near-edge portion of the X-ray absorption spectrum is comprised of transitions to bound states. For low-lying states, giving isolated peaks, the near edge can be very informative, but higher levels in many cases may overlap and obscure details of the spectrum. Nonetheless, near-edge X-ray spectroscopy has many advantages, and even a noisy spectrum of a challenging sample can be highly informative. Here we discuss some of the factors to which near-edge spectra are sensitive, along with attempting to correct some common misconceptions. The use of near-edge spectra in quantitative analysis of complex mixtures will be described in Chapter 11.

9.2 Electric dipole and electric quadrupole transitions

As we have previously discussed (see eq. (8.4)), the X-ray absorption $\mu(E)$ can be described by Fermi's golden rule:

$$\mu(E) \propto \sum \left| \left\langle \psi_f | (\mathbf{e} \cdot \mathbf{p}) \exp i(\mathbf{k} \cdot \mathbf{r}) | \psi_i \right\rangle \right|^2 \tag{9.1}$$

Here, $|\psi_i\rangle$ specifies the initial quantum state and $|\psi_f\rangle$ the final quantum state, $\mathbf{e}$ is the X-ray electric vector, $\mathbf{p}$ the electron momentum vector, $\mathbf{k}$ the X-ray forward propagation vector and $\mathbf{r}$ the electron coordinate. The summation is over all final states that contribute to $\mu(E)$. We can use the first two terms of a series expansion of the exponential to generate

$$\mu(E) \propto \sum \left| \left\langle \psi_f | (\mathbf{e} \cdot \mathbf{p}) + i(\mathbf{e} \cdot \mathbf{p})(\mathbf{k} \cdot \mathbf{r}) | \psi_i \right\rangle \right|^2 \tag{9.2}$$

With the first term, we obtain what are called the electric dipole-allowed transitions, with $\Delta l = \pm 1$, as discussed in Section 4.4, and which we can call $\mu_D(E)$:

$$\mu_D(E) \propto \sum \left| \left\langle \psi_f | (\mathbf{e} \cdot \mathbf{p}) | \psi_i \right\rangle \right|^2 \tag{9.3}$$

These are the most intense transitions observed and can be thought of as being stimulated by an oscillating electric field; they depend upon the orientation of the X-ray $\mathbf{e}$ vector, but not the direction vector $\mathbf{k}$. The next term in the series expansion gives smaller contributions, which are electric quadrupole-allowed, but electric dipole-forbidden transitions with $\Delta l = \pm 2$, which we can call $\mu_Q(E)$:

https://doi.org/10.1515/9783110570441-009

$$\mu_Q(E) \propto \sum \left| \left\langle \psi_f | i(\mathbf{e} \cdot \mathbf{p})(\mathbf{k} \cdot \mathbf{r}) | \psi_i \right\rangle \right|^2 \tag{9.4}$$

Quadrupole transitions are generally of low intensity and can be thought of as being stimulated by the electric field gradient, which for X-rays becomes increasingly significant at high X-ray energies due to the short wavelength.

For K-edges, therefore, with 1s core excitation (having $l = 0$), we expect the spectrum to be dominated by intense dipole-allowed transitions to levels with substantial p-character (having $l = 1$), i.e. $\Delta l = +1$, such as 1s→4p, along with weaker quadrupole transitions $\Delta l = +2$ such as 1s→3d. Figure 9.1 shows the spectrum of a binuclear Mn(IV) complex with approximately octahedral Mn site geometry. The most intense features of the spectrum are observed in the dipole-allowed 1s→4p region with a weak quadrupole-allowed 1s→3d pre-edge feature.

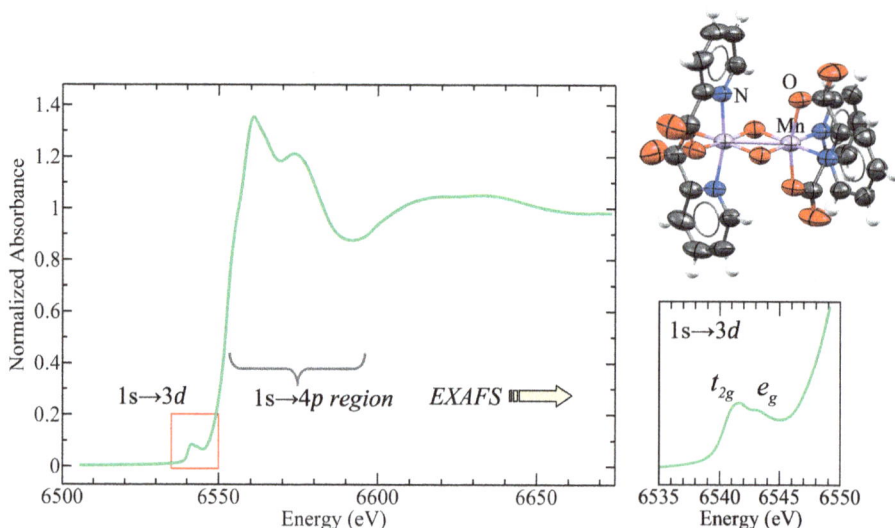

Figure 9.1: Mn K near-edge spectrum of a binuclear Mn(IV) complex exhibiting approximately octahedral site geometry for both Mn atoms. The lower inset shows the 1s→3d region expanded and the upper inset shows the structure of the complex using 75% probability thermal ellipsoids.

The intensity of the quadrupole transitions μ_Q is related to that of the dipole transitions μ_D by $\mu_Q/\mu_D = (Z\alpha)^2$ where α is the fine structure constant (Sommerfeld's constant)[1] with a value $\alpha \approx 1/137$. Hence, for the K-edge of calcium the 1s→3d quadrupole transitions are about 2.1% of the dipole, for the copper K-edge about 4.5% and for the molybdenum K-edge (1s→4d) about 9.4%. For p-block elements, such as selenium and

---

[1] The fine structure constant α is a dimensionless quantity that describes the strength of the electromagnetic interaction between charged particles.

especially for high-valent compounds such as Se(VI), which have a formally unoccupied 4p manifold, there is often an intense peak on top of the edge corresponding to dipole-allowed $\Delta l = +1$ 1s→4p transitions. Such features are often called white lines.[2] Similar white lines are observed for the L_{III}- and L_{II}-edges of the transition metal ions and of the rare earths. Quadrupole-allowed transitions and in particular the 1s→3d are of special interest, and we will return to a more detailed discussion of them below.

The above treatment neglects magnetic dipole transitions, which are governed by terms containing the cross-product $(\mathbf{k} \times \mathbf{e})$. In general, the magnetic dipole terms ought to be of the same order as the electric quadrupole, but with core-level spectroscopy for various reasons the matrix elements become vanishingly small, at least with conventional XAS, and we will not consider magnetic dipole transitions further.

Table 9.1: Major dipole and quadrupole transitions for different absorption edges.

Edge	K	L_I	L_{II}	L_{III}	M_I	M_{II}	M_{III}	M_{IV}	M_V
Core level	1s	2s	$2p_{1/2}$	$2p_{3/2}$	3s	$3p_{1/2}$	$3p_{3/2}$	$3d_{3/2}$	$3d_{5/2}$
Dipole	p	p	d	d	p	d	d	f	f
Quadrupole	d	d	f	f	d	f	f	g	g

Table 9.1 summarizes the dipole-allowed and quadrupole-allowed transitions of different edges, and Figure 9.2 shows the XAS of sodium tungstate Na_2WO_4, which is a formal $5d^0$ W(VI) compound showing all three L-edges, plus the M-edges of uranyl *tris*-acetate, $[UO_2(CH_3CO_2)_3]^-$. With tungstate the L_{III} and L_{II} edge show an intense peak at the absorption edge, corresponding to dipole-allowed transitions of the 2p electrons to the vacant 5d manifold. As expected, this intense peak is lacking for the L_I edge. Similarly, for the U M-edges of uranyl *tris*-acetate, the M_V and M_{IV} edges show intense dipole-allowed transitions to the formally vacant 5f manifold of U(VI), while these transitions are absent for the M_{III}, M_{II} and M_I absorption edges. As mentioned in Section 8.2, even the most intense features in X-ray absorption are weak compared to pronounced electronic transitions in the ultraviolet and visible parts of the electromagnetic spectrum. This is because the overlap between the wavefunctions of the core level initial state and those of the final states are small, with a very compact initial state and more diffuse final states. This same lack of overlap is responsible for the close correspondence between the features of the near-edge and the simple density of states that we have just discussed. If the wavefunctions of the initial state core levels and the valence levels overlap significantly, as is the case with first transition element L-edges, then this simple picture becomes more complicated.

---

2 The term "white line" refers to an intense absorption in the near-edge spectrum. The nomenclature dates from the days when spectra were recorded on strips of photographic film, and such intense absorption peaks showed up as a heavily exposed line on the developed film, appearing white in the photographic negative.

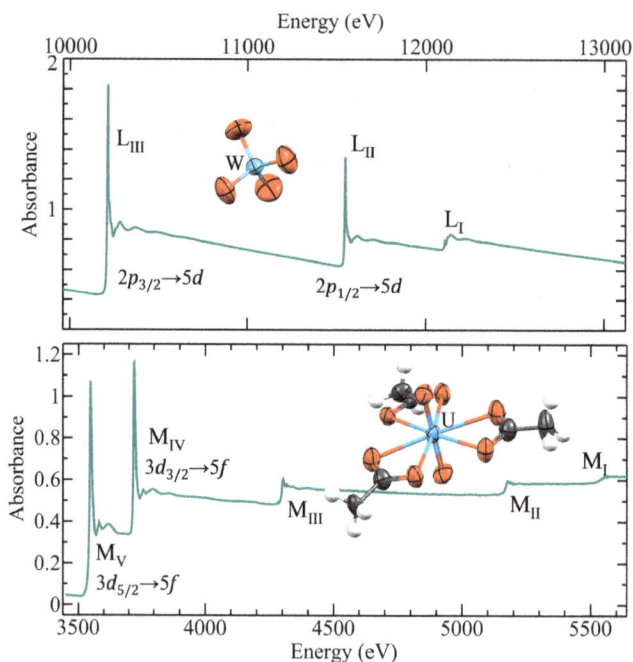

Figure 9.2: Tungsten L- and uranium M- X-ray absorption edges. The upper panel shows all three tungsten L-edges for solid sodium tungstate (Na_2WO_4) with the inset showing the structure of the tetrahedral $[WO_4]^{2-}$ anion (75% thermal ellipsoids), a formally $5d^0$ W(VI) complex. The L_{III} and L_{II} near-edge spectra can be seen to be dominated by intense dipole-allowed $2p_{3/2} \rightarrow 5d$ and $2p_{1/2} \rightarrow 5d$ transitions, respectively, which are absent for the L_I edge. The lower panel shows all five uranium M-edges for uranyl *tris*-acetate $[UO_2(CH_3CO_2)_3]^-$. Here, the M_V and M_{IV} show intense dipole-allowed transitions predominantly to the formally vacant 5f manifold, transitions which are mostly lacking for the M_{III}, M_{II} and M_I edges.

Returning to considerations relevant to the hard X-ray case, it probably serves to summarize some of what would be expected for near-edge spectra. As we have discussed, we expect to observe intense features due to dipole-allowed $\Delta l = \pm 1$ transitions plus somewhat weaker features due to quadrupole-allowed $\Delta l = \pm 2$ transitions. As we will discuss below, if there is mixing, for example p-d mixing in non-centrosymmetric transition metal complexes, then some nominally quadrupole-allowed transitions will gain dipole-allowed intensity. In the hard X-ray regime, the core hole that results from photoexcitation lies deep within the atom, and consequentially the loss of a negatively charged core electron in the final state of an absorber with atomic number Z approximates to the ground state of $Z + 1$, the next element along the period in the periodic table, corresponding to a +1 increase in nuclear charge. This can be important when comparing splittings measured from electronic spectroscopy with those observed from X-ray near-edge spectra. For example, 1s→3d splitting for a K near-edge spectrum of a Co(II) complex would approximately correspond to the optical splittings observed for an isostructural Fe(II) complex. The core hole will also cause the outer

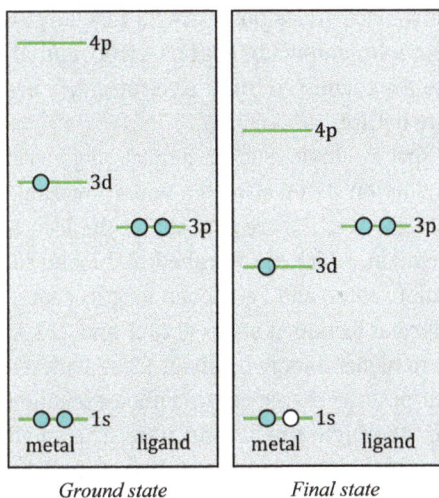

Figure 9.3: Schematic diagram (not to scale) of the energetic effects of a metal ion core electron excitation, in which the core hole causes the relaxation of the metal-based orbitals to lower energies.

orbitals to relax to lower energies, as shown schematically in Figure 9.3, for example, by close to 10 eV for Cu K-edge XAS. This in turn causes a shrinkage of their wavefunctions, and hence a reduction in overlaps for molecular orbitals involving ligand atoms. Consequentially, we expect hard X-ray near-edge spectra to be very atomic in some of their properties.

9.3 Chemical sensitivity of near-edge spectra

9.3.1 Oxidation state sensitivity

A commonly held notion is that near-edge spectra allow determination of oxidation state. Before going on to discuss this oversimplification, we briefly review what is meant by oxidation state. The oxidation state of an atom (sometimes referred to as its oxidation number) can be defined as the charge that the atom would bear if all of its bonds were completely ionic. Given this definition, and because bonding typically has some degree of covalency, we can see immediately that this might be unrealistic. We consider the sulfate anion $[SO_4]^{2-}$ as an example; with completely ionic bonds each of the four oxygen atoms would bear a -2 charge, totalling -8, and sulfate possesses an overall -2 charge, which would mean that the oxidation state of sulfur is $+6$. This might be written S^{6+}, S^{VI} or S(VI), all of which are correct. In fact, the bonds between oxygen and the central sulfur are highly covalent, and for an isolated $[SO_4]^{2-}$ in a vacuum, the charge on each oxygen would be approximately -0.68, and that on the sulfur would be approximately $+0.72$. Going down the group, the corresponding oxy-anions are selenate, tellurate and polonate, all of which are known. The charge both on oxygen and chalcogenide changes down the group, which is expected, but never gets above +1.41 (for $[TeO_4]^{2-}$),[3] despite all of the chalcogenides possessing a formally +6 oxidation state. Oxidation states are thus just a human-made characteristic, and although they are convenient in that they allow the chemist to count electrons, they are not real in that they have no simple relation to nature.

Because the near-edge X-ray absorption spectra do depend on nature, in general they do <u>not</u> a priori allow determination of oxidation states, although under favorable circumstances they can reflect oxidation state changes. Figure 9.4a shows the iron K near-edge spectra of the metalloprotein rubredoxin, which has tetrahedral thiolate coordination in formally Fe(III) and Fe(II) oxidation states, and Fe–S bond lengths changing only slightly between ferric and ferrous formal oxidation states (2.29 Å and 2.33 Å, respectively). In this case, the near-edge shifts to higher energy by about 1.3 eV with the higher oxidation state, reflecting the increased positive charge causing photoexcitation to be slightly more energetically difficult for the higher oxidation state. The small $1s \rightarrow 3d$ transition, which we will discuss in more detail below, is also shifted by

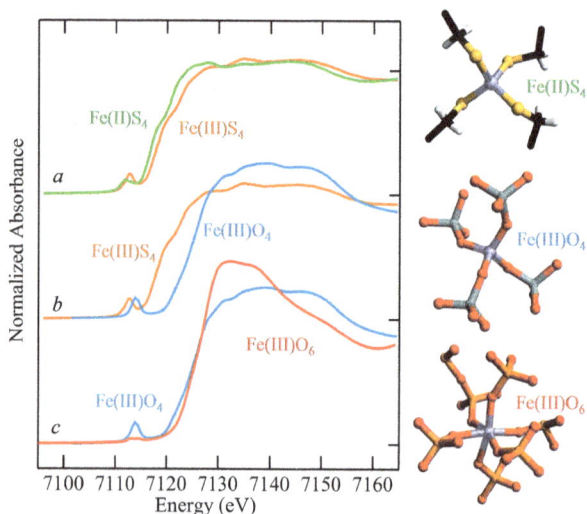

Figure 9.4: Factors influencing the near-edge spectrum. In *a* the effects of oxidation state change are compared for *Pyrococcus furiosus* rubredoxin in formally ferrous (green) and ferric (orange) oxidation states. The structure is shown in the inset on the right, with the iron exhibiting nearly tetrahedral local site geometry. The effects of ligand type are compared in *b* for two tetrahedral ferric sites, those of *P. furiosus* rubredoxin and of sodalite with iron doped into the zeolitic aluminosilicate framework. The site structure is shown to the right. The effects of coordination geometry are illustrated by *c*, which compares octahedral (ferric pyrophosphate, structure to right) and tetrahedral (iron-sodalite) coordination geometries.

---

3 Mulliken partial fitted charges were calculated using DFT (DMol[3] GGA/PBE/all-electron-relativistic cores) assuming Td point group symmetry.

about 0.9 eV to higher energy with the Fe(III) spectrum. Additional differences between the spectra of Figure 9.4a are relatively minor.

9.3.2 Sensitivity to chemical bonding

Figure 9.4b compares the effects of ligand type for a tetrahedral ferric site. The two spectra compared are those of ferric rubredoxin and Fe(III) framework-substituted sodalite (a zeolite[4]), with the iron occupying one of the normal aluminium sites. Both samples thus contain iron formally in the Fe^{3+} oxidation state with almost perfect tetrahedral coordination, differing only in the nature of the ligands to iron (sulfur versus oxygen). The near-edge spectra are dramatically different, with the oxygen-coordinated spectrum shifted to higher energy by approximately 5 eV, and the 1s→3d transition by about 1.4 eV. Hence, the effects of an altered ligand type (Figure 9.4b) upon the Fe K near-edge spectra are much greater than those of changing oxidation state (Figure 9.4a). A second example is provided in Figure 9.5, which compares the phosphorus K near-edge spectra of P_4S_{10} and P_4O_{10}. Both species contain phosphorus formally in the pentavalent oxidation state, and have similar structures, with an adamantane $P_4(O/S)_6$ core exhibiting Td point group symmetry, and one terminal and three bridging O/S ligands on each phosphorus. The oxygen compound is shifted by 5 eV to higher energy relative to the sulfur compound.

9.3.3 Sensitivity to coordination geometry

Figure 9.4c shows the effects of altered coordination geometry on the spectrum. Here we compare the spectrum of tetrahedral ferric substituted sodalite from Figure 9.4b with that of ferric pyrophosphate, which contains close to octahedrally coordinated ferric iron. While the spectra do not show large relative shifts in energy comparing iron sites coordinated by 4 or 6 oxygens, the shapes of the spectra are dramatically different, with the intensity of the 1s→3d transition being much diminished for the octahedral complex, and the upper 1s→4p portion of the spectrum at about 7,132 eV being more intense for the octahedral complex.

Thus, of the three factors considered in Figure 9.4, the effects of a change in oxidation state are considerably smaller than the other factors considered here, coordination geometry and ligand type. Our comments on oxidation state are underlined by Figure 9.6, which shows the sulfur K near-edge spectrum of the sulfate anion $[SO_4]^{2-}$ (+6 formal oxidation state) and compares the spectra of other compounds with formally S(0) and S(–II), and exhibiting entirely dissimilar spectra. Figure 9.6 demonstrates that spectra from com-

---

4 Zeolites are aluminosilicate entities containing pores and channels of molecular dimensions and have substantial industrial and academic importance. Sodalite is the simplest of the zeolites.

pounds having the same oxidation state can be dramatically different, and that spectra from compounds containing formally higher oxidation state sulfur can fall either lower in energy or higher in energy than spectra from compounds of lower formal oxidation state sulfur. We hope that this demonstrates the point that multiple factors contribute to the near-edge spectra and the inadequacy of the frequently heard statement that "XANES gives us the oxidation state".

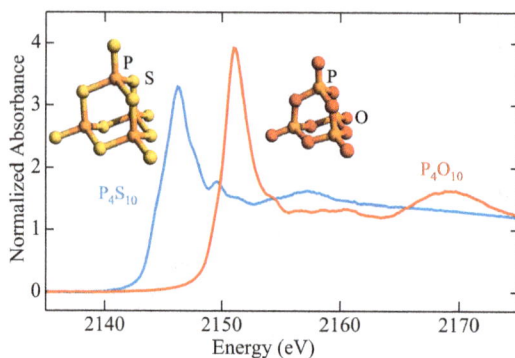

Figure 9.5: Effect of ligand type on the near-edge spectrum. Comparison of the phosphorus K-edge spectra for two isostructural formally P(V) compounds, P_4S_{10} and P_4O_{10}, showing a substantial (~5 eV) shift to higher energies for the oxygen compound.

9.3.4 Site symmetry and the near-edge spectrum

The transitions observed in the near-edge spectrum are highly dependent upon the coordination environment and upon the symmetry of the system. We now consider what is known as **parity**, which is an operation by which all Cartesian coordinates change sign, so that $\mathbf{r} = -\mathbf{r}$. Orbital parity is positive if the parity operation gives no change in sign of the orbital wavefunction, such as would be the case with an s orbital. Orbital parity is negative if there **is** a sign change with $\mathbf{r} = -\mathbf{r}$, such as would be the case with a p orbital, which has lobes with wavefunctions of differing signs (Figure 9.7); we note that orbitals are typically drawn with different colors or shading corresponding to the sign of the wavefunction. Atomic orbital parity is given by $(-1)^l$ where l is the azimuthal quantum number. The symmetry notation u and g is sometimes used in describing orbitals (both atomic and molecular); u and g respectively stand for *ungerade* (from the German, meaning odd or uneven) and *gerade* (meaning even).

In molecular-type systems the coordination geometry will be determined by the molecular orbitals that give rise to the chemical bonding that is characteristic of the system. The molecular orbitals can be regarded as being comprised of two or more atomic orbitals, which combine to make the molecular orbital. This means that we must consider the effects of orbital mixing. We have discussed above the 1s→3d feature that is fre-

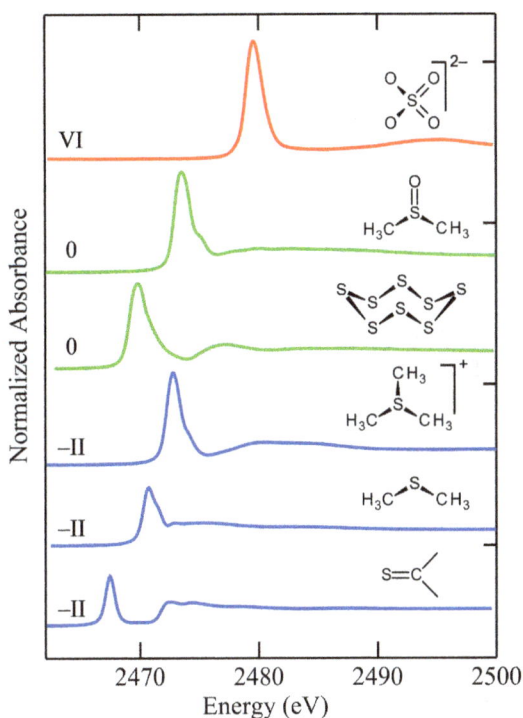

Figure 9.6: Spectral shifts within formal oxidation states. Comparison of sulfur K-edge spectra for sulfate (S(VI), top, red) and compounds with the same formal sulfur oxidation states (S(0), green; S(-II), blue) showing remarkably different spectra for the molecules shown.

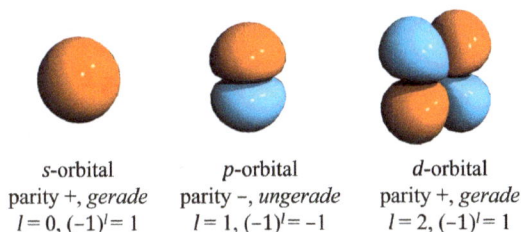

s-orbital	p-orbital	d-orbital
parity +, *gerade*	parity −, *ungerade*	parity +, *gerade*
$l = 0, (-1)^l = 1$	$l = 1, (-1)^l = -1$	$l = 2, (-1)^l = 1$

Figure 9.7: Parity of s, p and d hydrogen-like atomic orbitals. The colors indicate the sign of the wavefunction. The parity operation $r = -r$ would take any point on the orbital and translate through the nucleus ($r = 0$) to the opposite side; if the wavefunction sign is unchanged then the orbital is *gerade* and if it is changed the orbital is *ungerade*.

quently observed in the XAS of transition metal complexes, which is formally dipole-forbidden and quadrupole-allowed. The 1s→4p transitions are, however, dipole-allowed. If there were p-d mixing, for example 4p and 3d of a first-row transition metal ion, then the 1s→3d would take on some dipole-allowed character and become much more intense. Now, as we have discussed, p and d atomic orbitals have different parities; a consequence of this is that the site symmetry becomes very important in determining whether p-d mixing can occur. If there is inversion symmetry in the coordination environment, then orbitals of differing parity cannot mix. Thus, for octahedral coordination environments, which are inversion-symmetric, we would not expect p-d mixing and the 1s→3d would be only quadrupole-allowed, with no dipole-allowed contribution. Conversely, tetrahedral coordination environments are not inversion symmetric and consequently 4p-3d mixing may occur. Therefore, we can predict that tetrahedral complexes with allowed 4p-3d mixing would show a substantially more intense 1s→3d feature than would octahedral complexes, because of dipole-allowed character in the tetrahedral case. A comparison of the 1s→3d feature for tetrahedral- and octahedral-type coordination geometries and using the example of Figure 9.4c is shown in Figure 9.8.

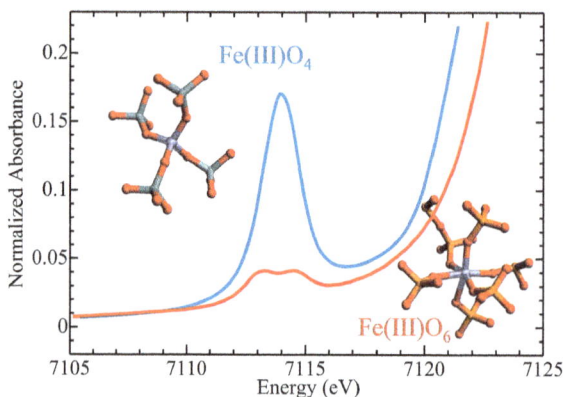

Figure 9.8: Comparison of the 1s→3d feature of the iron K-edge XAS for octahedral- and tetrahedral-type coordination ferric ion geometries.

Beyond the large difference in intensity of the 1s→3d feature in Figure 9.8, the 1s→3d feature of the octahedral complex is comprised of two discrete peaks. This is due to the coordination geometry causing a splitting of the 3d manifold. Two theoretical approaches have sought to understand these effects, which are well-known in physical chemistry. The first was crystal field theory, which considered only entirely the electrostatic interactions between a central metal ion and the ligand atoms. Subsequently, ligand field theory, a modification of crystal field theory additionally considered covalent bonding between the metal and its ligand. The basic idea of both can be understood from Figure 9.9. Here, we first consider an isolated transition metal ion possessing no ligands. Here, all the 3d orbitals have the same energy, and we say that the manifold is degenerate. We now consider six negatively charged ligand ions approaching the central metal ion in an octahedral arrangement, from $\pm x$, $\pm y$ and $\pm z$. Two of the five d-orbitals have lobes pointing at the approaching ligand ions ($d_{x^2-y^2}$ and $d_{z^2-r^2}$), while the other three have lobes that point between them (d_{xy}, d_{xz} and d_{yz}). According to crystal field theory, the electrostatic repulsion between the d-orbitals and the ligand atoms will split the energies of the d-orbitals into two, with a splitting Δ, so that $d_{x^2-y^2}$ and $d_{z^2-r^2}$ go up in energy by $\frac{3}{5}\Delta$ while the d_{xy}, d_{xz} and d_{yz} go down in energy by $\frac{2}{5}\Delta$ (Figure 9.9). The splitting Δ, or Δ_o for an octahedral complex, sometimes referred to as $10Dq$,[5] is called the **ligand field splitting**. The two different degenerate levels of d orbitals are often called t_{2g} (d_{xy}, d_{xz} and d_{yz}) and e_g ($d_{x^2-y^2}$ and $d_{z^2-r^2}$) orbitals, or sometimes d_ε and d_γ, respectively. Different coordination geometries have different ligand field splittings, as illustrated in Figure 9.10.

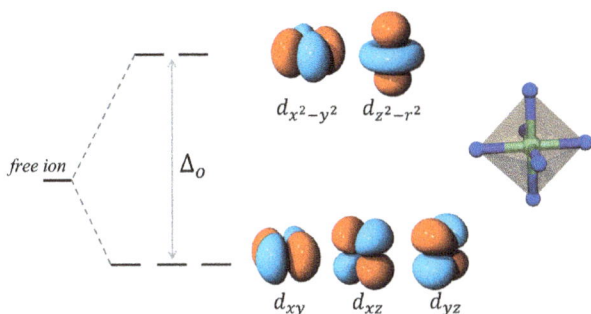

Figure 9.9: Metal ion d-orbital splitting by an octahedral ligand field.

We now consider a hypothetical metal ion with octahedral coordination and, in a thought experiment, add successive electrons to it. The lower energy orbitals are filled first, so that filling begins with the t_{2g} orbitals. The first three electrons to be added would occupy one each of the d_{xy}, d_{xz} and d_{yz} orbitals. This is because there is an energetic penalty for double occupancy, called the pairing energy, P. What happens upon adding two more electrons would depend upon whether Δ_o is larger than P. With $\Delta_o > P$ the electrons would enter the t_{2g} orbitals to form what is called a low-spin metal ion, in this case a d^5 configuration, which would have an overall spin S of ½. Conversely, with $\Delta_o < P$ the two electrons would enter the e_g orbitals to give a spin S = $^5/_2$ metal ion. Adding more electrons still would have paired electrons, irrespective of whether Δ_o is larger than P.[6] Figure 9.11 shows electronic configurations for low-spin and high-spin and Fe(III).

---

5 From 10 times the differential of quanta.
6 Hund's rule indicates that for double occupancy the spins of the two electrons must be opposed, such that they can only be ↑↓ and not ↑↑ or ↓↓.

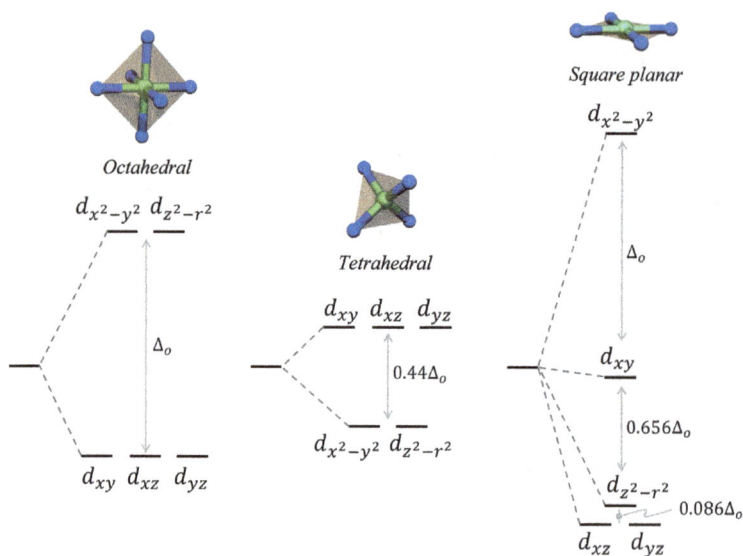

Figure 9.10: Ligand field splitting for different coordination geometries.

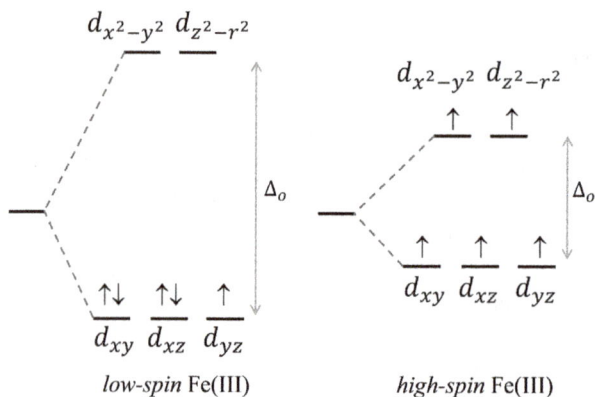

Figure 9.11: Electron configurations for low-spin ($\Delta_o > P$) and high-spin ($\Delta_o < P$) ferric in octahedral ligand fields.

The size of the ligand field splitting depends on the so-called **spectrochemical series**, which, arranged in order of increasing ligand field effect, is as follows:

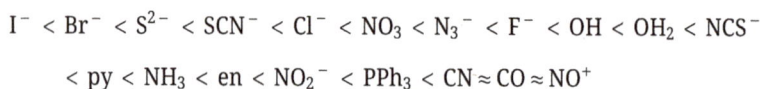

$$I^- < Br^- < S^{2-} < SCN^- < Cl^- < NO_3 < N_3^- < F^- < OH < OH_2 < NCS^-$$

$$< py < NH_3 < en < NO_2^- < PPh_3 < CN \approx CO \approx NO^+$$

where py is pyridyl (C_5H_5N) and Ph is phenyl (C_6H_5). Weak-field ligands of first transition elements typically give rise to high-spin complexes and strong-field ligands to low-spin complexes. The higher the charge on the metal ion the greater the splitting Δ. For the metal ions themselves there is a general sequence of Δ_0 values, which increase in the order:

$$Mn^{2+} < Ni^{2+} < Co^{2+} < Fe^{2+} < V^{2+} < Fe^{3+} < Cr^{3+} < V^{3+} < Co^{3+} < Mo^{4+}$$

Second and third row transition metals almost never form high-spin complexes. This is because the increased nuclear charge arising from larger atoms gives rise to a larger splitting Δ, so that in almost all cases $\Delta > P$ and the complex is nearly invariably low spin. This is not a rigid rule, and there are examples of high-spin metal ions beyond the first transition series, but the vast majority of high-spin species are first transition metal complexes. High-spin complexes tend to have longer metal-ligand bond lengths than analogous low-spin complexes, typically by about 0.2 Å for hard ligands such as oxygen or nitrogen, with smaller changes for soft ligands such as sulfur (ca. $\leq$ 0.1 Å). Some compounds can exist in either high-spin or low-spin states depending on temperature or pressure, with higher temperatures or lower pressures tending to favor a high-spin complex. One example is ferric *tris*-diethyldithiocarbamate[7] which is low-spin below ~250 K and high-spin above.

Observed splittings in the 1s→3d feature thus correspond to ligand field splittings in the 3d manifold containing information about site symmetry and coordination. They are excited state splittings, which correspond approximately to an increased nuclear charge, and remembering our discussion above, approximate to the ground state corresponding to $Z+1$. For $3d^6$ and higher occupancy, high-spin complexes can show two or more 1s→3d peaks, whereas low-spin complexes often show just a single feature.

What if the same system is observed using two different absorption edges: for example, the K-edge of a metal ion, and the K-edge of a ligand atom? Here we consider $CuCl_2\cdot 2H_2O$ as an example [1]. This formally $3d^9$ Cu(II) compound has two symmetry related Cu–Cl ligands (D_{2h} point group symmetry) and we will consider the near-edge spectra at both the Cu K-edge and the Cl K-edge (Figure 9.12). All four closely bound ligands lie in the same plane, hence we would predict a pseudo square planar ligand field with the single vacancy in the 3d manifold being in the $d_{x^2-y^2}$ orbital (e.g. see Figure 9.10). The lobes of the half-occupied $d_{x^2-y^2}$ orbital point directly along the Cu–Cl bonds, for which we can anticipate significant covalency, hence the singly filled

---

7 This complex caused temporary confusion for one of the authors (George) during early EXAFS measurements, as it was measured as a standard using low temperature EXAFS, which gave a bond length that was inconsistent with the room temperature crystal structure. Subsequent room-temperature EXAFS confirmed the crystallographic bond lengths and it was concluded that the compound converted to the low-spin form at cryogenic temperatures.

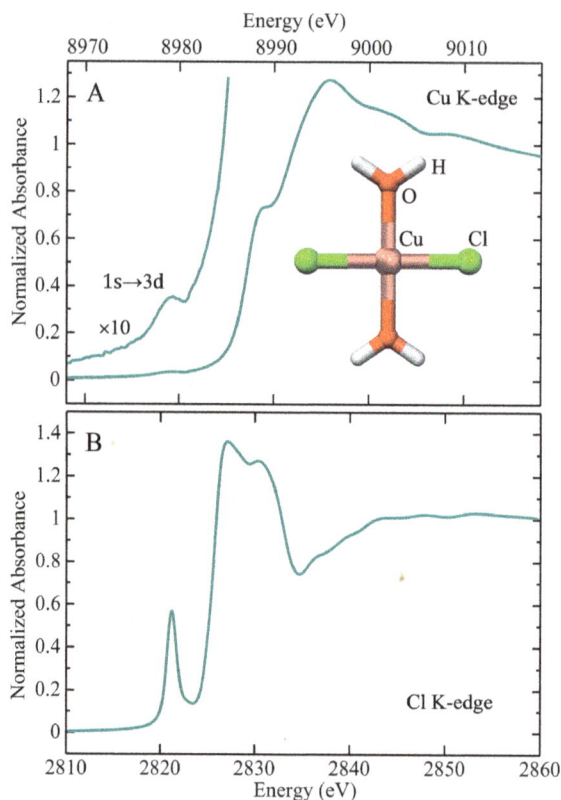

Figure 9.12: Cu K-edge (A) and Cl K-edge (B) near-edge spectra of solid $CuCl_2 \cdot 2H_2O$. The inset shows the planar structure of the $CuCl_2 \cdot 2H_2O$ molecule.

anti-bonding orbital should have both Cu $3d$ and Cl $3p$ character. Consider first the Cl K-edge XAS in Figure 9.12B. Formally speaking, Cl^- has a completely filled $3p$ shell, with no pre-edge transitions expected; however, this seems at odds with the presence of the intense feature at 2821.2 eV. If, however, there is covalency in the Cu–Cl bond, then a fraction of the hole in the $d_{x^2-y^2}$ orbital is shared with the Cl $3p$ manifold, leading to the observation of what might be called a [Cl]$1s\rightarrow$[Cu]$3d$ transition as part of the Cl K-edge XAS, consistent with the 2821.2 eV peak in Figure 9.12B. The relationship to the Cu K-edge $1s\rightarrow3d$ seems clear, but why then, is the intensity of the Cu K-edge so much less? Remembering that XAS is a very fast phenomenon (Section 4.7), Figure 9.13 provides a qualitative explanation of this observation.

The ground state is shown in the left panel, with the partial shading of the $3d$ and $3p$ electrons suggesting covalency. The middle panel in Figure 9.13 shows the effect of a metal core hole following photoexcitation. As we discussed above, this causes the outer orbitals to relax to lower energies, which in turn causes a shrinkage of their wavefunctions, and hence a decrease in molecular orbital overlap with ligand atoms.

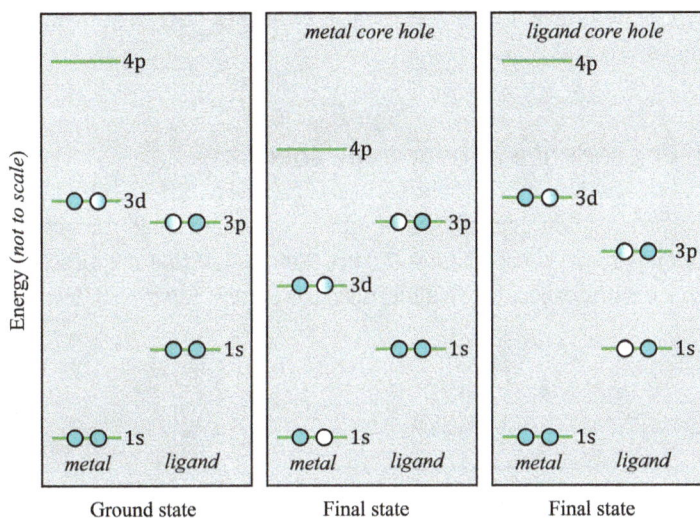

Figure 9.13: Schematic representation of the energetic effects of metal and ligand core holes.

The Cu K-edge XAS will thus be more atomic in character, and because the complex is centrosymmetric the major contribution to the 1s→3d will be quadrupolar, hence the low intensity of this feature. For the Cl K-edge, the core hole has a similar effect on the final state, but in this case the atom-like XAS transition will primarily involve the Cl 3p manifold, which will have dipole-allowed character and will thus be much more intense. The singly occupied molecular orbital will have contributions from all ligands to copper, but for the purposes of this review we will ignore all but the copper and one chloride. In this case the singly occupied anti-bonding orbital involved $|\varphi\rangle$ will be given as follows:

$$|\varphi\rangle = \left(1 - \alpha^2\right)^{\frac{1}{2}} \left|\text{Cu } 3d_{x^2 - y^2}\right\rangle - \alpha \left|\text{Cl } 3p\right\rangle \qquad (9.5)$$

in which α^2 is the covalency of the Cu–Cl bond. The intensity of the ligand K-edge transition $\mu(\text{Cl } 1s \rightarrow \varphi)$ is given as follows:

$$\mu(\text{Cl } 1s \rightarrow \varphi) = \alpha^2 c \left|\langle \varphi | H' | \text{Cl } 1s \rangle\right|^2 \qquad (9.6)$$

Here, c is a constant of proportionality that can be determined from experimental measurements. The intensity of the ligand pre-edge peak is thus directly proportional to the covalency α^2. This relationship between the intensity of ligand pre-edge features and covalency has been used as a tool to determine the covalency for a number of different bioinorganic systems by Solomon, Hodgson, Hedman and co-workers [2, 3]. We note here that in addition to the Cu K-edge and the Cl K-edge, the oxygen K-edge XAS of CuCl$_2$·2H$_2$O would have been very interesting but it proved problematic to acquire because under the vacuum conditions typical of oxygen K-edge XAS capable beamlines $CuCl_2 \cdot 2H_2O$ tends to lose water forming anhydrous $CuCl_2$.

9.4 Polarization dependence of near-edge spectra

Here we consider the polarization dependance of dipole and quadrupole transitions for a K-edge. We will employ polar coordinates (Figure 9.14) to describe the direction cosines of a unit vector **e** with respect to the molecular axis system x, y, z as follows:

$$\mathbf{e} = \begin{pmatrix} \sin\theta\cos\phi \\ \sin\theta\sin\phi \\ \cos\theta \end{pmatrix} \tag{9.7}$$

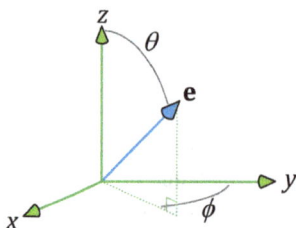

Figure 9.14: The polar coordinate system.

From eq. (9.3) for a K-edge (1s excitation) and using the angular wavefunctions of hydrogen-like orbitals we can derive an expression for the angular dependence of the dipole-allowed transitions to the three p-orbitals:

$$\mu_D(E)(1s \rightarrow p_y) \propto e_x^2 = \sin^2\theta\sin^2\phi \tag{9.8}$$

$$\mu_D(E)(1s \rightarrow p_x) \propto e_y^2 = \sin^2\theta\cos^2\phi \tag{9.9}$$

$$\mu_D(E)(1s \rightarrow p_z) \propto e_z^2 = \cos^2\theta \tag{9.10}$$

Thus, the transition polarization is aligned with the major direction of the orbital, as shown in Figure 9.15.

The results of a polarized single-crystal XAS experiment are shown in Figure 9.16 for dibenzyldisulfide. This molecule crystallizes with the S–S bonds almost aligned with the crystallographic c axis. When the crystal is oriented so that the **e**-vector points along the c-axis, the observed intensity of the S1s→(S–S)σ^* transition is maximal. When the crystal is rotated so that the **e**-vector rotates from the c axis to the b-axis in the b-c plane, the S1s→(S–C)σ^* becomes more intense and the S1s→(S–S)σ^* transition diminishes, as expected from the simple dipole transition behavior.

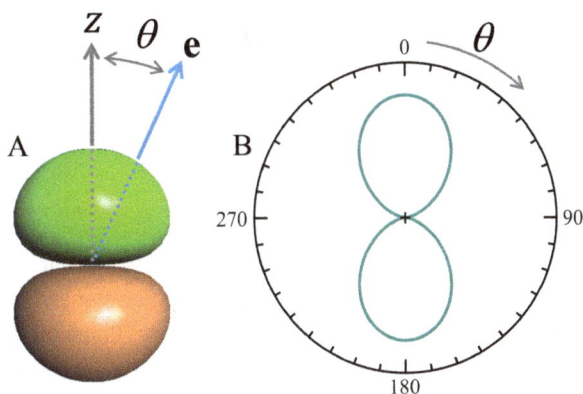

Figure 9.15: Polarization dependance of a transition to a *p* orbital. (A) The angle θ defined relative to a p_z hydrogen-like atomic orbital. (B) A polar plot of the predicted polarization dependence of the 1s→p_z transition.

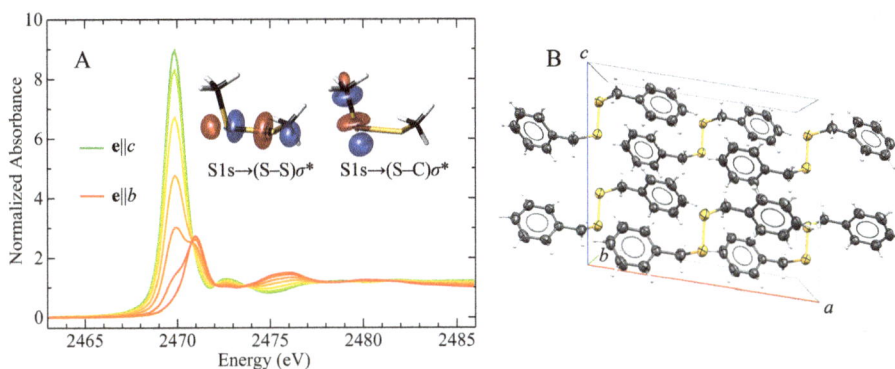

Figure 9.16: (A) Polarized single-crystal sulfur K-edge XAS of dibenzyldisulfide. Two prominent transitions in the near-edge spectra are clear, one a low-energy feature attributable to an S1s→(S–S)σ^* transition, maximal when the **e**-vector is parallel with the crystallographically aligned S–S bonds, and the other attributable to an S1s→(S–C)σ^*. The inset in (A) shows the excited state molecular orbitals corresponding to these transitions for dimethyldisulfide. (B) The crystal structure viewing along the crystallographic *b*-axis to highlight the S–S bond orientation, which is almost aligned with the crystallographic *c*-axis.

Similarly to the transitions to *p*-orbitals, for the quadrupole-allowed transitions we can use eq. (9.4) and the angular wavefunctions for the five hydrogen-like *d*-orbitals to show that the angular dependence is given as follows:

$$\mu_Q(E)\left(1s \rightarrow d_{xy}\right) \propto \left(e_x k_y + e_y k_x\right)^2 \tag{9.11}$$

$$\mu_Q(E)(1s \rightarrow d_{xz}) \propto \left(e_x k_z + e_z k_x\right)^2 \tag{9.12}$$

$$\mu_Q(E)\left(1s \rightarrow d_{yz}\right) \propto \left(e_y k_z + e_z k_y\right)^2 \tag{9.13}$$

$$\mu_Q(E)\left(1s \rightarrow d_{x^2-y^2}\right) \propto \left(e_x k_x - e_y k_y\right)^2 \tag{9.14}$$

$$\mu_Q(E)\left(1s \rightarrow d_{z^2-r^2}\right) \propto \left(2e_z k_z - e_x k_x - e_y k_y\right)^2 \tag{9.15}$$

Brouder [4] has provided a less superficial treatment. In eqs. (9.11)–(9.15) the angular intensity depends on both the X-ray electric vector **e** and the X-ray propagation vector **k** (with direction cosines k_x, k_y and k_z), similarly to eq. (9.7). This causes the somewhat more complex orientation dependence than for the dipole transitions. Consider a transition to a $d_{x^2-y^2}$ orbital with both the **e**-vector and the **k**-vector in the $x-y$ plane, so that $e_x = \cos\phi$, $e_y = \sin\phi$, $k_x = -\sin\phi$ and $k_y = \cos\phi$. Thus, from eq. (9.14), we can simplify the orientation dependence as follows:

$$\mu_Q(E)\left(1s \rightarrow d_{x^2-y^2}\right) \propto \left(-\cos\phi\sin\phi - \sin\phi\cos\phi\right)^2 = \left(-2\sin\phi\cos\phi\right)^2 = \sin^2 2\phi \tag{9.16}$$

This is shown in Figure 9.17, displaying the predicted cloverleaf-shaped orientation dependence.

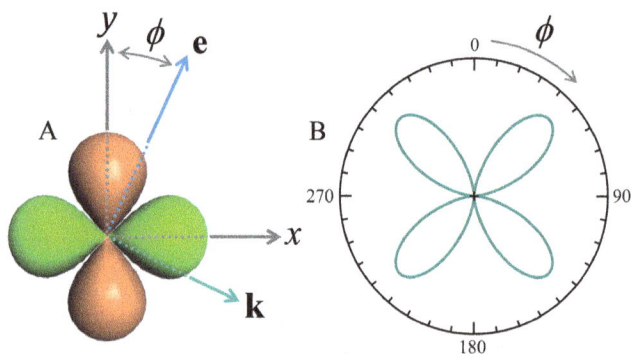

Figure 9.17: Predicted orientation dependence for a $1s \rightarrow d_{x^2-y^2}$ transition with both the **e**-vector and the **k**-vector in the $x-y$ plane. (A) The orbital and relevant axes and angles, and (B) a polar plot of the orientation dependence of the transition amplitude.

As an experimental example of the orientation dependence of a 1s→3d transition [5] we turn again to CuCl$_2$·2H$_2$O (see Figures 8.10 and 9.12) [1]. CuCl$_2$·2H$_2$O crystallizes in the orthorhombic space group *Pmna* with unit cell parameters $a = 8.05$, $b = 3.72$ and $c = 7.43$ Å. The structure shows CuCl$_2$·2H$_2$O molecules with *trans*-ligation, having D_{2h} point group symmetry, with the O–Cu–O aligned with crystallographic a-axis, and the Cl–Cu–Cl inclined at $\varepsilon = 38.7°$ from the crystallographic c-axis. The Cu–Cl and Cu–O bond lengths are 2.28 and 1.94 Å, respectively. The planar CuCl$_2$·2H$_2$O molecules exhibit two very long axial Cu····Cl at 2.94 Å and can therefore be essentially considered four-coordinate.

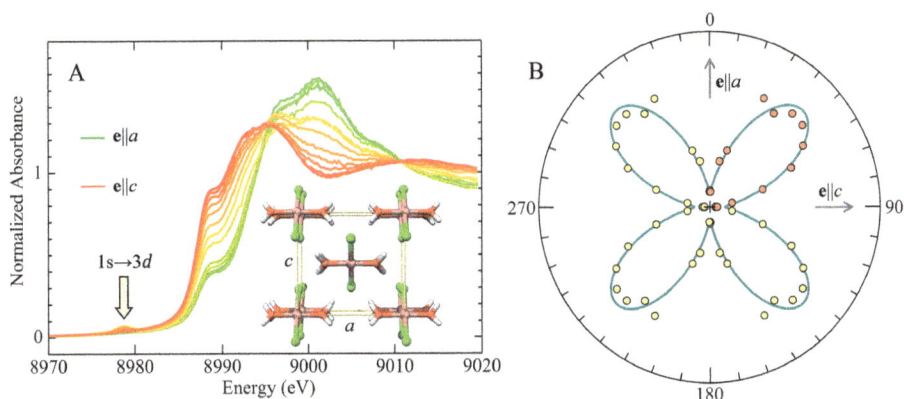

Figure 9.18: Polarized Cu K-edge XAS of a single crystal of $CuCl_2 \cdot 2H_2O$ rotated so that both **e** and **k** rotate in the crystallographic a–c plane. (A) The superimposed spectra, which are dominated by dipole-allowed transitions between 8,983 and 9,010 eV, with the inset showing the crystal structure viewed along the b-axis. (B) A polar plot of the 1s→3d orientation dependence. The orange points are experimentally measured 1s→3d amplitudes with the yellow points generated from the orange by symmetry.

Figure 9.18 shows a set of spectra obtained by rotating a single crystal of $CuCl_2 \cdot 2H_2O$ so that both **e** and **k** rotate in the crystallographic a–c plane. The predicted four-lobed cloverleaf angular dependence is clearly observed.

9.5 Shake-down and shake-up transitions

The K-edge spectra of first transition metal complexes often show an intense dipole-allowed feature that occurs at an energy, which seems too low to be a simple 1s→4p transition. An example of this is shown in Figure 9.19, for the strongly Jahn-Teller distorted Cu(II) complex **tetrakis**(imidazole)-**bis**(perchlorato) copper(II) which contains four imidazole ligands to copper with a Cu–N bond length of 2.01 Å along with a pair of axial distant oxygens at 2.60 Å from the two perchlorate anions. The coordination is thus close to square planar geometry, and typical of Cu(II). The XAS shows a well-resolved 1s→3d transition with the low intensity expected for a centrosymmetric complex and a strong and relatively isolated peak at 8,988 eV. This peak is perhaps 8.5 eV lower in energy than the expected location of a formal 1s→4p transition. It provides us with an example of what is known as a **shake-down transition**, which occurs when there is significant covalency in the bonding to the metal ion. To understand the nature of shake-down transitions, we recall from Section 9.1 (Figure 9.3) that the presence of a core hole causes the energies of the outer orbitals to relax to lower values.

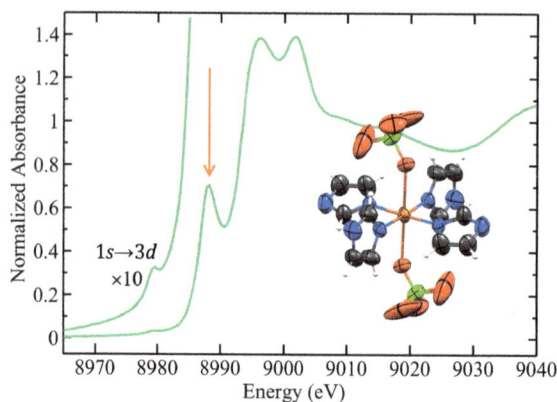

Figure 9.19: Cu K-edge XAS spectrum of **tetrakis**(imidazole)-**bis**(perchlorato) copper(II), the structure of which is shown in the inset (70% thermal ellipsoids). The feature marked with the orange arrow is the $1s{\rightarrow}4p_z$ + ligand-to-metal charge transfer shake-down transition discussed in the text.

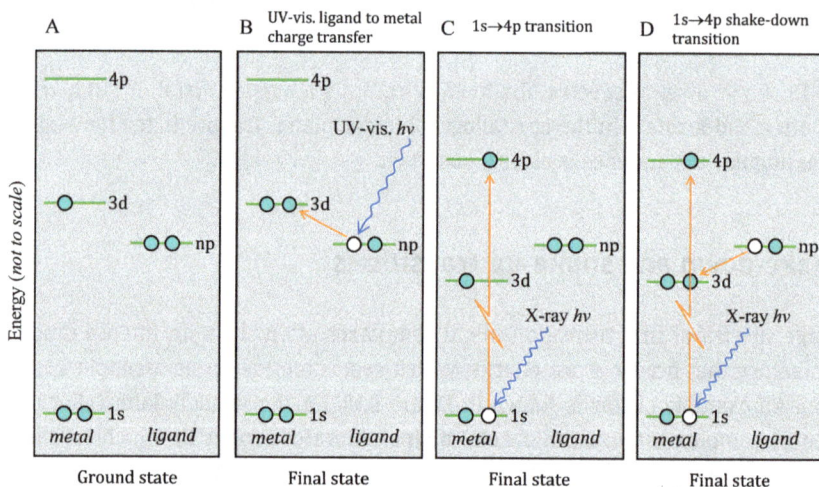

Figure 9.20: Schematic diagrams (not to scale) of processes relevant to $1s{\rightarrow}4p_z$ + ligand-to-metal charge transfer shake-down transition.

Figure 9.20 shows the relative energies of some of relevant metal and ligand orbitals. We note that these are not shown to scale – if they were, then the metal $1s$ level would be very much deeper than all of the other levels, which would appear almost superimposed at the top of the picture. Panel A shows the ground state and indicates filled metal 1s and ligand np levels, with a vacancy in the 3d. A schematic of a UV-visible ligand-to-metal transfer transition final state is shown in panel B. Here, a photon close to the visible energy range excites a ligand p-electron, which is promoted to the metal 3d level. Panel C shows a schematic of the final state for a 1s→4p transition. Here, the presence of the core hole causes the outer metal orbitals to relax to lower energies, but the energies of the ligand orbitals are relatively unchanged. Finally, panel D shows a schematic of the final state for a 1s→4p plus ligand-to-metal charge transfer transition. This is a one-photon two-electron situation, in which the X-ray excites a 1s electron to a 4p, with the core hole causing the metal orbitals to relax to lower energies. The ligand-to-metal charge transfer transition, previously "uphill" (requiring energy) in the absence of a core hole (Figure 9.20B), is now downhill due to the change in orbital energies, and occurs in a concerted manner along with the 1s→4p transition. The overall final state is energetically lower than that of the pure 1s→4p by the difference between the final state np and 3d levels, and consequently the shake-down transition, which is dipole-allowed, falls to lower energy by this amount.

As we have noted, shake-down transitions are frequently observed in the XAS of first transition element K-edges. A related type of transition is the **shake-up transition**, shown schematically in Figure 9.21A. Like shake-down transitions, these are dipole-allowed two-electron one-photon events, with dipole-allowed promotion of a core electron to a bound state, and with the core hole bringing a metal orbital in energetic proximity to a ligand orbital, but this time "uphill" (i.e. requiring energy). The resulting concerted two-electron one-photon transition requires more energy than the corresponding one-electron transition and thus occurs at higher incident X-ray energies than the corresponding metal-only based one-electron transition.

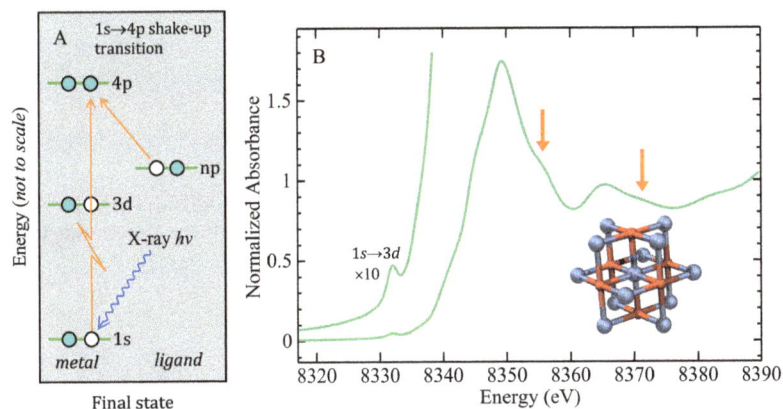

Figure 9.21: (A) A schematic of the shake-up transition. (B) The Ni K-edge XAS of NiO, with the inset showing a fragment of the structure. The broad features marked with the orange arrows are thought to be shake-up transitions, with other nearby features attributable to single electron events.

Shake-up features are expected to be more prominent in L_{III} and L_{II} edge and less prominent in K-edge XAS. This is because the dipole-allowed transitions with L_{III} and L_{II} edges are to d-orbital states, which are more susceptible to the effects of correlation than are p-orbital states. Because shake-up transitions typically are found above the most intense features in the absorption edge they are often broad, especially for K-edge XAS (Figure 9.21B). The simplistic view of shake-down and shake-up transitions presented here is not a comprehensive description of the possible X-ray correlative X-ray absorption processes, but these are the most common types, especially for hard X-ray spectra.

9.6 Electron spin effects

In our discussion thus far, we have neglected the effects of electron spin; however, these can often be important, as we will see.

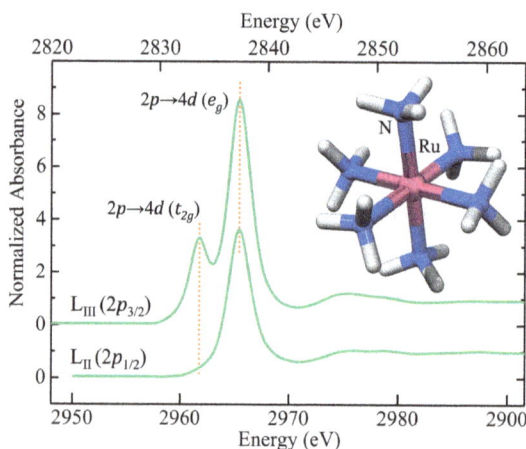

Figure 9.22: Ru L_{III} and L_{II} XAS spectra of the Ru(III) hexammine complex $[Ru(NH_3)_6][BF_4]_3$ showing dipole-allowed transitions to both the t_{2g} and e_g levels with the L_{III} edge, but only to the e_g levels with the L_{II} edge.

9.6.1 Spin-orbit coupling

A classic example of the importance of electron spin is provided by the Ru(III) hexammine compex $[Ru(NH_3)_6]^{3+}$. Early work by Sham [6] examined the Ru L-edges of the chloride salt and showed that there were significant differences between the L_{III} and L_{II} edges. These early measurements were complicated by the presence of the nearby chlorine K-edge at just below the Ru L_{III} edge energy, which gave rise to some distortions in the data. Figure 9.22 compares the Ru L_{III} and L_{II} edges of the tetrafluoroborate salt $[Ru(NH_3)_6][BF_4]_3$ which has no other absorption edges in the vicinity of the Ru L_{III} and L_{II} edges. The $[Ru(NH_3)_6]^{3+}$ has close to perfect octahedral symmetry with six uniform Ru–N bond lengths of 2.104 Å. Ru(III) has a formal $4d^5$ electronic configuration; the octahedral ligand field will split the $4d$ manifold into t_{2g} and e_g levels. The complex will be low-spin since Ru is a second transition metal ion and nitrogen is a strong-field ligand, resulting in a single vacancy in the t_{2g} in a manner similar to that shown schematically for low-spin Fe(III) in Figure 9.11. Figure 9.22 shows essentially the same results as Sham's early study [6], with the L_{III} showing clear transitions to both the t_{2g} and e_g levels, while the L_{II} edge only shows transitions to the e_g levels. Sham's analysis introduced spin-orbit coupling as a perturbation, lifting the degeneracy of the t_{2g} orbitals, and showed that the vacancy in the t_{2g}-related orbitals had $4d_{5/2}$ character. In Chapter 4 and in this chapter, we have discussed the dipole selection rule $\Delta l = \pm 1$; with consideration of the total angular momentum j, we have another selection rule $\Delta j = 0, \pm 1$. Hence, inclusion of the spin-orbit effects showed that the dipole transition of the $2p_{1/2}$ core level with $j = \frac{1}{2}$ to the vacancy in the t_{2g}-related orbital with the $j = \frac{5}{2}$ was forbidden, whereas the $2p_{3/2}$ core excitation transitions to both the t_{2g} and e_g levels are allowed. While this phenomenon will be rigorously true for octahedral symmetry, if the symmetry is broken then two peaks may be observed at the L_{II} edge. Close inspection of the L_{II} data of Figure 9.22 shows a very weak $2p_{1/2} \rightarrow e_g$ transition present as a shoulder, which might be due to slight departures from perfect octahedral symmetry or a quadrupole contribution to the transition.

In the tender X-ray regime, with $[Ru(NH_3)_6]^{3+}$ being something of an exception, the differences between L_{III} and L_{II} edges tend to be subtle [7], and in the hard X-ray regime they can be difficult to observe. With soft X-rays, however, such as the L-edges of the first transition elements, substantial differences are typically observed. Figure 9.23 compares the K-edge and L_{III}/L_{II}-edges of goethite, one of two polymorphs of Fe(III)O(OH). The iron in goethite is coordinated by three $\mu_3{}^8$ hydroxide ligands (average Fe–O 2.04 Å) and three μ_3 oxo ligands (average Fe–O 1.95 Å) in a distorted octahedral geometry. Figure 9.23 shows that the L_{III}/L_{II}-edges are considerably more detailed than the K-edge $1s \rightarrow 3d$ transition.

9.6.2 X-ray magnetic circular dichroism

Almost all applications of X-ray magnetic circular dichroism (XMCD) [8] have used the L-edges of the first transition elements, in which XMCD is used to provide a detailed understanding of the electronics of the 3d manifold. XMCD uses left and right circu-

---

8 Ligands that bridge two or more metal ions are often described using the notation μ_n where n is the the number of metal ions bridged. If $n = 2$, then the subscript may be omitted.

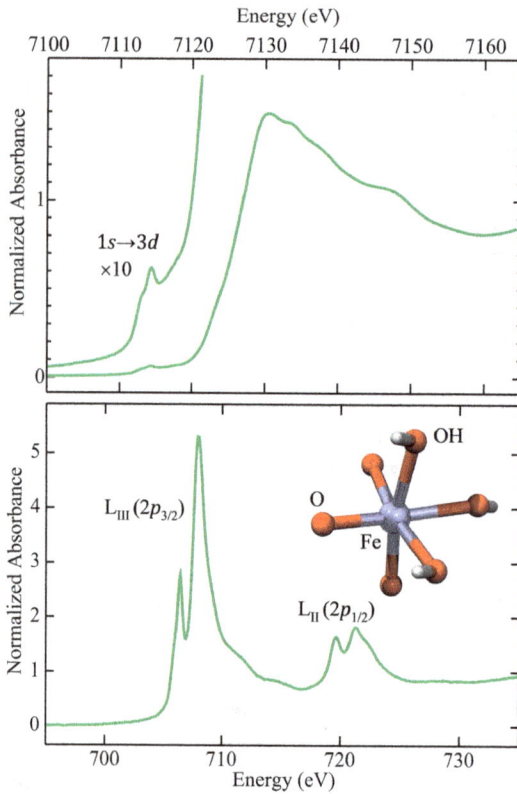

Figure 9.23: Comparison of Fe K-edge and Fe L_{III}/L_{II}-edge XAS of goethite. The inset in the L-edge plot shows the local structure around one Fe(III) atom.

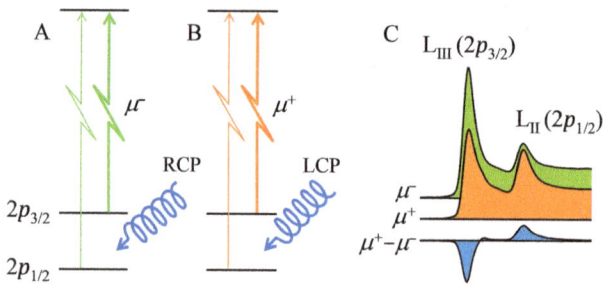

Figure 9.24: Schematic of the XMCD measurement. A and B respectively show the absorption of RCP and LCP photons, and C shows a schematic of L_{III}/L_{II}-edge XMCD result.

larly polarized X-ray light, which can be obtained either from a helical undulator (see Section 2.9) or, at much lower intensity, from a bend magnet source by setting an aperture above and below the plane of the storage ring by $\pm 1/\gamma$.

We first recall some essentials for circularly polarized light. If we consider two beams of in-phase plane-polarized light of wavelength λ and inclined at 90° to each other, then the result will be a polarized beam of light with the direction of polarization inclined at 45°. Now if we were to retard one of these plane-polarized beams beam by $\lambda/4$ – the result would describe a left-handed helix in which the electric and magnetic vectors rotate about the propagation vector, defined as left-handed circularly polarized light. Conversely, a $\lambda/4$ retardation of the other beam would yield right-handed circularly polarized light. Just as plane-polarized light would cause a free charge to oscillate, a free charge interacting with circularly polarized light will be driven in a circular trajectory. With circularly polarized light each photon carries a spin angular momentum of $\pm\hbar$, which would be $+\hbar$ for left circular photons (LCP) and $-\hbar$ for right circular photons (RCP). XMCD thus provides a direct probe relevant to electron spin effects.

To understand XMCD in a little more detail we again consider spin-orbit $(l-s)$ coupling. An external applied magnetic field will change the energy levels of unpaired electrons, since the electrons align with the magnetic field leaving hole states with spins of the opposing spin symmetry. For L_{III} excitation we probe $2p_{3/2}{\rightarrow}3d$ with $j=l+s$, so that X-rays with ↑ circular polarization are more likely to excite spin ↓ electrons into the hole states. With the L_{II} edge the situation is reversed; we probe $2p_{1/2}{\rightarrow}3d$ with $j=l-s$, so that X-rays with ↓ circular polarization are more likely to excite spin ↓ electrons into the hole states. Figure 9.24 shows a schematic diagram of the processes giving rise to XMCD. If we define the X-ray absorption with helicity parallel to the magnetization as μ^+ and that with helicity antiparallel as μ^- then the XMCD is given by $\Delta\mu=\mu^+-\mu^-$, with the conventional XAS being given by $\mu=(\mu^++\mu^-)/2$. A variety of different sources of magnetic field are possible, and in some cases, samples can be externally magnetized. XMCD has even been used to microscopically and directly visualize the magnetized domains in magnetic disk computer storage media. Ferromagnetic materials are also inherently simple to study.

With paramagnetic systems the size of the XMCD signal will depend strongly upon the magnetic field strength and the sample temperature. The splitting of the spin energy levels ΔE (the Zeeman splitting) will be directly proportional to the magnetic field, with the size of the signal dependent on the population difference between these split levels. There is a well-defined relationship of this with electron paramagnetic resonance (EPR), which uses microwave radiation to stimulate transitions between the upper and lower halves of the electronic Zeeman split level. Transitions can be stimulated from the lower to the upper level with absorption of a microwave photon, and from the upper to the lower level with coherent emission of a second microwave photon. Hence, the EPR signal size depends upon the population difference between the upper and lower electronic Zeeman split levels. Likewise, for magnetic circular dichroism in the UV-visible regime, LCP and RCP, respectively, stimulate promotion of electrons to an excited state preferentially from the lower and upper halves of the electronic Zeeman split level. A schematic summarizing both EPR and UV-visible MCD is shown in Figure 9.25.

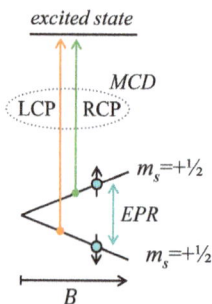

Figure 9.25: Relationship between EPR and UV-visible MCD.

The signal size in all these cases will thus be governed by a Boltzmann-type distribution describing the population difference, and which will vary as $\exp(-\Delta E/k_B T)$, where k_B is Boltzmann's constant and T is temperature. For an $S = 1/2$ system the XMCD will vary as follows:

$$\Delta\mu(E) \propto \left(\frac{n_{\text{upper}}}{n_{\text{lower}}}\right) = \exp\left(-\frac{gB\beta_e}{k_B T}\right) \tag{9.17}$$

where n_{upper} and n_{lower} are the populations in the upper and lower levels, respectively, g is the electron g-factor (usually ~2.0 for $S = 1/2$ systems), B is the magnetic field and β_e is the Bohr magneton. Hence, the lower the temperature the bigger the XMCD, and the higher the applied field the bigger the XMCD. Some experimental setups have used superconducting solenoids to generate very high fields, combined with ^{3}He refrigerators to enhance signal size by using temperatures in the milli-Kelvin regime.

While some XMCD can be subtle, for strongly magnetic systems such as complexes of rare earth metal ions (e.g. Gd(III), which is $S = 7/2$) it can be most pronounced. Rare earth L_{III} and L_{II}-edges tend to be dominated by the dipole-allowed $2p \rightarrow 5d$ transitions, and because the electronic configuration is always $5d^0$ the form of these spectra are relatively invariant along the period. The M_V and M_{IV} edges, conversely, respectively exhibit intense dipole-allowed $3d_{5/2} \rightarrow 4f$ and $3d_{3/2} \rightarrow 4f$ transitions, which are expected to be chemically sensitive and which, like transition metal L-edges, will show large XMCD. Figure 9.26 shows the M_V and M_{IV} XMCD of a solution of the Gd(III)-containing MRI contrast agent known as Magnevist, clearly showing its quite remarkable XMCD.

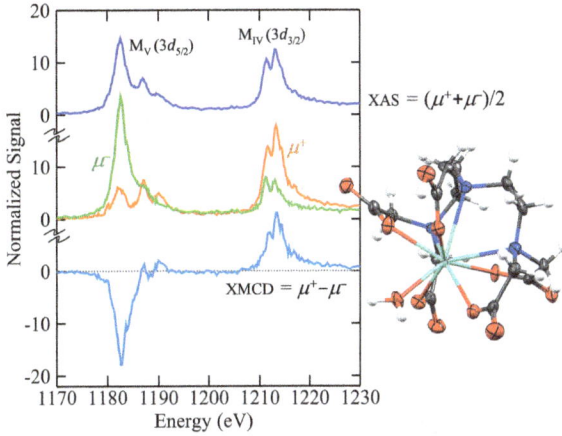

Figure 9.26: Gd M_V and M_{IV} XMCD of a solution of Magnevist, also known as gadopentetate, which is an $S = 7/2$ Gd(III) complex with the chelator diethylenetriaminepentacetate (DTPA). These data were measured using an applied magnetic field of 6 T and a sample temperature of 5 K (data courtesy of Simon George of Simon Scientific). The inset shows the crystal structure.

9.7 Spectral linewidths

We mentioned in Section 4.7 that there are two major components to the line broadening in near-edge spectra. As we will see below this is something of an oversimplification, but before discussing this further we will recap our previous discussion. One source of broadening is due to the beamline optics, discussed in Chapter 3, which contributes a Gaussian component to the overall experimental lineshape. A second source of broadening is what is known as lifetime broadening. This refers to the lifetime of the core hole that is created by the primary X-ray photoexcitation event. Here we refer to Heisenberg's uncertainty principle, which can be written in a number of ways, for example position and velocity, or time and energy. Using the latter we can write as follows:

$$\Delta E_L \Delta t \geq \frac{\hbar}{2} \tag{9.18}$$

where ΔE_L is the lifetime broadening and Δt the core-hole lifetime. As we have previously discussed, in many hard X-ray XAS experiments the core-hole lifetime is a fraction of a femtosecond (10^{-15} s). We have also discussed that the core hole will decay by various radiative and non-radiative processes, such as X-ray fluorescence and Auger emission, respectively. The core-hole lifetime thus corresponds to the inverse of the sum of various contributing transition rate constants. To a first approximation, however, the core hole will decay over time in a close to exponential fashion, with Δt representing the mean lifetime, which is related to the half-life $t_{1/2}$ of the exponential decay $\Delta t = t_{1/2}/\ln 2$. The exponential nature of the decay with time gives rise to the Lorentzian line-broadening function in energy since the Fourier transform of an exponential function is a Lorentzian, alternatively called Cauchy and Briet-Wigner distributions. The overall broadening of spectroscopic features is thus a convolution of a Gaussian from the beamline optics and a Lorentzian from core-hole lifetime effects, which gives what is known as a Voigt peak shape function. Since the Voigt peak shape function can be computationally intensive, in calculations it is common practice to approximate the Voigt as a Gaussian + Lorentzian sum, which is known as a pseudo-Voigt. At first thought one would expect a core hole associated with an absorption edge lying at a high energy to have a shorter lifetime, and conversely those at lower energies to have longer lifetimes, but the situation is complicated by various non-radiative decay processes. An example is shown in Figure 9.27, which compares the experimental linewidths of the K, L_I, L_{II} and L_{III} edges of an aqueous solution of molybdate $[MoO_4]^{2-}$ obtained by fitting the lowest energy feature in the spectra using peak-deconvolution. Table 9.2 compares the experimental and calculated [9] linewidths for the lowest energy XAS feature in each edge.

Table 9.2: Experimental linewidths for a series of absorption edges of $[MoO_4]^{2-}$.

Edge	Γ_{expt} (FWHM) (eV)	% Gaussian	Γ_{calc} (FWHM) (eV)	ΔE_{min}(eV) (mono.)
K	5.38	44	4.52	1.15
L_I	4.16	9	4.25	0.37
L_{II}	1.67	7	1.97	0.34
L_{III}	1.74	5	1.78	0.33

Different monochromator crystals were used to access the L-edges and the K-edge – Si(111) and Si(220), respectively – which gives a different instrumental broadening ΔE; ΔE_{min}, the minimum value for each edge, is given in Table 9.2. The L_{III} and L_{II} edges will show some multiplet splitting (Section 9.8), which is not accounted for in this analysis, but this is expected to be much smaller than for the first transition elements. As expected, the energy resolution is highest in the L_{III} and L_{II} edges, with close to pure Lorentzian peak shapes, suggesting that the lifetime broadening is dominant. The same is true for the L_I edge, except that the linewidth is more than double that of the L_{III} and L_{II}, and close to that observed for the K-edge, which at first sight would seem anomalous. However, these experimental results are in good agreement with theoretical predictions (Table 9.2) [7], with the apparently anomalous L_I energy resolution due to additional broadening mechanisms occurring at the L_I, and in particular Coster-Kronig emission [7].

This comparison serves as an illustration that spectral linewidths can be quite complex. Additional non-lifetime broadening can result from a range of phenomena,

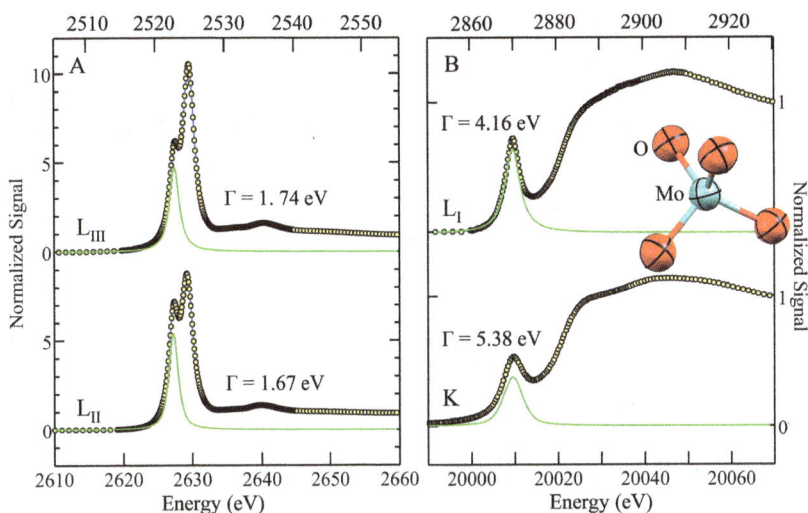

Figure 9.27: Spectroscopic energy resolution at different X-ray absorption edges for an aqueous solution of sodium molybdate, containing the $[MoO_4]^{2-}$ anion. (A) Comparison of the L_{III} and L_{II} edges and (B) the L_I and K edges. Yellow points show experimental data and the underlying blue lines the peak deconvolution fits. The green lines show the lowest energy peak from the fits, from which the Γ_{expt} and % Gaussian values in Table 9.2 are derived. The inset in (B) shows the structure of the $[MoO_4]^{2-}$ anion.

of which exchange coupling, or multiplet splittings, is conspicuous in lighter elements, such as the first transition elements. While we discuss multiplet splitting in the next section, here we note that if the energy separation of multiplet components is less than the lifetime broadening, substantial spectroscopic broadening can result.

9.8 Multiplet splittings

Up to this point we have mostly considered X-ray absorption as a one-electron process, with shake-up and shake-down transitions as special cases. The one electron picture works well for the K-edges, for which computation of spectra can be relatively simple. However, as we have already noted, the L-edges of the first transition metal complexes may exhibit a rich structure (e.g. Figure 9.23). In these cases the one electron model can work less well, necessitating consideration of multiplet splittings. Multiplet splittings arise from overlap of the wavefunction of the core hole created by photoexcitation and the wavefunctions of the valence electrons; if we consider a first transition metal L_{III}/L_{II}-edge and again taking Fe(III) as an example, the initial state will be $2p^6 3d^5$ and the final state will be $2p^5 3d^6$, remembering that the 2p and 3d radial wavefunctions overlap significantly. The final state has an unpaired spin in the 2p level, which couples with the valence electrons to effect what is called multiplet splitting. While orbital overlap is of course present for the initial state, it will not be subject to multiplet splitting because the $2p$ level is completely filled. This picture is compounded by the ligand field effects, which split the 3d manifold (e.g. Figures 9.9 and 9.10) and by the effects of spin-orbit coupling. We can think of the general case for n d-electrons and transitions within the L_{III} and L_{II} XAS as $2p^6 3d^n \rightarrow 2p^5 3d^{n+1}$ transitions. With multiplet splitting we can thus expect that there will be a large number of final states, which for a $3d^n$ configuration is given as follows [10]:

$$6 \times \frac{10!}{(10-n)!n!} \tag{9.19}$$

Among these numerous final states will be many that give rise to XAS transitions with appreciable intensity, depending upon the selection rules. The resulting spectra can be complex and, unlike both the K-edge 1s→3d transitions (discussed above) and the L-edges of the second transition elements [11], may be less than intuitive.

Multiplet effects of even greater complexity are expected for the M_V, M_{IV} XAS for compounds of the rare earths, also known as the lanthanides (e.g. Figure 9.26). Relative to the 3d orbitals, the 4f orbitals are more spread-out in nature, with angular wavefunctions possessing six or eight lobes, rather than the three or four present in the 3d. Consequently, the ligand field splitting of the 4f manifold tends to be smaller. For example, an octahedral ligand field will split the seven 4f orbitals into three levels with degeneracies of 1, 3 and 3, called respectively, a_{2g}, t_{2g} and t_{1g}. In addition the metals have a tendency towards high coordination numbers, such as the 9-coordinate species shown in Figure 9.26. In many cases, therefore, the ligand field splitting of rare earth complexes will be smaller than the lifetime broadening (Section 9.6), so that the end result is a poorer spectroscopic resolution.

9.9 Calculation of near-edge spectra

A number of years ago we might have written in this section that although the calculation of the EXAFS was well-developed, the calculation of near-edge spectra was still far from mature. Thankfully this is no longer the case, as there are now a large number of codes that can successfully calculate spectra, each with its own particular strengths and limitations. In this section we will not attempt to present an exhaustive review of all the different codes that are available. Any such attempt would become rapidly out of date and would almost inevitably omit one or other important code. Instead, we will discuss examples of various codes which employ different approaches. In the last section we discussed multiplet effects, and thus we will begin with multiplet calculations of near-edge spectra, followed by the multiple scattering approach and quantum mechanical theoretical approaches.

9.9.1 Multiplet calculations

One emerging *de facto* standard code for simulation of first transition metal L-edges is known as TT-multiplets or CTM4XAS. This code was originally an atomic code and was modified from early versions to incorporate crystal field splittings and charge transfer parameters. The input required includes the initial and final state electronic configurations (e.g. for Fe(III) a $2p^63d^5$ initial state and a $2p^53d^6$ final state), and initial and final state charge-transfer configurations (e.g. again for Fe(III), a $2p^63d^6$–L initial and $2p^53d^7$–L final) and a handful of other parameters, such as symmetry and crystal field splitting Δ_0.

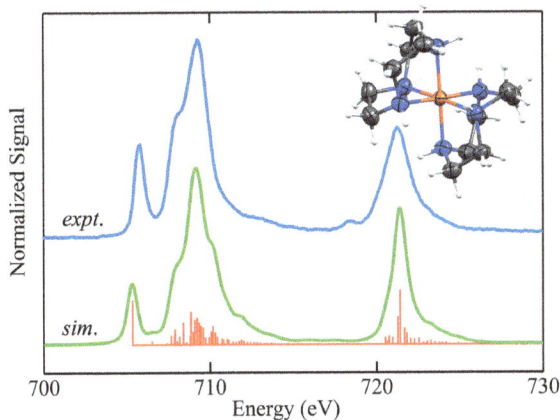

Figure 9.28: Example of a simple TT-multiplet calculation of the L_{III}/L_{II} spectrum of an Fe(III) octahedral complex (the structure is shown in the inset). The input assumed O_h site symmetry with a value of Δ_0 of 3.15 eV. The upper blue line shows an experimental spectrum and the lower green line the simulation, which has a line-shape function convoluted with the computed transitions. The red line shows a stick spectrum, with bars indicating individual transition intensities.

Such calculations require no atomic coordinates and given the sparsity of parameters that are input it is remarkably successful in calculation of spectra. Of course, the code only calculates transition energy positions and intensities; to obtain a realistic looking simulation one must convolute with a peak-shape function that may be Gaussian, Lorentzian, Voigt or pseudo-Voigt. Figure 9.28 shows an example of a typical calculation, that of an octahedral ferric species.

9.9.2 Multiple scattering calculations

We have discussed multiple scattering in the context of EXAFS in Section 8.7. At low k values the multiple scattering paths will proliferate considerably; early approaches to near-edge simulations summed very large numbers (thousands) [12] of individually computed multiple scattering paths to approach simulation of near-edge spectra, with much recent work focusing on methods to increase the efficiency of computation [13]. The input required for a multiple scattering calculation typically consists simply of atomic coordinates with Cartesian values. Only atoms that provide significant scattering need to be included, and there is no requirement for satisfying the strictures of chemical bonding (one can neglect hydrogens, for example). The most successful multiple scattering calculations of near-edge spectra typically include a large cluster of scattering atoms around the central absorber, although if the atoms are symmetrically disposed then calculations can be very effective.

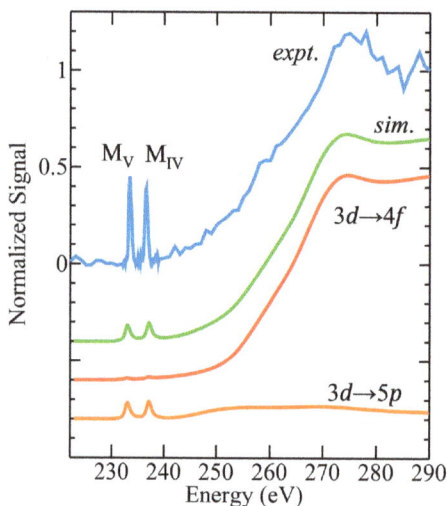

Figure 9.29: FEFF calculation of the M_V/M_{IV} near-edge spectrum of molybdate [12], showing the experimental data for Na_2MoO_4 (blue line), the total FEFF-10.0 simulation for $[MoO_4]^{2-}$ (green line) and contributions from $3d{\to}4f$ and $3d{\to}5p$ transitions (red and orange lines, respectively). The calculated lines have been offset vertically for clarity.

Figure 9.29 shows a FEFF full multiple scattering calculation of the five-atom cluster of the molybdate oxy-anion $[MoO_4]^{2-}$ compared to an experimental M_V/M_{IV} near-edge spectrum [14]. The only exogenous inputs needed for the calculation are the coordinates of the five-atom cluster $[MoO_4]^{2-}$. The sharp M_V, M_{IV} features are well reproduced by the TT-multiplets code [12] (not illustrated), which confirms that these features are transitions from the $3d$ to the $5p$ levels, and the presence of a single sharp peak for each of the M_V and M_{IV} is consistent with a lack of any discernible ligand field splitting of the $5p$ manifold from the tetrahedral arrangement of oxygens around molybdenum. Notably, the FEFF full multiple scattering calculations better reproduces the $3d{\to}4f$ absorption, a broad structure, which is lacking in both TT-Multiplet [12] and TD-DFT ORCA simulations of the spectra (not illustrated). Figure 9.29 thus provides an example where FEFF succeeds while other methods fail.

9.9.3 Density functional theory

For both density functional theory (DFT) and time-dependent density function theory (TD-DFT) (considered in Section 9.8.4), two major factors are important in generating accurate spectral simulations. These are the quality of the basis sets used, and the nature of the core-hole potential used in modelling the excited state. Calculations usually employ a relatively large basis set for the photoexcited atom with more compact basis sets for other atoms in the calculation. An important but obvious difference between programs such as FEFF and DFT (or TD-DFT) codes is that every atom and electron is important and that there can be no unsatisfied bonding requirements in any DFT calculation. An early and successful DFT approach was that adopted within the StoBe-deMon code.[9] This code can employ what is called the half-core-hole approximation in which half an electron is removed from the core level to generate the potential that is used to compute the exited states. This approach provides a good compromise between accurately modelling initial and final state effects and generates realistic-looking spectral simulations, which resemble experimental spectra. We have already commented upon convolution of computed transition intensities with a peak shape function for TT-multiplets. StoBe-deMon uses what might be called a ramping function-width convolution method where the width of the peak-shape function varies linearly between minimum and maximum values at arbitrarily chosen energies. This is a pragmatic method, which is designed to compensate for discrete continuum sampling by smearing out unrealistic sharp features that would otherwise occur at higher energies. One advantage of the DFT method, in common with the TD-DFT method, is that the excited state molecular orbitals that are involved in the excited states can be visualized, which can give insights into the nature of the various transitions involved. If a molecule contains more than one type of atom whose near-edge spectrum is being investigated, it is often straightforward to compute the spectra expected from individual sites, as in Figure 9.16 which shows excited state molecular orbitals for one of two sulfurs in a molecule, obtained from a StoBe-deMon calculation.

9.9.4 Time-dependent density functional theory

TD-DFT is an extension of the DFT method, which is very useful for calculation of time-dependent phenomena such as the interactions with electromagnetic fields (e.g. photons), giving transition energies and probabilities. The TD-DFT method is increasingly popular for computing excited states in general, as well as for our purposes here, which is the computation of core-level spectra. A number of modern codes such as ORCA and

---

9 StoBe-deMon is a version of deMon (density of Montréal) originating from Stockholm (Sto) and Berlin (Be).

ADF provide capabilities to compute near-edge spectra, and the many advantages include all of those discussed for DFT above. Modern codes supporting TD-DFT are typically relativistic codes, meaning that spectra from large atoms in which relativistic effects become important can be computed. As an added bonus, many of the codes are now quite easy to use, having evolved from being more or less the exclusive domain of the specialist to be accessible by a wide variety of users. Transition energies computed with TD-DFT require shifts in the computed energies to align with experimental spectra as transition energies are typically underestimated (we note that XES energies are typically overestimated and require shifts in the opposite direction). This is because what is known as the self-interaction error gives core orbital energies that are too high, resulting in an underestimation of the energy difference between core and the orbital for the transition. Energy shifts are also required for DFT transition energies, although these are typically smaller.

9.10 Peak deconvolution of near-edge spectra

The process of peak deconvolution of near-edge spectra can be useful when it is desired to obtain accurate peak positions and areas. The approach is to model the spectrum using the sum of a series of discrete peaks, together with an edge-step function. Overall, the absorption edge function $c(E)$, after pre-edge background subtraction, is modelled as follows:

$$c(E) = a_s s(E) + \sum_{i=1}^{n} a_i V_i(E) \tag{9.20}$$

where a_s is the amplitude (step height) of the edge-step function, which for normalized spectra will be unity, and a_i are the amplitudes of the individual peaks, with the summation proceeding over all n peaks. The edge-step function $s(E)$ is described below. Ideally, the peak shape would be described by a Voigt peak shape function, which is a convolution of a Gaussian with a Lorentzian, the former arising from optical components in the beamline and latter from core-hole lifetime broadening as described in Section 9.7. In almost all cases the pseudo-Voigt peak shape $V(E)$ will suffice, which is a simple sum of Gaussian $G(E)$ and Lorentzian $L(E)$ functions, which is used because it is quicker to compute:

$$V_i(E) = m_i G_i(E) + (1 - m_i) L_i(E) \tag{9.21}$$

in eq. (9.21) where subscript i denotes the peak number in the summation from eq. (9.20), m_i is the mixing parameter, and the functions $G_i(E)$ and $L_i(E)$ are defined as follows:

$$G_i(E) = \exp\left(-\ln 2\, \frac{(E-E_i)^2}{\Gamma_i^{\,2}}\right) \tag{9.22}$$

$$L_i(E) = \frac{\Gamma_i^2}{\Gamma_i^2 + (E-E_i)^2} \tag{9.23}$$

where Γ_i is the half-width at half-maximum of the peak i and E_i specifies the peak's energy position (mid-point).

The areas A_i of the individual peaks can be computed using the following equation:

$$A_i = \Gamma_i\left(m_i\sqrt{\frac{\pi}{\ln 2}} + (1-m_i)\pi\right) \tag{9.24}$$

In practice a peak-skew is often needed to fit the experimental spectrum, and some codes provide for different widths between $G(E)$ and $L(E)$, so that

$$G'_i(E) = \exp\left(-\ln 2\, \frac{(E-E_i)^2}{(\Gamma_i + (E-E_i)\eta_i)^2 \xi^2}\right) \tag{9.25}$$

$$L'_i(E) = \frac{(\Gamma_i + (E-E_i)\eta_i)^2}{(\Gamma_i + (E-E_i)\eta_i)^2 + (E-E_i)^2} \tag{9.26}$$

where η_i is a peak skew and ξ is the ratio of Gaussian to Lorentzian linewidths.

The edge-step function $s(E)$ is approximated as the sum of an erf function (the integral of a Gaussian) and an arctangent, (the integral of a Lorentzian). The edge-step function $s(E)$ is defined by eq. (9.24), in which E_s is the mid-point energy of the edge-step and Γ_s is the half-width, which is simply the integral of eq. (9.21):

$$s(E) = \left(\frac{\pi + \sqrt{\pi \ln 2}}{4}\right)\left[m\frac{1}{2\sqrt{\pi \ln 2}}\left(1 + \mathrm{erf}\left(\sqrt{\ln 2}\,\frac{(E-E_s)}{\Gamma_s}\right)\right)\right.$$
$$\left. + (1-m)\left(\frac{1}{2} + \frac{1}{\pi}\tan^{-1}\left(\frac{(E-E_s)}{\Gamma_s}\right)\right)\right] \tag{9.27}$$

The method of peak deconvolution typically uses non-linear optimization, refining parameters of peak positions, amplitudes and widths, to minimize the sum of the squares of differences between the experimental points and the calculated points from eq. (9.20). Very often the same mixing m is used for all peaks, or at least for all sharp peaks at lower energies. We have already presented this type of analysis in obtaining estimates of linewidths through linear combination fits in Figure 9.27. An additional example is shown in Figure 9.30, that of the sulfur K-edge spectrum of p-toluene-sulfonic acid, where the lowest energy feature is present as a barely resolved shoulder on the major feature. The exact position and area of this peak would be difficult to determine without peak deconvolution analysis.

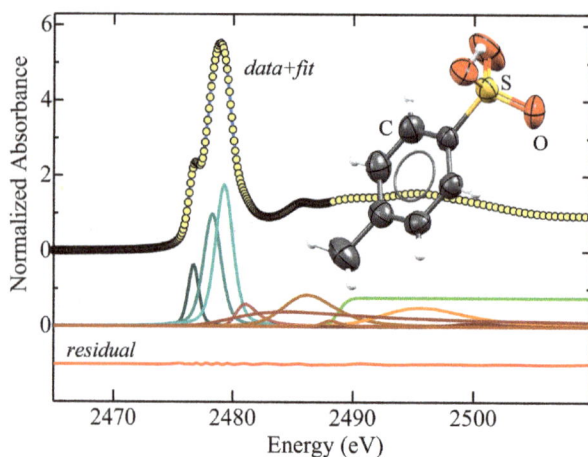

Figure 9.30: Example of a peak-deconvolution analysis. The yellow points show the experimental sulfur K-edge X-ray absorption spectroscopic data for a 100-mM toluene solution of *p*-toluene-sulfonic acid, with the blue line underlying the points the peak-deconvolution fit. The individual peak profiles given by the fit are shown below the data, with major transitions shown in blue, minor transitions shown in brown-orange, and the edge step in green. The lowest red line shows the residual (experiment minus fit). The crystal structure is shown in the inset.

References

[1]   Pickering, I. J.; George, G. N. Polarized X-ray absorption spectroscopy of cupric chloride dihydrate. *Inorg. Chem.* **1995**, *34*, 3142–3152.

[2]   Solomon, E. I.; Hedman, B.; Hodgson, K. O.; Dey, A.; Szilagy, R. K. Ligand K-edge X-ray absorption spectroscopy: Covalency of metal–ligand bonds. *Coord. Chem. Rev.* **2005**, *249*, 97–129.

[3]   Ha, Y.; Dille, S. A.; Braun, A.; Colston, K.; Hedman, B.; Hodgson, K. O.; Basu, P.; Solomon, E. I. S K-edege XAS of Cu^{II}, Cu^{I}, and Zn^{II} oxidized dithiolene complexes: Covalent contributions to structure and the Jahn–Teller effect. *J. Inorg. Biochem.* **2022**, *230*, 111752/1–7.

[4]   Brouder, C. Angular dependence of X-ray absorption spectra. *J. Phys.: Condens. Matter.* **1990**, *2*, 701–738.

[5]   Hahn, J. E.; Scott, R. A.; Hodgson, K. O.; Doniach, S.; Desjardins, S. R.; Solomon, E. I. Observation of an electric quadrupole transition in the X-ray absorption spectrum of a Cu(II) complex. *Chem. Phys. Lett.* **1982**, *88*, 595–598.

[6]   Sham, T. K. X-ray absorption spectra of ruthenium L-edges in $Ru(NH_3)_6Cl_3$. *J. Am. Chem. Soc.* **1983**, *103*, 2269–2273.

[7]   George, G. N.; Cleland, W. E.; Enemark, J. E.; Smith, B. E.; Kipke, C. A.; Roberts, S. A.; Cramer, S. P. L-edge spectroscopy of molybdenum compounds and enzymes. *J. Am. Chem. Soc.* **1990**, *112*, 2541–2548.

[8]   Funk, T.; Deb, A.; George, S. J.; Wang, H.; Cramer, S. P. X-ray magnetic circular dichroism – A high-energy probe of magnetic properties. *Coord. Chem. Rev.* **2005**, *249*, 3–30.

[9]   Krause, M. O.; Oliver, J. H. Natural widths of atomic K and L levels, Kα X-ray lines and several KLL Auger lines. *J. Phys. Chem. Ref. Data.* **1979**, *8*, 329–338.

[10] De Groot, F. Multiplet effects in X-ray spectroscopy. *Coord. Chem. Rev.* **2005**, *249*, 31–63.

[11] George, G. N.; Cleland, W. E.; Enemark, J. E.; Smith, B. E.; Kipke, C. A.; Roberts, S. A.; Cramer, S. P. L-edge spectroscopy of molybdenum compounds and enzymes. *J. Am. Chem. Soc.* **1990**, *112*, 2541–2548.

[12] Farges, F.; Brown, G. E., Jr; Rehr, J. J. Ti K-edge XANES studies of Ti coordination and disorder in oxide compounds: Comparison between theory and experiment. *Phys. Rev. B.* **1997**, *56*, 1809–1819.

[13] Rehr, J. J.; Ankudinov, A. L. Progress in the theory and interpretation of XANES. *Coord. Chem. Rev.* **2005**, *249*, 131–140.

[14] George, S. J.; Drury, O. B.; Fu, J.; Friedrich, S.; Doonan, C. J.; George, G. N.; White, J. M.; Young, C. G.; Cramer, S. P. Molybdenum X-ray absorption edges from 200 to 20,000 eV: The benefits of soft X-ray spectroscopy for chemical speciation. *J. Inorg. Biochem.* **2009**, *103*, 157–167.

10 Analysis I – EXAFS data reduction and analysis

This chapter focuses on the methods of XAS data reduction and EXAFS analysis, from how to treat raw data collected at the beamline to determining a structure from the extracted EXAFS. While much of what will be discussed here is built into analysis software in a more or less semi-automated manner, we describe the various steps in detail to help enable the use of such software to be less like a "black box" and more informed. The methods also include the vital step of incident X-ray energy recalibration, but as this has already been discussed in Chapter 7, we will not revisit this here.

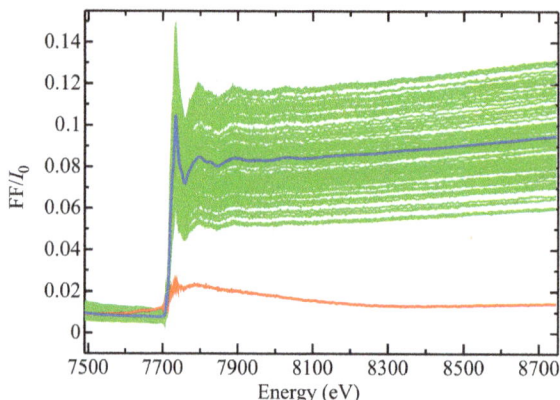

Figure 10.1: Automated screening for problematic channels in a single sweep of an XAS data set measured using a 100-pixel germanium monolithic detector of a sample of a dilute cobalamin-containing metalloprotein. The algorithm compares the shapes described by each channel with the average of all the others, with iteration until no more bad channels are identified. The green lines show the "good" channels, the blue line shows the average of these and the red line shows a "bad" channel that has been automatically identified.

10.1 Signal averaging and screening

In many cases, XAS data from more than one individual sweep is averaged. For concentrated solid samples measured in transmittance, the rule is that two or more sweeps must superimpose. Sweep-to-sweep – it is normal to compare individual sweeps to check for beam-induced or other chemical changes in the sample with time, which in most cases is undesirable. Ideally, this type of screening should be done at the beamline, at which time some kind of intervention following the observation of changes might be possible. For example, the sample might be moved slightly so that the beam interrogates a fresh area of sample to minimize photo damage. Multiple sweeps, beyond the two needed to check reproducibility, may be needed to improve signal to noise, for example for lower concentrations or higher k-range.

https://doi.org/10.1515/9783110570441-010

With X-ray fluorescence detection of dilute samples, in addition to multiple sweeps, each sweep may contain data from multiple detector elements, since many modern fluorescence-detected XAS experiments use multi-element solid-state detector arrays (e.g. Section 5.3.8). Here, the signal averaging process should ideally involve screening for and rejecting from the average, what might be called bad channels or even entire bad sweeps that may have some problem. With multi-element array detectors, a bad channel might arise because of some issue with the detector system itself, such as an improperly connected cable on the pre-amplifier. Alternatively, one or more array elements of the detector may have previously been count-rate saturated, which can cause baseline determination errors in some digital signal processing firmware. Other possible sources might include ice diffraction within the sample itself. Irrespective of the origins, bad channels may need to be eliminated, and this is ideally done using a somewhat laborious scan-by-scan channel-by-channel screening process. Given that the detector might be a 100-pixel monolithic detector, it may be tempting to just average all of the channels without examining them, but this is ill-advised as serious problems can be introduced by doing this. However, modern software can at least, in part, relieve the user of this chore, with tools to compare individual channels with all the others, and with those from other sweeps. Figure 10.1 shows an example of this process with a single sweep of a dilute metalloprotein. Once screening is complete, all the good fluorescence detector channels are summed for each sweep and ratioed with the incident intensity, I_0. Sweeps are then also summed.

Incident X-ray energy re-calibration is also conducted at this point. Even if the energy has been calibrated at the beamline, it is a good practice to perform energy recalibration. As we have discussed in Section 7.9, the energy calibration comprises measuring a standard, which in hard X-ray experiments is most usually a foil placed downstream of the sample, with spectra shifted so that the foil absorption edge corresponds to the tabulated value. This important process is described in Section 7.9. At the risk of being repetitive, we again remind the reader that for extended range data sets, such as are needed for EXAFS, re-calibration should never be done using a simple energy shift because this can introduce errors in the bond lengths that will be later determined using EXAFS curve-fitting analysis (see Section 10.8).

10.2 Background removal and normalization – removing the pre-edge

The first step in data reduction is background removal. This is a routine but very important step, since if background removal and normalization are poorly done, then this can significantly impact everything that follows. The process of background removal can be considerably simpler with transmittance XAS data, which will be discussed first, before moving on to consider more complex examples.

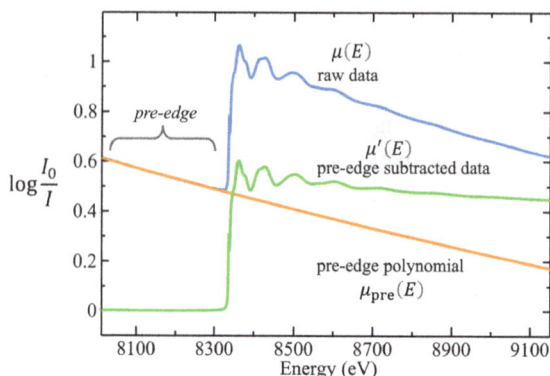

Figure 10.2: Pre-edge subtraction for the transmittance spectrum of [(n-butyl)$_4$N][Ni(mnt)$_2$].

Figure 10.2 shows such a simple transmittance example, that of a nickel K-edge XAS measurement on a solid sample of the approximately square planar nickel tetrathiolate complex **tetra**-n-butyl ammonium **bis**-maleonitriledithiolate-nickel(III), or [(n-butyl)$_4$N] [Ni(mnt)$_2$], appropriately diluted with boron nitride. With transmittance data, one would in most cases fit a polynomial $\mu_{\mathrm{pre}}(E)$ to the unstructured pre-edge region of the data (Figure 10.2), and extrapolate this through the entire data set to obtain an absorption spectrum corresponding to only the absorption edge of interest $\mu'(E)$:

$$\mu'(E) = \mu(E) - \mu_{\mathrm{pre}}(E) \tag{10.1}$$

With dilute samples, there may be significant signal contamination from a number of sources; for example, if a helium cryostat is used, then small amounts of zinc are usually present in the mylar windows of the cryostat.

Figure 10.3 shows an example of a dilute zinc-containing metalloprotein from which a background spectrum showing a small zinc contamination signal has been subtracted. The background spectrum would be recorded on a sample with essentially the same composition (buffer, glassing agent etc.) as the experiment, but lacking the zinc-containing entity, here, the metalloprotein. The disadvantage here is that the background spectrum must be recorded with good signal to noise, and so at least, the same number of sweeps needs to be averaged as for the experimental data set. If signal contamination is negligible, then a polynomial can be fitted to the background spectrum to obtain a noise-free approximation to the background, with this polynomial then being subtracted from the raw data; in which case, the stricture of good signal to noise for the background can be relaxed. The background in the pre-edge region of Figure 10.3 can be seen to rise in a curving way, which differs from transmission data (e.g. Figure 10.2). We will next discuss the origins of this and common ways of pre-edge subtraction. Figure 10.4 shows the origins of the rising background, which is frequently observed in fluorescence data.

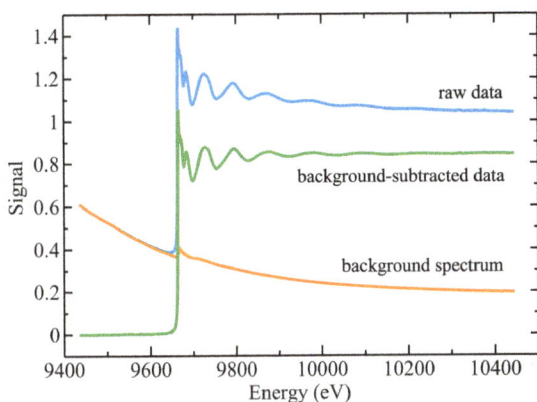

Figure 10.3: Background subtraction of a zinc-containing metalloprotein with minor contamination. The raw data, which is a zinc-containing metalloprotein (ca. 0.35 mM Zn), is shown by the blue line. The background spectrum is obtained by scanning a "blank" sample, which is the same as the protein sample but without the protein. The difference (green line) removes minor contamination observed in the blank and also provides very effective pre-edge subtraction.

In Figure 10.4, when the incident X-ray energy is well above the absorption edge (e.g. $E = 9,400$ eV), an emission spectrum (Figure 10.4B, orange line) shows three peaks, a strong clear Cu $K\alpha_{12}$ and minor $K\beta$ fluorescence, in addition to the strong scatter near the incident energy. The electronic window, shown in Figure 10.4B as a green box, effectively discriminates these signals, only registering the Cu $K\alpha_{12}$. When the incident X-ray energy lies in the pre-edge part of the spectrum (e.g. $E = 8,600$ eV), no X-ray fluorescence is produced but the wings of the X-ray scatter peak now fall within the electronic window for Cu $K\alpha_{12}$ and give a rising pre-edge signal, corresponding to the filled green area in Figure 10.4B. In the absorption spectrum shown in Figure 10.4A, the $E = 9,400$ eV gives rise to near-pure fluorescence in the post-edge region, whereas the $E = 8,600$ eV point below the edge, instead, shows significant scatter, whose intensity decreases as energy increases towards the edge. In example absorption spectrum, the X-ray energy has been deliberately scanned through a wider-than-normal range to include the energies of the Cu $K\alpha_{12}$ electronic window in order to show how the scatter peak gives rise to a large peak in the pre-edge. Normally, the pre-edge would not be extended to such low energies and only the high-energy tail of the peak would be observed.

The X-ray scatter peak will exhibit some asymmetry due to inelastic X-ray scattering, which will change subtly as the incident energy is swept, and the pre-edge function will be a convolution of the scatter peak and the shape of the X-ray window function. If we approximate the window as a rect function and the X-ray scatter peak as a Gaussian, then pre-edge shape can be modelled by two half-Gaussians separated by a plateau. If the widths of the Gaussian and the rect are similar, then this approximates to a Gaussian function, and it is common practice to model the pre-edge $\mu_{pre}(E)$ as a Gaussian function,

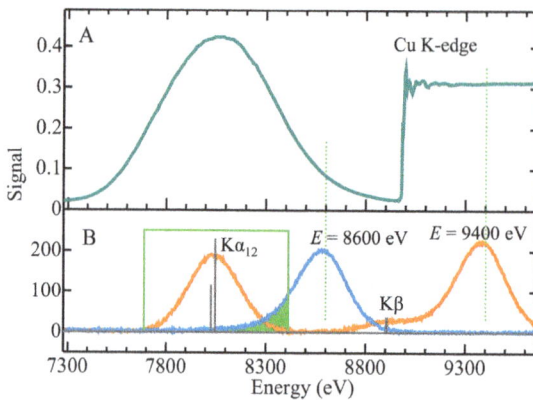

Figure 10.4: Origins of rising pre-edge in X-ray fluorescence XAS. A shows the XAS spectrum of a copper-containing sample, with the incident X-ray energy scanned over a wider-than-normal range. This is measured using an energy-dispersive detector, with the signal corresponding to the windowed Cu Kα fluorescence line. B superimposes the X-ray fluorescence emission spectrum from the solid-state detector at two different incident excitation energies, above (E = 9,400 eV, orange line) and below (E = 8,600 eV, blue line) the Cu K-absorption edge. The electronic window used for discriminating fluorescence is shown by the green line and with the grey sticks showing the X-ray fluorescence line positions for copper.

often combined with a low-order polynomial. That this works reasonably well for the data of Figure 10.4 is shown in Figure 10.5. In the particular case of Figure 10.5, the overall spectrum has been fitted to match the Cu K-edge X-ray absorption cross-section obtained from tabulated values [1], to give a somewhat better background representation. As we have already noted, the data shown in Figures 10.4 and 10.5 extends to lower incident X-ray energies than normal, and in most cases, only the high-energy tail would be fitted.

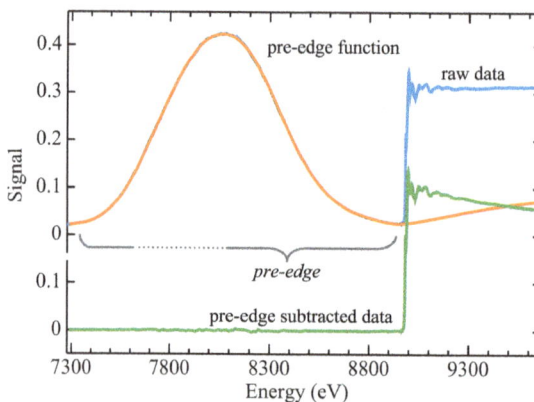

Figure 10.5: Modelling the pre-edge function of X-ray fluorescence XAS with a Gaussian. Raw data (blue) from Figure 10.4A is fitted with a pre-edge function (orange) comprising a Gaussian and a polynomial, which is subtracted to result in the pre-edge subtracted function (green).

10.3 Background removal and normalization – the spline

Following successful pre-edge subtraction, the next step in data reduction is subtraction of a spline from the EXAFS region of the data to isolate the EXAFS oscillations. The fitted spline is also used to define the size of the edge jump in the raw data in order to normalize the data. In the mathematical context, the spline is a stiff but flexible curve that runs through the data. In XAS, it is used to extract the EXAFS oscillations from a smoothly varying background without following the EXAFS wiggles themselves. The spline is named for a traditional drafting tool made of a readily bendable yet springy wood such as pine, with locations along the curve secured by lead weights called ducks. The mathematical spline is actually very similar, comprising a series of polynomials mathematically connected by what are called knot points, the equivalent of the ducks of the draftsman's spline. The mathematical spline is fitted to the pre-edge subtracted data $\mu'(E)$, including only the region above the absorption edge to obtain $\mu_s(E)$. Because only data above the absorption edge are included, we can express the absorption as a function of k (see eq. (4.17)), which we will restate here as follows:

$$k = \sqrt{\frac{2m_e}{\hbar^2}(E - E_0)} = \sqrt{0.262468(E - E_0)} \qquad (10.2)$$

Thus, if the background removal was perfect and there were no other distortions present, then the EXAFS would be given as follows:

$$\chi(k) = \frac{\mu'(k) - \mu_s(k)}{\mu_s(k)} \qquad (10.3)$$

In practice, the background subtraction might be less than perfect, so a better denominator for eq. (10.3) might be a calculated function $\mu_c(k)$, which is available from tabulated X-ray absorption cross-sections, such as those provided by McMaster et al. [1] or Victoreen [2], as follows:

$$\chi(k) = \frac{\mu'(k) - \mu_s(k)}{\mu_c(k)} \qquad (10.4)$$

The process of spline removal for the data of Figure 10.2 is shown in Figure 10.6.

A close inspection of the trajectory of the spline through the pre-edge subtracted data is often a good idea to make certain that the spline is not being influenced by the EXAFS oscillations, which would be the case when using too flexible a spline. When using early analysis codes, a lot of effort typically went into getting a "good" spline, and while some of this may have been directed at removal of AXAFS (Section 8.6), modern codes possess smart default values or other mechanisms for defining the best spline, so that obtaining an appropriate spline is less of an issue.

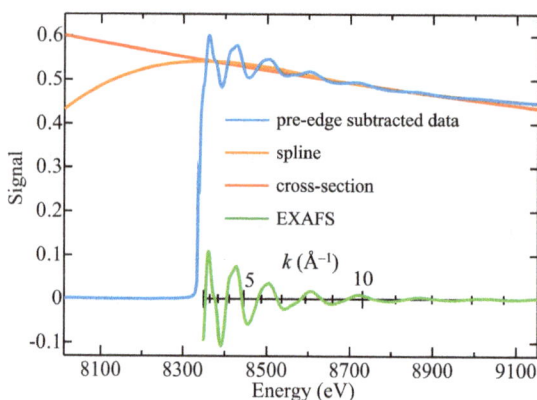

Figure 10.6: Extraction of the EXAFS for [(n-butyl)$_4$N][Ni(mnt)$_2$] using a spline. The pre-edge subtracted data (blue line) μ' corresponds to the pre-edge result of Figure 10.2, the spline (orange line) μ_s was calculated automatically and the cross-section (red line) corresponds to the calculated μ_c discussed in the text.

As we have previously noted in Chapter 8, with analysis going forward from this point, the EXAFS $\chi(k)$ is usually weighted by k^3. There are two reasons for this. First, this weighting helps counteract the inherent damping at high k of $\chi(k)$ due to smaller backscattering amplitudes at high k, the Debye-Waller factor and the $1/k$ term in the EXAFS equation. Second, the theory used to describe the EXAFS tends to be more challenged at low k and improves approximately linearly as a function of k.

The **spline point** would be the energy where μ_s and μ_c cross (μ_c is scaled to μ_s at this point) and this should be defined as a point above but close to the rise of the absorption edge, and may be used to define E_0, the origin for k. There are different strategies in use to determine E_0; one strategy adopted by some is to use the foil calibration point for E_0 or the first inflection point of the experimental spectrum. While this is easy to define experimentally, the main issue is that it may not be physically reasonable. Figure 10.7 shows a close up of the near-edge region of [(n-butyl)$_4$N][Ni(mnt)$_2$], superimposed with the spectrum of a standard nickel foil. As we discussed in Chapter 8, the near-edge is made up of transitions to bound states superimposed upon the edge step at E_0, which corresponds to excitation to the continuum, often called the threshold energy. The real issue is that the true value of E_0 is not readily determined, but placing E_0 below the major features of the spectrum perhaps makes less sense than defining a somewhat arbitrary position above the absorption edge. In the author's analysis code (called EXAFSPAK), the arbitrary assignment of E_0 is preferred, with the default value being computed relative to the calibration energy +15 eV and rounded to the nearest 5 eV.

The spline point: the value of μ_s at E_0, provides a quick and easy way to normalize the near-edge spectrum as the spline will ideally run through any structure that is present.

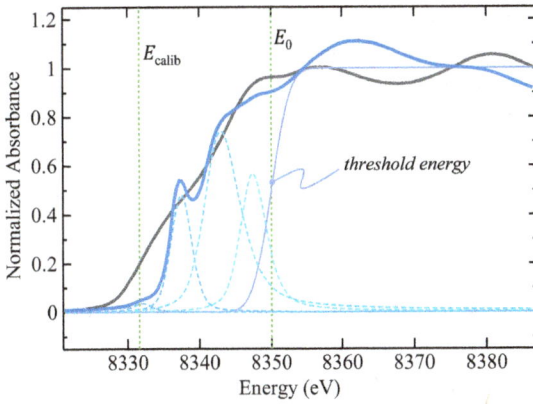

Figure 10.7: Defining the value for E_0. The near-edge region of the K-edge XAS of [(n-butyl)$_4$N][Ni(mnt)$_2$] is shown as the bold blue solid line, superimposed with the spectrum of a standard nickel foil (solid grey line).The broken blue lines show various peak deconvolution contributions to the spectrum, corresponding to different bond-state transitions. The green broken lines show the calibration point (E_{calib}) corresponding to the first inflection of the Ni metal foil at 8,331.6 eV, and the selected threshold energy, E_0, which, here is 8,350 eV.

At this point, the background subtraction procedure might be regarded as complete, and we will now consider the EXAFS in more detail.

10.4 The EXAFS

The k^3-weighted EXAFS of [(n-butyl)$_4$N][Ni(mnt)$_2$], obtained from the spline subtraction shown in Figure 10.6 and computed using eq. (10.4), is shown in Figure 10.8.

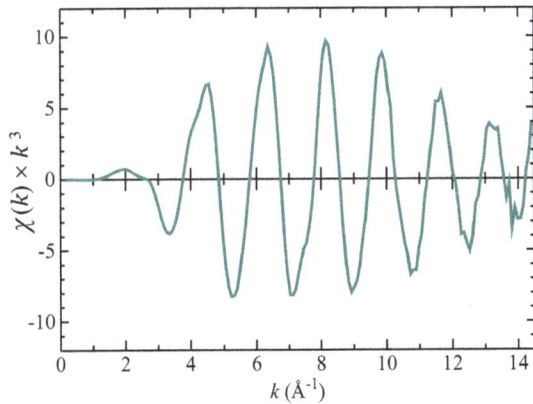

Figure 10.8: The k^3-weighted EXAFS of [(n-butyl)$_4$N][Ni(mnt)$_2$].

In this particular case, the EXAFS resembles a single sine wave dominated by the four thiolate donors, two from each of the two maleonitriledithiolate ligands. Before proceeding further with the approach to analysis, we will revisit the EXAFS expression using the combined amplitude function $A_i(k)$ of eq. (8.33):

$$\chi(k) = \sum_{i=1}^{n} \underbrace{\frac{N_i A_i(k)}{kR_i^2} e^{-2k^2\sigma_i^2} e^{-2R_i/\lambda(k)}}_{\text{Amplitude}} \underbrace{\sin(2kR_i + \varphi'_i(k))}_{\text{phase}} \tag{10.5}$$

Here, $\varphi'_i(k) = \varphi_i(k) + \phi_c(k)$, where $\varphi_i(k)$ is the backscatterer phase function for backscatterer i and $\phi_c(k)$ is the central atom phase function. All other terms are as previously defined and discussed in more depth in Chapter 8, and recapped here for convenience: R_i is the interatomic distance between the absorber atom and backscatterer atom i with mean square deviation due to static disorder and thermal effects σ_i^2, N_i is the coordination number and $\lambda(k)$ is the electron mean free paths.

Our task with analysis of the EXAFS is to extract the structurally relevant parameters R, N and σ^2 from the data. How these contribute to the EXAFS can be seen from eq. (10.5); those variables under **Amplitude** will affect the EXAFS amplitude (all of R, N and σ^2), while R, under **Phase**, will affect the phase of the EXAFS. In practice, an additional parameter must also be included due to the inevitable mismatch between the theoretical and assumed values for E_0 (Figure 10.7), which we call ΔE_0. Modern theoretical approaches to the EXAFS amplitude and phase functions allow for a common ΔE_0 for all backscatterers included in the refinement, which reduces the number of parameters considerably. The effects of changing the threshold energy from E_0 to E_0' by ΔE_0 involve a change in the abscissa from k to k', eq. (10.6). There are two ways to adjust this, either by shifting the experimental k-scale or by shifting the k-scale for the calculated EXAFS. Some data analysis codes allow the user to choose between these alternatives, but in general, keeping the experimental data invariant is considered preferable:

$$k' = \sqrt{k^2 - \frac{2m_e}{\hbar^2}\Delta E_0} \tag{10.6}$$

Assuming that the structure is unknown, the process of deriving these structural parameters is one of **model building**, in which the user will postulate certain ligands to a metal (for example), calculate the EXAFS, and then determine values for N, R, σ^2 and ΔE_0 by minimizing the sum of squares of differences between the experimental and calculated EXAFS. This will be discussed in more detail below. The effect of ΔE_0 will be to shift the EXAFS oscillations more at low k than at high k.

10.5 Sensitivities of the EXAFS

Before proceeding to discuss specific methods of EXAFS analysis, we will revisit aspects of the sensitivities of the EXAFS that have already been approached in Chapter 8, but here discussed from a more practical point of view.

10.5.1 Sensitivity to structural parameters and threshold energy

The effects of varying N, R, σ^2 and ΔE_0 are shown in Figure 10.9A. The effects of coordination number are straightforward because N only appears in the numerator of the amplitude part of the EXAFS expression, eq. (10.5). If all the other structural parameters are the same, then changing the EXAFS coordination number simply scales the amplitude, for example doubling it going from $N = 1$ to 2. This is shown in Figure 10.9A for a Ni–S interaction at 2.15 Å. The interatomic distance R occurs in both phase and amplitude parts of eq. (10.5); as expected, the EXAFS phase is the most sensitive, allowing the detection of small changes in R, with longer interatomic distances leading to increases in the frequency of the EXAFS (Figure 10.9A). The effects of σ^2 are confined to the amplitude, changing the k-dependent damping of the EXAFS, while as we have discussed above, the effects of ΔE_0 are to shift the EXAFS oscillations more at low k than at high k (Figure 10.9A).

10.5.2 Sensitivity to backscatterer type

Figure 10.9B compares the calculated EXAFS for Ni–S at 2.15 Å ($Z = 16$) with three different backscatterer types. In (a), Ni–O ($Z = 8$) at the same distance illustrates that the EXAFS amplitude is approximately half that for Ni–S due to the lower Z backscatterer, with a distinctive phase difference of close to 180°, demonstrating that these backscatterers should be readily distinguished from each other in an analysis. Plot (b) shows the comparison of Ni–Cl ($Z = 17$) again at 2.15 Å, and plot (c) compares the EXAFS for Ni–Se ($Z = 34$). The combination of phase and amplitude differences makes Ni–Se like Ni–O, easy to discriminate from Ni–S. Conversely, in the Ni–Cl case, the close similarity of the EXAFS from the two different backscatterers means that these would probably be impossible to distinguish without taking into account additional information. Like Ni–S versus Ni–Cl, other backscatterers with similar atomic numbers would also be difficult to distinguish, such as nitrogen and oxygen ligands, or arsenic and selenium.

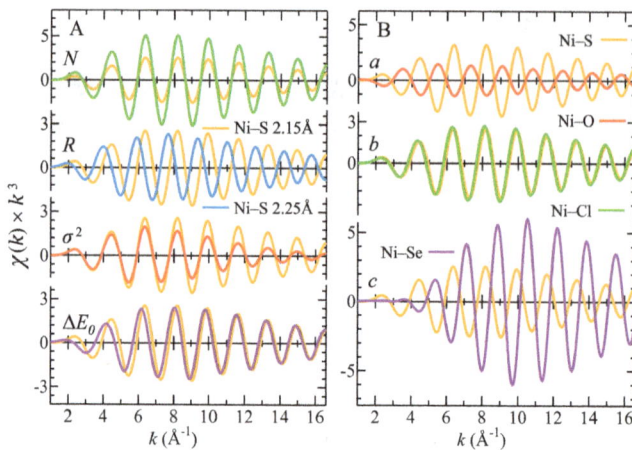

Figure 10.9: (A) Sensitivity of the EXAFS to structural parameters, N, R, σ^2 and ΔE_0. In all cases, the orange line shows the calculated EXAFS for Ni–S with $N = 1$, $R = 2.15$ Å, and $\sigma^2 = 0.0025$ Å^2. The uppermost pair of traces, labelled N, compares $N = 1$ with $N = 2$ while other parameters remaining the same (note that ordinate scale is more compressed in this case, compared to the others in (A)). The pair of traces, labelled R, compares $R = 2.15$ and 2.25 Å, that, labelled σ^2, compares $\sigma^2 = 0.0025$ and 0.0055 Å^2, and those labelled ΔE_0 show the effects of a -10 eV shift in E_0. (B) Sensitivity of the EXAFS to backscatterer type. In each of **a**, **b** and **c**, the orange lines show the same computed EXAFS as in (A), and the red, green and magenta lines, respectively, compare Ni–O, Ni–Cl and Ni–Se EXAFS with Ni–S, all at the same distance of $R = 2.15$ Å with $N = 1$ and $\sigma^2 = 0.0025$ Å^2.

10.6 The EXAFS Fourier transform

After the extraction of the EXAFS using the spline, the typical next step in data reduction and analysis would be to compute the EXAFS Fourier transform. The use of Fourier transforms is central to the field of EXAFS analysis. Here, we will assume familiarity with the basics of Fourier theory; for those who are not, or who would like a refresher, a primer is included in Appendix B.

The EXAFS Fourier transform, which we will call $\rho(R)$, can be computed using eq. (10.7), which assumes that the usual k^3 weighting of the EXAFS is used:

$$\rho(R) = \frac{1}{\sqrt{2\pi}} \int_{k_{min}}^{k_{max}} \chi(k)k^3 \exp(i2kR)dk \tag{10.7}$$

Because there is a phase-shift term $\varphi_i'(k)$ in our expression for the EXAFS, eq. (10.5), the EXAFS cannot be regarded as a simple sum of sine waves. As a result of this phase-shift term, an uncorrected Fourier transform such as that of eq. (10.7) will show peaks that are asymmetric in shape and shifted in R by half the total phase-shift slope, averaged across the k-range of the EXAFS Fourier transform:

$$\delta R \approx \frac{1}{2}\left(\overline{\frac{\partial \varphi'(k)}{\partial k}}\right) \tag{10.8}$$

A better strategy is to use a **phase-corrected EXAFS Fourier transform**, sometimes called an optical transform, which uses $\varphi'(k)$ chosen for one backscatterer, as in eq. (10.9). This has the effect of making the Fourier transform peaks more symmetrical, and more closely locating them at the R value of their respective shell:

$$\rho(R) = \frac{1}{\sqrt{2\pi}} \int\limits_{k_{min}}^{k_{max}} \chi(k)k^3 \exp(i2kR + i\varphi'(k))dk \tag{10.9}$$

In most cases, the **magnitude of the Fourier transform** is viewed (also known as the power spectrum), which is given by eq. (10.10) (see eq. (B.12)). While this lacks any phase information, it serves the purpose of visualizing the radial structure. In some software packages, the imaginary part of the transform may be viewed, but this is not an established standard:

$$|\rho(R)| = \sqrt{\mathrm{Re}[\rho(R)]^2 + \mathrm{Im}[\rho(R)]^2} \tag{10.10}$$

Figure 10.10 compares the phase-corrected and un-corrected Fourier transforms for the EXAFS data of $[(n\text{-butyl})_4\text{N}][\text{Ni(mnt)}_2]$ shown in Figure 10.8. The phase-corrected Fourier transform shows clear peaks corresponding to the different shells of atoms around the central nickel atom. The EXAFS is dominated by the first-shell backscattering from the four thiolate donors. The phase-corrected Fourier transform shows a peak position of 2.149 Å, which is very close to the crystallographically determined Ni–S bond-length of 2.146 Å. The transform lacking phase correction, on the other hand, shows a slightly asymmetric peak that is shifted to lower R ($\delta R \approx -0.41$ Å), relative to the Ni–S bond-length.

The use of phase correction in EXAFS Fourier transforms has been argued to be a matter of personal preference, with some groups routinely avoiding it, while others always use it. Arguments levelled against the practice of phase correction usually suggest that it is an additional and unnecessary manipulation of the data, or point out that it is only correct for one kind of backscatterer. The first of these arguments is specious because one is already manipulating the data by Fourier transforming it; the use of phase correction just replaces one manipulation that is more incorrect with one that is closer to being correct. The argument that phase correction is only valid for one type of backscatterer is valid, but as we have discussed in Chapter 8, in most cases, the total phase-shift function $\varphi'(k)$ is dominated by the central atom phase shift, $\phi_c(k)$. For this reason, at least for the transition metal ions having backscatterer with $Z>17$, using an incorrect backscatterer to phase-correct the transform will not distort the peaks or change the EXAFS Fourier transform peak positions very much at all. In our opinion, the use of phase correction is preferred. Although much less-used

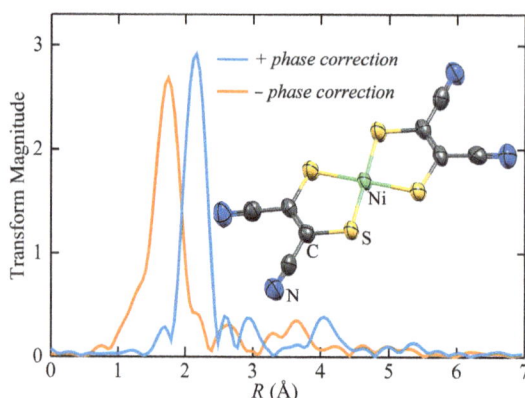

Figure 10.10: Comparison of EXAFS Fourier transforms with and without phase correction. Fourier transforms for [(n-butyl)$_4$N][Ni(mnt)$_2$] with (blue line) and without (orange line) phase correction for Ni–S backscattering. The inset shows the crystal structure of the complex. As expected, the EXAFS is dominated by the first shell Ni–S coordination, with a bond length of 2.146 Å.

than phase correction, amplitude correction of the Fourier transform is also a simplifying method. This can be computed using eq. (10.9), which would calculate a combined phase- and amplitude-corrected Fourier transform, in which $A(k)$ is the total backscattering amplitude function, as described in eq. (8.33), and possibly other terms such as the electron mean free path term. The disadvantage of application of amplitude correction is that this tends to reverse the natural envelope of the EXAFS in which the low k- and high k-ends of the k^3-weighted spectrum have lower amplitudes than the middle of the k-range. This means that series termination artefacts are usually much more pronounced in amplitude-corrected Fourier transforms, which limits any advantages given by using it:

$$\rho(R) = \frac{1}{\sqrt{2\pi}} \int_{k_{min}}^{k_{max}} \frac{\chi(k)}{A(k)} k^3 \exp(i2kR + i\varphi'(k)) dk \qquad (10.11)$$

The EXAFS Fourier transform gives the experimenter a first look at the information that the EXAFS data holds. When conducting experiments at the beamline, viewing the Fourier transform after the first few scans can sometimes provide surprises for the experimenter.[1] An example of a multi-shell EXAFS Fourier transform is shown in Figure 10.11. The complex in question is a model of a biological iron-sulfur cluster,

---

[1] For example, in bio-inorganic chemistry, sometimes XAS will show that the active site of a protein is completely different from what is expected. One example of this was a protein, initially called **red tungsten protein**, from hyperthermophilic archaeon *Pyrococcus furiosus*, which was initially thought to contain tungsten in a W-Fe-S cluster. Early XAS provided the first evidence that this protein did not contain such a cluster, and instead should be classified along with the mononuclear Mo enzymes.

which contains a $[(PhS)_4Fe_4S_4]^{2-}$ cluster. The cluster core comprises a cube-like arrangement of four sulfides and four iron atoms, with mixed Fe(II)/Fe(III) oxidation state having approximately tetrahedral coordination, with an average core Fe–S bond length of 2.293 and 2.270 Å for sulfide and pendant thiolate sulfur groups, respectively. The Fe···Fe separation in the core is 2.720 Å. The EXAFS Fourier transform shows two clear peaks, one at ~2.3 Å from the Fe–S interaction, and the other at ~2.7 Å from the Fe···Fe interaction.

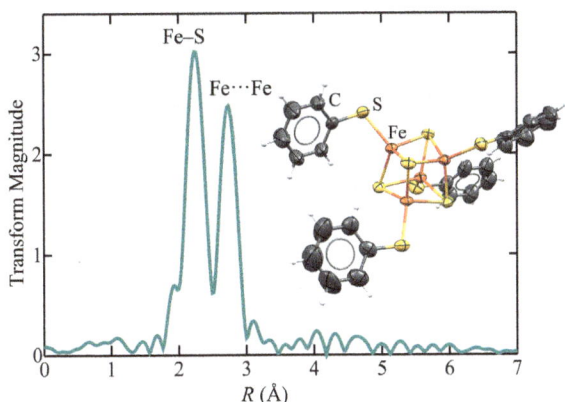

Figure 10.11: EXAFS Fourier transform of a multi-shell data set. The data is the Fe K-edge EXAFS of $[Fe_4S_4(SPh_4)_4]^{2-}$ with a k-range of 0–17 $Å^{-1}$, with the Fourier transform phase-corrected for Fe–S. The EXAFS clearly shows backscattering from first-shell Fe–S bonds and from outer shell Fe···Fe interactions. The inset shows the crystal structure of the complex.

10.6.1 k-Windowing

If the k-range of the data terminates, as it must, but with an abrupt change in the signal $\chi(k) \times k^3$ at the two ends of the data (specified here as k_{min} and k_{max}), then series termination artefacts (Appendix B) will occur in the Fourier transform. These have the effect of convoluting a sinc function (Appendix B) with the transform, resulting in small peaks or ripples appearing on either side of each transform peak, progressively decreasing in intensity away from the peak centre. Since such series termination artefacts are sometimes considered undesirable in EXAFS analysis, to help remove them, a window function $w(k)$ is sometimes applied to the k-space data, with the Fourier transform being computed using the windowed EXAFS $\chi_w(k) = w(k)\chi(k)$. Because the presence of even a small step or sill at the ends of the data can result in series termination ripples, what is ideally desired for $w(k)$ is a function that smoothly adjusts the $\chi(k)$ data to zero at the ends of the data range to be transformed, with a plateau in between, in the region we define as $(k_{min} + k_w)$ to $(k_{max} - k_w)$, in which k_w defines the k-range over which the window taper extends. Popular window functions include the Hanning window (10.10) and the half-Gaussian (10.11):

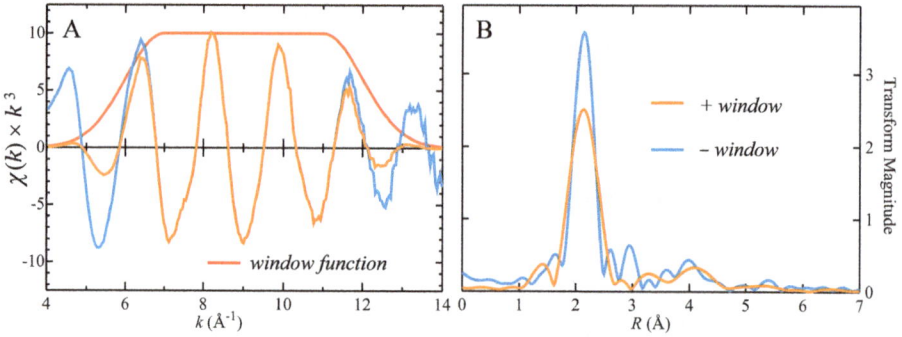

Figure 10.12: Example of k-space window function to minimize EXAFS Fourier transform data truncation artefacts. The example data set is the Ni K-edge EXAFS data for [(n-butyl)$_4$N][Ni(mnt)$_2$]. (A) The original EXAFS (blue line) without the window function, while the red curve shows a Gaussian window function (10.13), with k_w = 3 Å^{-1} computed between 4 and 14 Å^{-1}, a range deliberately selected to show sharp sills without the window function. The EXAFS Fourier transforms, computed with and without the Gaussian window function, both with Ni–S phase-correction, are compared in (B).

$$w(k) = \begin{cases} \sin^2\left(\frac{\pi}{2}\frac{k-k_{\min}}{k_w}\right), & k_{\min} < k < (k_{\min} + k_w) \\ 1, & (k_{\min} + k_w) < k < (k_{\max} - k_w) \\ \cos^2\left(\frac{\pi}{2}\frac{k-(k_{\max}-k_w)}{k_w}\right), & (k_{\max} - k_w) < k < k_{\max} \end{cases} \quad (10.12)$$

$$w(k) = \begin{cases} \exp\left(-c\frac{(k-k_{\min}-k_w)^2}{k_w^2}\right), & k_{\min} < k < (k_{\min} + k_w) \\ 1, & (k_{\min} + k_w) < k < (k_{\max} - k_w) \\ \exp\left(-c\frac{(k-k_{\max}+k_w)^2}{k_w^2}\right), & (k_{\max} - k_w) < k < k_{\max} \end{cases} \quad (10.13)$$

In eq. (10.13), the constant c is set to an arbitrary value of 4.60517019 such that $w(k)$ will be ~0.01 at both $k_{\min}$ and $k_{\max}$. The advantage of the Hanning function is that it will always go to zero at the ends of the taper, whereas with a Gaussian window, small 1% sills will be present.

Other window functions include Parzen, which uses linear tapers to zero, Welch, with k^2 tapers to zero, and Kaiser-Bessel, in which a zero-order-modified Bessel function is used to describe the tapers. All of these window functions will effectively reduce the Fourier ripples, but their use also comes at a substantial cost, which is that the transform peaks will be significantly broadened (e.g. see Figure 10.12). Many experimenters, including ourselves, prefer to live with the series-termination artefacts, and simply Fourier transform the data without including a k-space window function at all.

10.6.2 R-windowing – Fourier filtering

Fourier filtering is a technique that can partially isolate the EXAFS from one or more Fourier transform peak, simplifying the overall data analysis by reducing complexity. In order to Fourier filter a peak, one isolates the peak in R-space by multiplication of both the real and imaginary parts of the transform with an R-window function, then computes the inverse Fourier transform to obtain what is called Fourier-filtered EXAFS (Figure 10.13). In the early days of EXAFS, Fourier filtering was a fairly standard part of the analysis procedure, in part because it makes the data (often noisy, in the early days) appear noise-free by removing the high-frequencies in the data. The problem is that Fourier filtering also removes the aforementioned structure due to Fourier series termination, hence the Fourier-filtered data is inevitably distorted.

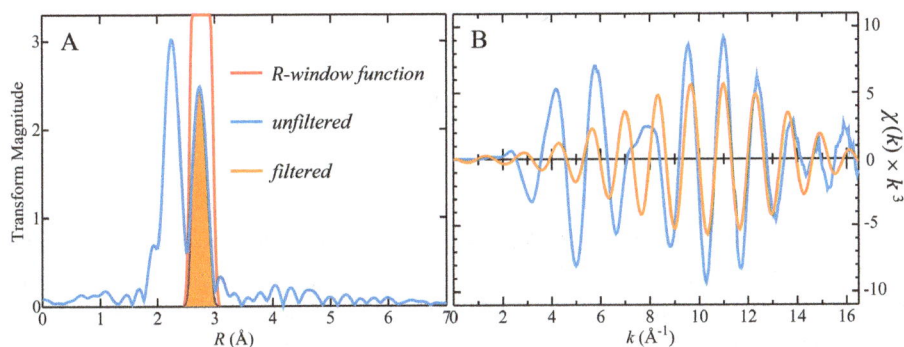

Figure 10.13: EXAFS Fourier filtering of the Fe···Fe interaction for $[Fe_4S_4(SPh_4)_4]^{2-}$. (A) The Transform magnitude of the original data (blue), the R-window (red) and the filtered Fe···Fe transform peak (orange). The corresponding original and filtered EXAFS are shown in (B).

In the modern context, it is widely agreed that fitting the raw rather than filtered EXAFS data is preferable. However, in cases where a small peak of unknown origins is observed in the Fourier transform, Fourier filtering can be invaluable for its identification. Here, Fourier filtering could be used to isolate the EXAFS and to provide a preliminary fit, from which the derived parameters could be then incorporated into a final refinement using the raw, unfiltered, EXAFS data.

10.6.3 Extraction of phase and amplitude functions using Fourier methods

For a single backscatterer and eq. (10.5), we can write the EXAFS as follows:

$$\chi(k) = a(k) \sin \Phi(k) \tag{10.14}$$

where $\Phi(k) = 2kR + \varphi'(k)$, $a(k) = NA(k) \exp(-2R/\lambda(k)) \exp(-2\sigma^2 k^2)/kR^2$ with all other symbols having meanings as previously defined. We can re-cast eq. (10.14) using complex numbers as follows:

$$\chi(k) = \frac{1}{2i} a(k) \exp(i\Phi(k)) - \frac{1}{2i} a(k) \exp(-i\Phi(k)) \tag{10.15}$$

If we now Fourier transform to obtain $\rho(R)$, the first term in eq. (10.15) corresponds to positive R values and the second, to negative R values (see Appendix B). For computational convenience, we can set the latter to zero and define a new function $z(k)$:

$$z(k) = \frac{1}{2i} a(k) \exp(i\Phi(k)) \tag{10.16}$$

From eq. (10.16), we can easily compute the total effective phase function (10.17) and the total amplitude function (10.18):

$$\Phi(k) = \mathrm{atan}\left(\frac{\mathrm{Im}[z(k)]}{\mathrm{Re}[z(k)]}\right) \tag{10.17}$$

$$a(k) = \sqrt{\mathrm{Re}[z(k)]^2 + \mathrm{Im}[z(k)]^2} \tag{10.18}$$

Prior to the availability of good *ab initio* theoretical EXAFS standards such as are available now from the FEFF code, such Fourier methods were used to extract phase and amplitude functions from standard compounds of known structure, in some cases with computation of σ^2 values from vibrational spectra [3], or refinement of a $\delta\sigma^2$, if these are not available.

With the availability of *ab initio* theoretical phase and amplitude functions, we can use the so-called linear method to estimate the bond length R by computing $\{\Phi(k) - \varphi'(k)\} = 2kR$, where ΔE_0 can be computed by adjusting its value until the function $\{\Phi(k) - \varphi'(k)\}$ plotted against k runs through the origin, with $2R$ being the slope of the plot. To extract N and σ^2, we can define an additional function $B(k)$ using the theoretically derived $A(k)$ and $\lambda(k)$:

$$B(k) = \frac{NA(k)}{kR^2} \exp\left(\frac{-2R}{\lambda(k)}\right) \tag{10.19}$$

Plotting the natural logarithm of the ratio $a(k)/B(k)$ versus k^2 will also give a straight line with slope $2\sigma^2$ and ordinate intercept of $\ln(N/R^2)$, from which N can be obtained using the value of R from the linear phase plot:

$$\ln\left(\frac{a(k)}{B(k)}\right) = \ln\left(\frac{N}{R^2}\right) + 2k^2\sigma^2 \tag{10.20}$$

The linear method provides a very quick alternative to curve fitting using non-linear optimization methods.

10.6.4 Wavelet transforms

The use of wavelets in EXAFS analysis has been gaining in popularity in recent years, although most users still employ the standard Fourier methods discussed above. While a Fourier transform gives a two-dimensional plot, typically with transform magnitude versus R, the wavelet transform gives a three-dimensional plot, with magnitude versus R versus k, showing the k-dependence of each feature in R. This can allow the experimenter to distinguish between lighter and heavier backscatterers, as the latter will show more intensity at high k. The wavelet transform, cast into a form that is appropriate for EXAFS analysis and using k^3 weighting of the EXAFS, is given as follows:

$$w(R, k) = \left(\frac{R}{R_0}\right)^{\frac{1}{2}} \int_{-\infty}^{+\infty} \chi(k')k'^3\psi\left[\left(\frac{R}{R_0}\right)(k' - k)\right]dk' \tag{10.21}$$

in which the function ψ is what is known as the mother-wavelet function, with the condition specified as follows:

$$\int_{-\infty}^{\infty} \psi(k)dk = 0 \tag{10.22}$$

Two major different types of wavelets have been used for the analysis of EXAFS, the Gaussian-based Morlet[2] wavelet and the Continuous Cauchy wavelet. It has been argued that the Morlet wavelet is more appropriate for analysis of EXAFS because its structure can be similar to the EXAFS signal, with a slowly varying amplitude term and a rapidly oscillating phase. The Morlet wavelet function is given as follows:

$$\psi(k) = \exp(-2iR_0k) \exp(-\sigma_0^2 k^2) \tag{10.23}$$

Here, R_0 is the characteristic frequency of the Morlet wavelet and this, along with a value for σ_0, must be selected by the user. These parameters control the resolution of

---

2 The Morlet wavelet was in fact introduced by Physicist Dennis Gabor in 1942, and it is alternatively known as the Gabor wavelet. It is more popularly named for French Geophysicist Jean Morlet, who introduced Gabor's work to the field of seismology in 1984.

the wavelet transform in both k and R. The wavelet transform can be conveniently calculated in Fourier space as the product of the EXAFS Fourier transform and the Fourier transform of the mother wavelet, which for Morlet can be shown analytically to be a simple Gaussian function. The Morlet wavelet transform of an EXAFS data set is thus numerically identical to computing Fourier-filtered data for all values of R, but using a Gaussian window function with a variable width. Figure 10.14 shows the wavelet transform of the data $[Fe_4S_4(SPh_4)_4]^{2-}$, illustrating the separation of the Fe–S and Fe···Fe interactions in both k and R.

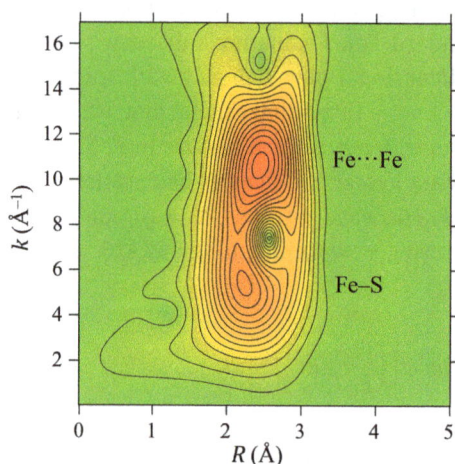

Figure 10.14: Morlet wavelet transform. The dataset is the EXAFS for $[Fe_4S_4(SPh_4)_4]^{2-}$ (Fe–S phase-corrected), showing clear separation of the Fe–S and Fe···Fe interactions in both k and R. The colour map indicates intensity, with green zero and red, highest intensity.

At the time of writing, the use of wavelet transforms in the analysis of EXAFS data is still relatively uncommon, with a total of several dozen papers, compared to tens of thousands employing conventional Fourier transforms. Nonetheless, the potential advantages of wavelets are clear – they are quick to compute and, at a glance, they allow the user to discriminate light and heavy backscatterers. Less common still is the use of maximum entropy transforms, and hence we will only briefly review these.

10.6.5 Maximum entropy transforms

The maximum entropy method provides a technique for producing high resolution power spectra-lacking truncation artefacts. The method is one of effectively using a series approximation to extrapolate the data to a significantly greater extent than the actual run of the data, thus the name "maximum entropy". We will not go into the mathematical details of the method here, but once again, there is no free ride. The difficulty with the maximum entropy transform and EXAFS analysis lies in knowing the correct value for the extent of the series approximation, also called the number of poles, usually called m. If too low a value for m is used, then the maximum entropy transform will show only a single peak, even when two are obvious in the conventional Fourier transform, and conversely too high a value for m leads to the appearance of spurious peaks, known as ghost peaks. Overall, the method provides some interesting possibilities, but its promise as a useful tool in the analysis of experimental EXAFS data has yet to be realized.

10.7 EXAFS curve fitting – goodness of fit, accuracies and precisions

Before discussing the procedure of EXAFS analysis itself, we first briefly visit some elementary statistical considerations.

10.7.1 Goodness of fit: metrics of fit quality

The quality of a fit can be judged using a range of different metrics, often referred to as goodness of fit indicators or error functions. We will name these metrics F here, rather than the more usual R, to avoid confusion with bond length R. For n data points, using k^3 weighting of the data, where $\chi(k)$ is the experimental EXAFS and $\chi_f(k)$ is the fitted (calculated) EXAFS, then the simplest error function would be the sum of the squares of the differences between the experimental and calculated EXAFS, normalized by the number of data points:

$$F = \frac{1}{n}\sum k^6 \left(\chi(k) - \chi_f(k)\right)^2 \tag{10.24}$$

in which, the summation is carried out over all n data points. If we have m variables in the refinement, what is called the reduced error function might be used, given as follows:

$$F = \frac{1}{n-m}\sum k^6 \left(\chi(k) - \chi_f(k)\right)^2 \tag{10.25}$$

Our preferred error function is the weighted F-factor, F', which is analogous to but slightly different from the crystallographic R-factor:

$$F' = \sqrt{\frac{\sum k^6 \left(\chi(k) - \chi_f(k)\right)^2}{\sum k^6 \chi(k)^2}} \tag{10.26}$$

For low-noise data, F' of 0.1 is an excellent fit, while 0.4 would be a poor fit. F' can also be expressed as a percentage, so, for example, an F' of 0.2 would be quoted as

20%. Also, often quoted, depending on the software used, is the expected weighted F-factor F_e' which is given as follows:

$$F_e' = \sqrt{\frac{n-m}{\sum k^6 x(k)^2}}$$

(10.27)

The value for F_e' is typically greater than F', and if this is not the case, then this may be a warning sign that there are problems with the fit.

10.7.2 Accuracy and precision

Two measures that are sometimes a source of confusion are accuracy and precision. **Accuracy** is a measure of the anticipated discrepancy with reality, or how good our estimate is compared to the true value. All sorts of factors contribute to the accuracy – such as how well the theory that we are using to simulate the EXAFS performs. Accuracy can be quite difficult to quantify as we need to know the true values. It must be guessed by comparing results from other methods (each with their own built-in inaccuracies) and is often unknowable on a case-by-case basis. **Precision** is an index of the error in a variable, primarily arising from uncertainties in the fitting process, and uncertainties (noise) in the data. This is relatively straightforward to quantify and the estimated standard deviations produced by fitting software are typically precisions. In general, uncertainties estimated by repeated measurements of the same quantity are precisions. Figure 10.15 shows a pictorial representation of these definitions.

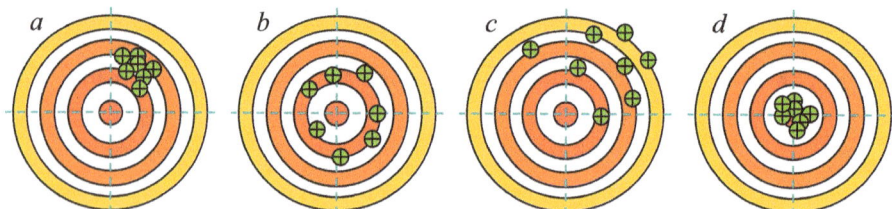

Figure 10.15: Accuracy and precision. The bullseye of the target represents the true value of a variable, with the green shots showing estimates of that value. Target *a* shows a collection of shots that is inaccurate but precise, target *b* shots that are accurate but imprecise, target *c* neither accurate nor precise, and target *d* both accurate and precise.

Curve-fitting analysis of the EXAFS data normally employs standard methods of non-linear optimization, such as the Levenberg-Marquardt (LM)-damped least-squares method. The LM method is a good choice for systems that have few variables, such as EXAFS with rapid convergence, and can also give estimated standard deviations directly from the diagonals of the variance-covariance matrix that is used during the non-linear optimization. As we have discussed, these are precisions, and are often quoted in parentheses following the least significant digits of the value, emulating crystallographic convention for reporting atomic coordinates. For example, a curve-fitting analysis of an EXAFS spectrum might give a refined value for R of 2.145 ± 0.001 Å, which could be written as $2.145(1)$ Å.

The typical accuracies of fitted parameters were determined during early EXAFS work, while the method was being validating as a structural tool [4], by comparison to good quality small-molecule X-ray crystal structures. Typically, for directly coordinated atoms, R can be determined to be better than ± 0.02 Å, while for N and σ^2, the accuracy is in the vicinity of $\pm 20\%$. The precisions from the fit are generally much smaller than these values.

10.8 EXAFS curve-fitting analysis

As we have previously mentioned, the curve fitting of the EXAFS can be done either in k-space or in R-space, and the two should be exactly equivalent. The advantage of R-space fitting is that convergence to final fitted values for variables can be more efficient, especially if the code employed uses a LM algorithm using numerical estimation of derivatives in the non-linear optimization procedure. The disadvantage of R-space fitting is that it requires rather more manipulation of the data than k-space fitting and keeping data manipulation at a minimum is always desirable.

As already noted, quantitative analysis of the EXAFS of a completely unknown species is essentially an exercise in model building. With a correct model, analysis of the EXAFS can provide important information, however with an incorrect model, it can be misleading. As we will see, both initial and final models may be informed by chemical knowledge. Once a model of the coordination environment has been postulated, it is usually a very simple matter to compute the EXAFS, and if the model is dramatically wrong, for example using oxygen donors in place of sulfurs, then the error will often be immediately obvious. To illustrate the process, we return to the EXAFS data for [(n-butyl)$_4$N][Ni(mnt)$_2$], and first consider a single-scattering analysis of the first coordination shell. Let us presume that the experimenter suspects that the complex has six Ni–O ligands with a bond length corresponding to the (phase-corrected) Fourier transform peak position of 2.15 Å. Notwithstanding the issue that this bond length would be chemically unlikely for six-coordinate Ni(II) and Ni(III), the EXAFS would appear close to 180° out of phase with the experimental data, as shown in Figure 10.9Ba. Analysis software allows the user to select variables for refinement, and if the experimenter were to start with this Ni–O model and simultaneously refine all of N, R, σ^2 and ΔE_0, then a reasonable match to the experimental EXAFS will be obtained. However, this will have large negative values for ΔE_0 (e.g. –30 eV) and impossibly large values for N (e.g. ~12), with R of $2.229(2)$ Å and σ^2 of $0.0035(2)$ Å^2, and F' of 0.2426, which can be compared to F' for a single-scattering fit to four Ni–S (described in more detail below) of 0.1905. For the erroneous Ni–O fit, the value obtained for σ^2 is chemically reasonable, and that for R, is a little large, but N and ΔE_0 are clearly outside of reasonable bounds. Hopefully, the experimenter would reject this model, think again, and postulate a more correct Ni–S model. The simultaneous fitting process of all of N, R, σ^2 and ΔE_0 that our hypothetical experimenter employed is also bad practice, and the reason for this is that some of the variables show a high degree of mutual correlation in the refinement. This is illustrated in Figure 10.16.

The search profiles in Figure 10.16 show F' as a function of R versus N, σ^2 versus N, and R versus ΔE_0. The R versus N plot shows that, as expected, the minimum for R is sharply defined, hence the small precisions for this variable, and that for N, is substantially broader, while the lack of any diagonal character to the valley shows that these variables are largely not correlated in the refinement, meaning a change in one is unlikely to influence a change in the other, and these two parameters can be refined together. The σ^2 versus N plot shows broad minima for both variables with a well-defined diagonal valley, clearly indicating a strong degree of mutual correlation. Because of this correlation, it is unwise to freely co-refine σ^2 and N, and a normal strategy is to compare fits for different integer values of N. Finally, the R versus ΔE_0 plot in Figure 10.16 also shows a clear diagonal valley, indicating that these two variables are correlated, however the minimum for R is very sharp and that for ΔE_0, broad, so there is usually little concern about co-refining R and ΔE_0. Notwithstanding the caution expressed against co-refining σ^2 and N, with our model data for [(n-butyl)$_4$N][Ni(mnt)$_2$], and using the FEFF code to compute single-scattering phases and amplitudes, if all four structural variables are co-refined, then the fit gives $N = 4.0(1)$, $R = 2.144(1)$ Å, $\sigma^2 = 0.0038(1)$ Å^2, and $\Delta E_0 = -0.4(4)$ eV (quoted relative to the nominal choice of E_0), with $F' = 0.1905$. Examination of the correlation matrix, shown in Table 10.1, is also informative, showing that both N and σ^2 and R and ΔE_0 show more than 90% correlation.

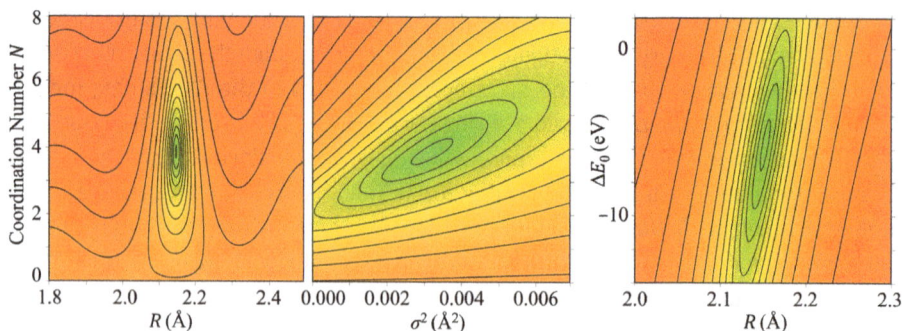

Figure 10.16: EXAFS curve-fitting search profiles, plotting goodness of fit (red worse, green best) for different variable combinations.

Table 10.1: Correlation matrix.

	N	R	σ^2	ΔE_0
N	1	0.311	0.931	0.300
R		1	0.218	0.929
σ^2			1	0.225
ΔE_0				1

10.8.1 EXAFS bond length resolution

While EXAFS has outstanding accuracy and precision for determination of average bond lengths, it has poor bond length resolution. The EXAFS resolution is defined as the minimum difference in interatomic distance that can be discerned for similar backscatterers. Thus, with our example of [(n-butyl)$_4$N][Ni(mnt)$_2$], there are four nearly equivalent Ni–S bonds. The high-resolution X-ray crystal structure shows that these have bond lengths of 2.143, 2.146, 2.147 and 2.146 Å, with individual bonds differing only by thousandths of an Ångström. These small differences would be impossible to resolve by EXAFS, with these four Ni–S bonds appearing as a single shell. The EXAFS resolution δR is approximately given as follows:

$$\delta R \approx \frac{\pi}{2k_{max}} \tag{10.28}$$

This relationship comes from a trivial treatment in which estimation is made when beats in the EXAFS are expected; two Ni–S backscatterers at sufficiently different interatomic distances will cause the EXAFS to have a beat. By beat, we mean a place along the run of the data in which the amplitude becomes temporarily diminished. Notwithstanding these simple origins, this relation works very well in practice. When similar backscatterers separated by less than δR, they are not resolved, and only the mean R can be determined, with the presence of different R values being betrayed only by an increase in the static Debye-Waller factor. Figure 10.17 shows a series of calculated EXAFS spectra with corresponding EXAFS Fourier transforms for two different Ni–S interactions, one at 2.15 Å, and the second separated by different δR, ranging from 0 to 0.2 Å. The beat can be observed propagating from the high-k end of the spectrum as δR increases, appearing first for a δR value of $\sim \pi/2k_{max} = 0.1$ Å. Simultaneously, the Fourier transform transitions from a single peak to two peaks. We note that many texts define the EXAFS resolution as $\delta R \approx \pi/2(k_{max} - k_{min})$; this is not really correct, as the beat will appear from the high-k end and for practical purposes, the resolution is approximately described by 10.18, although obviously, if the low-k end of the data is set high enough, then the beat may vanish.

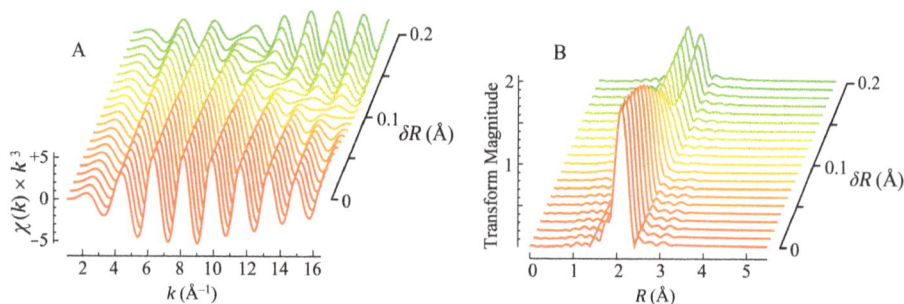

Figure 10.17: Resolution of Ni–S EXAFS using simulations with different δR: (A) the EXAFS oscillations and (B) the Ni–S phase-corrected EXAFS Fourier transforms. One Ni–S has $R = 2.15$ Å, and the other gets progressively longer by δR, with all other simulation parameters being the same and as described for Figure 10.9.

10.8.2 Number of relevant independent points in EXAFS analysis

Another important criterion is the number of relevant independent data points in the EXAFS N_{idp}, which is a value that the total number of variables in the fit should not exceed. This is available from what is called the Nyquist sampling theorem, sometimes called the Nyquist interval, which gives us the following equation:

$$N_{idp} \approx \frac{2\delta k \delta R}{\pi} \tag{10.29}$$

Here, δk is the k-range of the EXAFS being analysed ($k_{max} - k_{min}$), and δR, the R-space range. Stern has argued [5] that this relationship should really be written according to

$$N_{idp} \approx \frac{2\delta k \delta R}{\pi} + 2 \tag{10.30}$$

For example, if we assume that the data extends from 1 to 14 Å^{-1}, so that $\delta k = 13$Å^{-1}, and that we are fitting a single Fourier transform peak such as the Ni–S in [(n-butyl)$_4$N] [Ni(mnt)$_2$], then $\delta R \approx 1$ Å^{-1}, which gives us $N_{idp} \approx 10$. While the difficulty here is that there is no clearly defined value for δR, the point is clear – the number of independent data points in EXAFS tends to be quite small, and it would be easy to construct a complicated multi-shell model and exceed this limit. In our simple four-variable refinement for this data, we have clearly not exceeded the Nyquist limit, but there is clear potential for over-interpretation of data if appropriate caution is not used.

10.8.3 EXAFS cancellation

We have already observed that EXAFS from two different backscatterers can be close to 180° out of phase for all or part of the k-range of the data, for example for Ni–S and Ni–O at the same distance, in Figure 10.9Ba. If this is the case, and if the amplitudes of the two EXAFS contributions match, then the EXAFS will effectively cancel. EXAFS cancellation is a surprisingly common phenomenon, especially when partial cancellation is considered, and for example, it has been observed in biologically relevant cuprous thiolate clusters [6, 7], in fully reduced forms of the nitrogenase iron protein [8] and in Hg···Hg interactions in biologically relevant complexes [9]. Here, we will discuss one example – a mixed-metal cluster containing a $[Cu_2Mo_2S_4]^{4+}$ core, an industrially relevant sample. These experiments were carried out in 1987, at which time no crystal structures of related compounds were available, but the core cluster was hypothesized to be a cubane with the four metals and four sulfides on alternate corners. The hypothetical metal coordination was completed by two **di-n**-octyldithiocarbamates, each one providing two external sulfur donors to the two molybdenum atoms, and two chlorides, one on each copper. The expectation was for strong first-coordination sphere EXAFS for both the Cu and Mo perspective from the five Mo–S and three Cu–S plus one Cu–Cl, and for the observable outer shell EXAFS, resulting from backscattering from Mo and Cu second shell neighbours at about 2.7 Å. We first ran the Mo K-edge EXAFS and upon computing the Fourier transform, were surprised to observe significant intensity only from the directly coordinated sulfur ligands, as shown in Figure 10.18a. The Cu K-edge EXAFS Fourier transform, however, showed a clear outer shell peak, suggestive of metal-metal backscattering, as shown in Figure 10.18b. Curve-fitting analysis of the data confirmed that from the molybdenum perspective, there was one Mo and two Cu atoms, with the Mo having approximately twice the amplitude than the Cu, and with the EXAFS being close to 180° out of phase. Because there are two Cu and one Mo, the amplitudes match and consequently, the outer shell EXAFS mostly cancelled, giving only a shoulder rather than a well-defined peak in the Fourier transform. From the copper perspective, however, two outer Mo and one Cu are present, again with Mo backscatterers giving greater amplitude than Cu, and with EXAFS, that is also close to 180° out of phase, but only resulting in partial cancellation, so that a clear outer shell Fourier transform peak and a beat in the k-space data are observed with the Cu EXAFS. The isolated EXAFS of first and second shell is shown in Figure 10.18c and d, from the perspective of Mo and Cu, respectively. The derived structural model accurately predicted the core cluster structure, observed in the much later crystal structure of a related cubane complex, shown in Figure 10.18e.

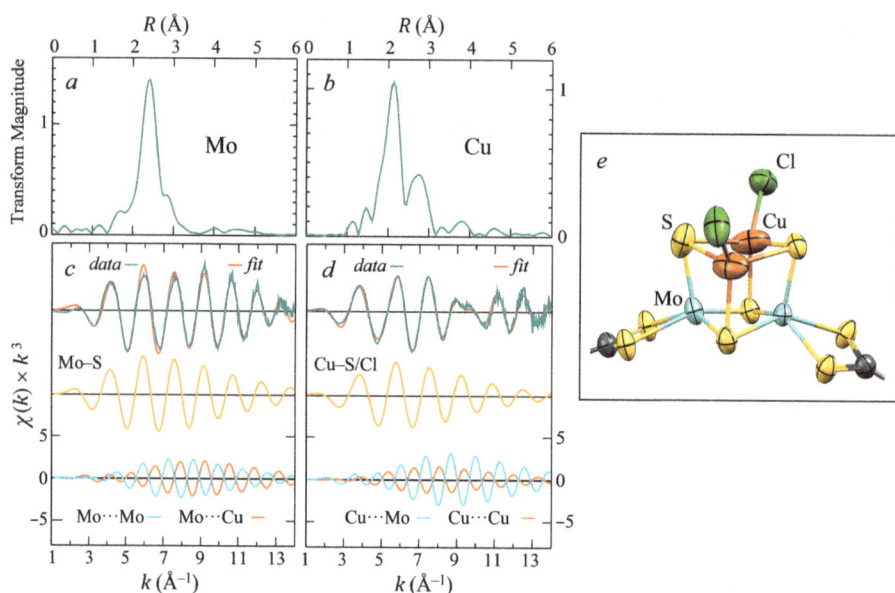

Figure 10.18: EXAFS cancellation in a $[Cu_2Mo_2S_4]^{4+}$ cluster. (a) and (b) The Fourier transforms of the Mo K-edge EXAFS and Cu K-edge EXAFS, respectively. (c) and (d) he EXAFS curve-fitting analysis for Mo and Cu, with the lower curves showing the EXAFS components, with cancellation and partial cancellation of outer-shell EXAFS. (e) The X-ray crystal structure of the cluster core of a related complex.

10.8.4 Physical and chemical checks – do the derived structural parameters make sense?

We have already hinted that the structural parameters derived from EXAFS analysis should make physical and chemical sense. If they do not, then this may be an indication that something is seriously wrong with an analysis. As we have discussed, the most accurately determined structural parameter is the interatomic distance, R, and the least accurately determined parameters are coordination number, N and σ^2, with the latter sometimes being treated as a kind of throw-away variable, mainly because R and N have more obvious relevance to physical structure. We will discuss the physical bounds for σ^2, but for now, we note that bond lengths can, however, be used to help determine coordination numbers as these change systematically with N, usually in an approximately linear manner. For example, Hg(II)-thiolate complexes are known with coordination numbers of 2, 3 and 4, with the average Hg–S bond lengths changing systematically as 2.35, 2.47 and 2.54 Å, respectively, and these trends have been used to help determine coordination numbers in the EXAFS of biologically relevant species [9, 10]. Figure 10.19 shows a more challenging example; oxygen-coordinated Zr(IV) complexes exist with coordination numbers of 4, 5, 6, 7, 8 and 9, and given the uncertainties of ~25% for determination of N by EXAFS, without considering bond length trends, coordination numbers would be challenging to determine. Again, these systematic trends of bond lengths with coordination numbers have been used with EXAFS to help identify the coordination numbers for solution complexes [11, 12].

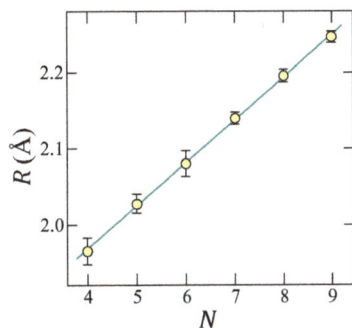

Figure 10.19: Average bond lengths R versus coordination number N for crystallographically characterized Zr(IV)–O species in the Cambridge Crystal Structure Database [13]. The error bars indicate the standard deviations of the Zr–O bond lengths.

EXAFS-derived bond lengths can also give clues about coordination geometry, and we will return to a discussion of how the various pieces of information might be put together in the last section in this chapter.

Similarly, ligand identity can be inferred from bond length trends. We have previously noted that oxygen and nitrogen backscatterers are almost impossible to distinguish by EXAFS. Using Cu(I) complexes as an example, two-coordinate digonal Cu(I) with oxygen donors have a characteristic bond length of 1.83 Å, whereas nitrogen ligands give a slightly longer characteristic bond length of 1.88 Å. There are a large number of trigonally coordinated Cu(I) species with nitrogen ligands (bond length 1.98 Å), but no analogous complexes with oxygen ligands. Thus, the bond length can give clues to ligand identity, and while these methods should always be used with caution, this too may be reinforced by examination of the near-edge spectra.

As we discussed in Chapter 8, the σ^2 values have physical meaning and can be estimated using codes such as FEFF, together with force constants (sometimes called spring constants) for each bond and angle to be considered. These force constants can be conveniently obtained from density functional theory calculations via the Hessian matrix of geometrical second partial derivatives.[3] Remembering from Chapter 8 that σ^2 values are composed of the sum of vibrational and static components ($\sigma^2 = \sigma^2_{vib} + \sigma^2_{stat}$, eq. (8.25)), and as elaborated in Chapter 8, Cotelesage et al. [14] have discussed methods by which physical bounds can be estimated for σ^2 values. Here, σ^2_{vib} can be computed as a lower bound using codes such as FEFF, with force constants from density func-

---

3 Care must be taken with units as the values required by FEFF are required to be in $N \cdot m^{-1}$ whereas most DFT packages use atomic units.

tional theory, and an upper bound for σ^2_{stat} can be estimated from the k-range of the experimental EXAFS.

Table 10.2, modified from Cotelesage et al. [14], gives a sampler of limits for σ^2 (upper and lower bounds) computed for four-coordinate transition metal species with R values from averages of crystallographic values obtained from the Cambridge Crystal Structure Database [13], and force constants obtained from the Hessian matrix of DMol^3 density functional theory geometry optimizations, giving FEFF-derived minimum values as σ^2_{vib} (lower bounds). Maximum values (upper bounds) were obtained from eq. (8.31), assuming $k_{max} = 12$ Å^{-1} and $n = 2$, giving $\sigma^2_{stat} = 0.0043$ Å^2, plus an arbitrary additional contribution of 0.0005 Å^2 to account for contributions not incorporated into our simple model.

Table 10.2: Limiting values for σ^2.

Bond	R (Å)	σ^2_{vib} (minimum) (Å^2)	σ^2 (maximum) (Å^2)
Fe–N	2.08	0.0028	0.0076
Fe–O	1.87	0.0021	0.0069
Fe–S	2.28	0.0027	0.0075
Zn–N	1.98	0.0023	0.0071
Zn–O	1.95	0.0022	0.0070
Zn–S	2.35	0.0027	0.0075
Mo = O	1.75	0.0015	0.0063
Mo–N	1.92	0.0025	0.0073
Mo–O	1.93	0.0023	0.0071
Mo = S	2.17	0.0018	0.0066
Mo–S	2.40	0.0029	0.0077

The σ^2_{vib} values will also depend upon the bond length because bond-stretch vibrational modes will tend to lower frequencies as R increases. This has been shown experimentally by Cramer et al. [3], for a range of different Mo–S complexes (Figure 10.20).

10.8.5 Getting fooled – confusion of backscatterers and lack of control of ΔE_0

We have already mentioned that wild variations in ΔE_0 can cause problems with data analysis. Ideally, values for ΔE_0 should be constrained to be similar to those of standard compounds of a known structure. If such precautions are not taken and if the EXAFS is of limited k-range, then it is possible to be effectively fooled in an analysis. For example, with a hypothetical data set arising from a four-coordinate Ni(III) species with 2 Ni–S at 2.145 Å and 2 Ni–O at 2.005 Å, with reasonable σ^2 for both (see Table 10.2) and ΔE_0 –0.36 eV, a nearly identical fit can be obtained with 3 Ni–S at 2.126 Å and ΔE_0 –10.69 eV, again with a chemically reasonable σ^2 (Figure 10.21). Valid

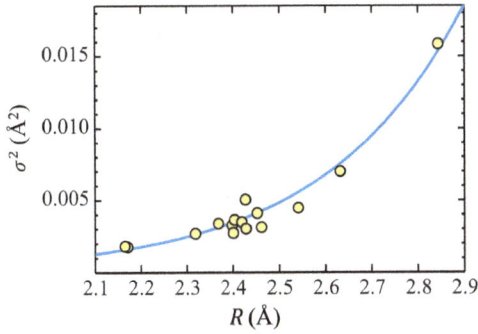

Figure 10.20: Dependency of room temperature σ^2 values (points) on bond length R for Mo–S EXAFS in a series of different coordination compounds (data from Cramer et al. [3]). The solid line is drawn to guide the eye.

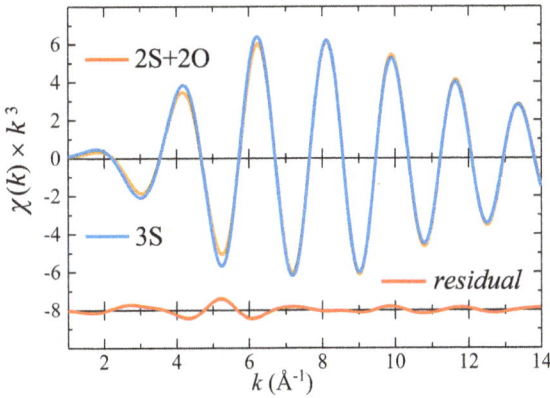

Figure 10.21: Getting fooled by ΔE_0. Computed curves are compared for 2Ni–S at 2.145 Å + 2Ni–O at 2.005 Å ($\sigma^2 = 0.0038$ Å^2) and ΔE_0 –0.36 eV (orange line), and 3Ni–S at 2.126 Å ($\sigma^2 = 0.0039$ Å^2) and ΔE_0 –10.69 eV (blue line), together with the residual (red line). Nearly identical EXAFS are obtained, illustrating potential errors resulting from indiscriminate refinement of ΔE_0.

values for ΔE_0 will vary between fits, especially if different oxidation states are present, but these changes should be small. A ΔE_0 that is different from the others may signal an erroneous analysis and should be taken as a warning flag.

10.8.6 EXAFS range

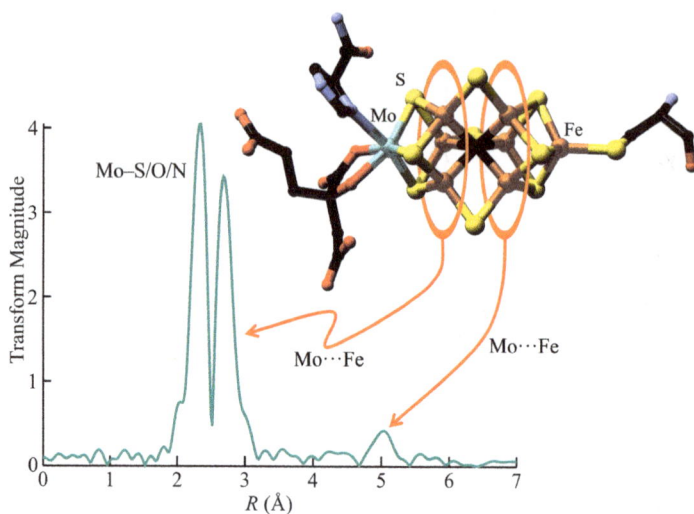

Figure 10.22: Distance limitations of EXAFS illustrated using the Mo K-edge EXAFS Fourier transform (Mo–S phase-corrected) of *Azotobacter vinelandii* nitrogenase. The inset shows the crystal structure of the active site, with the two different Mo⋯Fe outer shell transform peaks, each originating from three iron atoms, indicated.

The upper limit in distance that EXAFS might observe a backscatterer depends upon both the size of the backscatterer in question, the number of such backscatterers, and the *k*-range of the EXAFS. Remote backscatterers with higher atomic numbers are, in general, more readily observed than equivalent positioned lighter atoms. However, in general, EXAFS from distant backscatterers tends to be weak, both because of the R^{-2} term in eq. (10.5), and because distant backscatterers can have very large σ_{vib}^2. In well-ordered extended solids, especially those possessing co-linear atomic arrangements that would give substantial EXAFS multiple scattering, EXAFS from remote atoms can be observed (e.g. 10 Å), however in most biological samples, or in general species in solution, it is rare to observe backscatterers beyond about 4 Å. One example of an exception to this is the probable observation of an iodide bound within a positively charged pocket within the molybdenum enzyme sulfite oxidase at ~5 Å from Mo. In this case, the observation of iodine required very good EXAFS signal to noise and was supported both by density functional theory calculations and by experimental observation of well-resolved [127]I hyperfine coupling in the Mo(V) electron paramagnetic resonance spectrum [15]. Another biological example is the observation of a weak long-range interaction, observed as a Fourier transform peak, at about 5 Å in the Mo K-edge EXAFS of the enzyme nitrogenase.

The active site of this enzyme contains a novel prolate ellipsoidally shaped [MoS-$_9$Fe$_7$C]$^{2-}$ cluster[4], with Mo bound by two oxygens from an external homocitrate and one histidine ligand from the protein, and a central carbide bound to a core of six iron atoms. From the molybdenum perspective, there is a first-shell Fourier transform peak consisting of overlapping Mo–S, Mo–O and Mo–N interactions, an intense Fourier transform peak from the three nearest iron atoms at about 2.69 Å, and a much weaker peak from three more distant iron atoms at about 5.10 Å (Figure 10.22). The difference in intensities of the 2.69 Å and of 5.10 Å transform peaks clearly illustrates the distance limitations of EXAFS. We note that this is both an advantage and a limitation, as limiting the quantitative analysis to close atoms alone makes data analysis tractable.

10.8.7 Reporting the results of EXAFS curve-fitting analysis

Before moving on to a discussion of the analysis of multiple scattering EXAFS, we will review the preferred methods for reporting the results of EXAFS curve-fitting analysis. Almost all analysis codes give estimated standard deviations, or some other metric related to the precision for the fitted parameters, and these are typically provided in what is sometimes called structural notation. Here, the error is given in parentheses on the last decimal place quoted. For example, an interatomic distance, fitted as 2.7137 ± 0.0013 Å, would be quoted as 2.714(1) Å. Table 10.3 shows an example of a curve-fitting analysis for the EXAFS data of [Fe$_4$S$_4$(SPh$_4$)$_4$]$^{2-}$ cluster shown in Figures 10.11, 10.13 and 10.14.

Table 10.3: EXAFS curve-fitting analysis of [Fe$_4$S$_4$(SPh$_4$)$_4$]$^{2-}$.

Backscatterer	N	R (Å)	σ^2 (Å^2)	ΔE_0 (eV)	F'
Fe–S	4	2.270(1)	0.0032(1)	−12.0(2)	0.161
Fe···Fe	3	2.714(1)	0.0035(1)		

Here, a common ΔE_0 value, expressed relative to the arbitrary value assigned during data reduction, has been floated for both shells included in the fit. Most analysis software allow for various types of linking of refinable variables, which serves to reduce the overall number of variables floated. Common types of linking might be fixed ratio, fixed difference, fixed sum or defined according to some geometrical constraint. Both the X-ray energy assumed for incident X-ray energy calibration and that assumed for E_0 should be reported. Other important factors to note in reporting are the units of the parameters and the nature of the error parameter, which in our case is

---

4 The total charge here is still uncertain.

given by eq. (10.26). Some analysis packages report $2\sigma^2$ rather than σ^2, and this should be noted in the reporting. If what are normally considered undesirable practices have been followed, for example free floating N and σ^2, then the rationale for doing this must be included in the discussion of the results. The checklist below provides a list of good practices for reporting EXAFS curve-fitting analyses:

- Show both the EXAFS oscillations and the Fourier transform.
- Indicate the k-range used for fitting.
- If R-space fitting was used, indicate the k-weighting used to compute the transform.
- Was the raw (measured) data interpolated? If so, this should be stated.
- If R windowing was used, this should be stated, together with relevant parameters.
- If smoothing or filtering has been used, this should be stated with relevant parameters.
- Include a metric of the goodness of fit, which should be defined.

10.9 Multiple-scattering EXAFS analysis

Multiple scattering was introduced in Section 8.7, where it was emphasized that linear arrangements of atoms give rise to the most prominent multiple scattering. In this section, we consider two examples; the first will be an example where multiple scattering was essential for the determination of structural details of an intermediate of a dinitrogen cleavage reaction [16]. The formally triple-bonded N_2 molecule is often thought of as being close to chemically inert. The N≡N bond is very strong with a bond length of just less than 1.10 Å; moreover, the gap between the highest occupied and lowest unoccupied molecular orbitals is very large at close to 10 eV. This, together with the lack of a dipole moment, means that N_2 is chemically very difficult to activate. The starting molecule in this chemically novel series of reactions was a mononuclear three-coordinate red-orange coloured Mo(III) species Mo[N(R)Ar]$_3$. Mo(III) complexes have a tendency to dimerize forming metal-metal bonds, however with a ligand specially designed to prevent this through steric hindrance, the complex reacts with N_2, forming a gold-coloured Mo(VI) terminal nitrido complex N≡Mo[N(R)Ar]$_3$. This reaction proceeds via a novel purple-coloured intermediate, (μ-N$_2$)(Mo[N(R)Ar]$_3$)$_2$, which is stable at low temperatures; Figure 10.23 shows a schematic of the reaction, with the EXAFS Fourier transforms of the starting material, the intermediate and the product shown in Figure 10.24.

The EXAFS Fourier transform of the purple intermediate shows pronounced features attributable to long-range interactions through a linear Mo–N=N–Mo core, giving a Fourier transform peak at 4.95 Å. As an aside from the main thread of this section, we note that the 4.95 Å peak is only visible when the high k-portion of the EXAFS is included, illustrating the need for adequate k-range in experimental data. Including the high-frequency contributions, the EXAFS of the purple intermediate could only be interpreted using multiple scattering [16], and the four-atom inner core of Mo–N=N–Mo required inclusion of

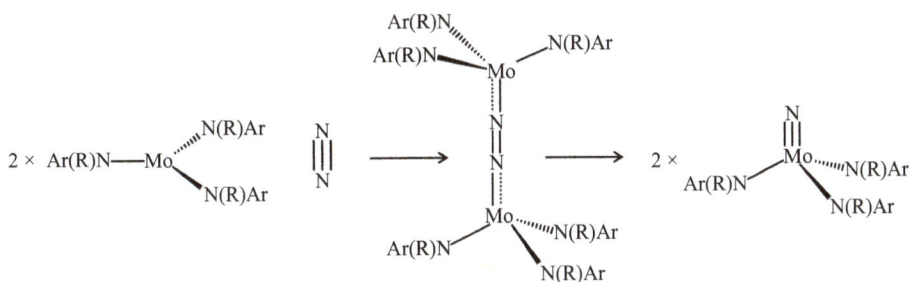

Figure 10.23: Reaction of three-coordinate Mo(III) species Mo[N(R)Ar]$_3$ with N$_2$ to form the Mo(VI) nitrido complex N≡Mo[N(R)Ar]$_3$ via a purple intermediate (μ-N$_2$)(Mo[N(R)Ar]$_3$)$_2$ (R is tertiary-butyl and Ar is 3,5-C$_6$H$_3$(CH$_3$)$_2$).

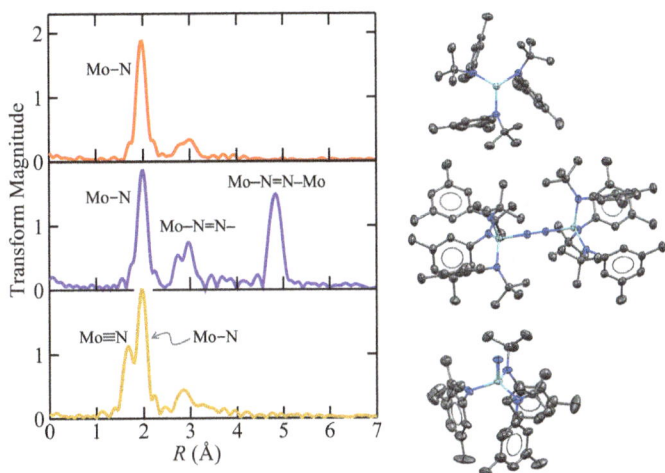

Figure 10.24: EXAFS Fourier transforms (Mo–N phase-corrected, k-range 0–21 Å^{-1}) of the starting material (red), the purple intermediate and the product of the reaction (gold) shown in Figure 10.23. The right panel shows the crystal structures (protons omitted for clarity) [17].

up to six multiple-scattering legs [16]. The use of multiple-scattering analysis allowed determination of the Mo–N bond length to the central dinitrogen, which would normally be difficult to resolve from the three N(R)Ar nitrogen ligands. It also allowed determination that the Mo–N=N–Mo core was essentially linear, and the –N=N– bond length, which was important for understanding the nature of the intermediate. Figure 10.25 displays plots of geometric searches involving both **cis** and **trans** distortions of the Mo–N=N–Mo core, which showed that the four-atom core arrangement was linear, and the –N=N– bond distances, which indicated a bond length of 1.19(2) Å. A subsequent crystal structure was in excellent agreement with the structural details derived from the EXAFS [17].

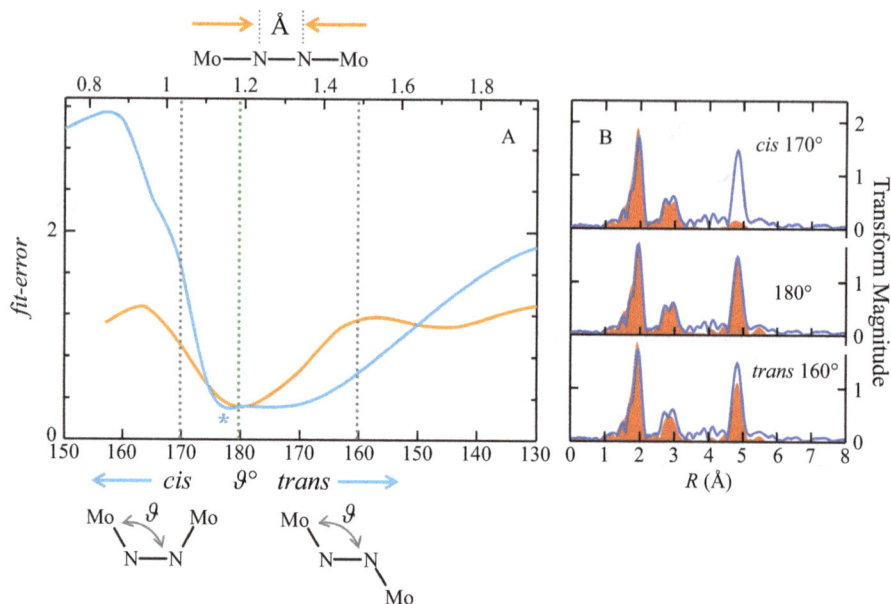

Figure 10.25: A: EXAFS curve-fitting search profiles investigating the geometry of the Mo–N=N–Mo core of (μ-N₂)(Mo[N(R)Ar]₃)₂. Profiles were calculated using a Mo···Mo distance that was constrained to be 4.94 Å. The angle search profile (blue line) used a fixed –N=N– distance of 1.19 Å, while the –N=N– distance search profile (orange line) assumed $\vartheta = 180°$ (linear Mo–N=N–Mo), and gave a –N=N– value of 1.19(2) Å. The other parameters were refined to obtain the best fits possible for each point on the curves. B: Fourier transforms of the fits (red filled curves) superimposed on the data (blue lines) for three different values of the angle ϑ. In the ϑ search profile, a subtle minimum at $\vartheta = 178°$ is evident (*), suggesting that in-solution distortions from linearity might be present.

Our second multiple-scattering example returns to the Ni K-edge EXAFS data for [(*n*-butyl)₄N][Ni(mnt)₂]. Here, and as we have discussed in Section 10.8, a close-to-adequate fit can be obtained with only the first coordination shell using four Ni–S interactions. However, fitting the outer shells requires the inclusion of several different multiple-scattering paths, including paths that span the central nickel atom, as shown in Figure 10.23. The observant reader may already have noted the Fourier transform peak at approximately twice that of the first-shell Ni–S peak. As we have discussed in Section 8.6 (e.g. Figure 8.16), such 2×R features are characteristic of back-and-forth-type multiple-scattering paths around the central absorber, which in this case involves the linear **trans** S–Ni–S arrangements within the [Ni(mnt)₂]⁻ core. Moreover, Fourier filtering this 2×R peak and examining its amplitude profile (not illustrated) shows the function peaking at a lower k than the single-scattering peak at 1×R. While this is expected because of bigger Debye-Waller terms for outer shell features, further damping of the high-k EXAFS is also expected for back-and-forth multiple scattering because these paths involve scattering angles β that are close to 180°, and for such β, $|f_{\text{eff}}(k,\beta)|$ decreases rapidly with increasing k (e.g. see

Figure 8.17 and associated discussion). In practice, inclusion of all these additional scatter-ing paths is a simple matter, given the coordinates of the metal site, because software such as FEFF will compute them automatically. Figure 10.26 shows the result of such com-putations, and represents a final fit to the EXAFS. We note that the back-and-forth multi-ple-scattering paths will also be present in complexes that have symmetrically bound linear groups such as cyanide ligands ($-C\equiv N$), but in this case, the smaller back-and-forth multiple-scattering EXAFS will be essentially overwhelmed by the intense multiple-scattering EXAFS from the linearly bound $-C\equiv N$ groups.

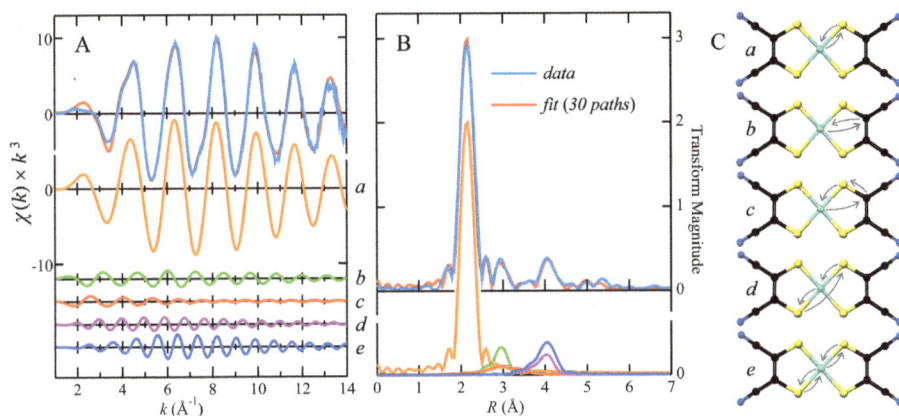

Figure 10.26: Multiple-scattering fit of the Ni K-edge EXAFS data for [(n-butyl)$_4$N][Ni(mnt)$_2$]. (A) The EXAFS oscillations, with the upper pair of curves the experimental data and the fit employing 30 scattering paths. The lower traces show the five most intense scattering paths, a through e. (B) The EXAFS Fourier transforms (Ni–S phase-corrected) of the data, fit and the same five prominent scattering paths, and (C) the five scattering paths, a through e, superimposed on the structure.

10.10 Putting it all together . . .

We close this chapter with a mention that EXAFS analysis generally should not be used in isolation of other methods. First, as we have discussed, a comparison of the EXAFS-derived bond lengths with the crystallographic literature can be very helpful. Thus, if the Cambridge Structure Database (CSD) [13] is searched for complexes with four thiolate donors and the search is restricted to complexes with square-planar-type geometry, then formally Ni(II) and Ni(III) complexes show average Ni–S bond lengths of 2.174 Å and 2.155 Å, respectively. The latter compares very well with our EXAFS-derived bond length of 2.147(1) Å (from the analysis shown in Figure 10.26). As an aside, we note that the EXAFS-derived bond length for the Ni(II) complex [(n-butyl)$_4$N]$_2$[Ni(mnt)$_2$]$_2$ is 2.175(1) Å, again in excellent agreement with the CSD.

If we were to consider other possible coordination geometries apart from effectively square planar, then this might include tetrahedral-type coordination. There are plenty of CSD entries for such complexes that possess formally Ni(II) oxidation state, but there are none that possess formally Ni(III). As might be expected, the Ni(II) tetrathiolate complexes with close to tetrahedral coordination geometry show a longer bond length than for square planar of 2.288 Å, which agrees very well with the EXAFS data, which gives 2.287(1) Å. We can complement the available information from the CSD with density functional theory (DFT) calculations,[5] which give Ni–S bond lengths of 2.148 Å for formally Ni(III) square-planar type coordination, again in excellent agreement with our EXAFS, and for the tetrahedral-type thiolate-coordinated Ni(III) that are missing from the CSD, we can come up with a value of 2.205 Å, clearly distinct from the EXAFS-derived bond length. Thus, with our [(n-butyl)$_4$N][Ni(mnt)$_2$] example EXAFS data, the bond length trends clearly indicate both the square-planar-type coordination geometry and a formal Ni(III) oxidation state.

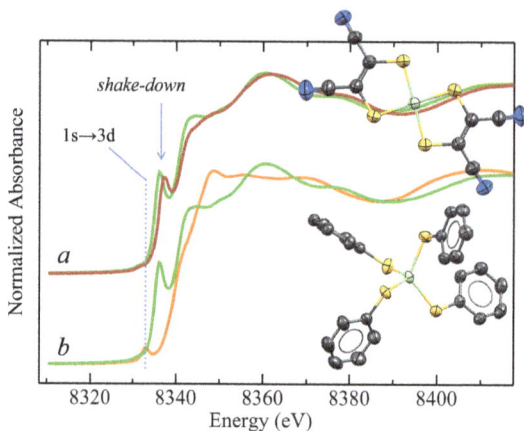

Figure 10.27: Ni K-edge X-ray absorption near-edge spectra of [(n-butyl)$_4$N][Ni(mnt)$_2$] and related compounds. Traces *a* compare the spectra of the formally Ni(III) [(n-butyl)$_4$N][Ni(mnt)$_2$] (dark red line) with that of the isostructural Ni(II) complex [(n-butyl)$_4$N]$_2$[Ni(mnt)$_2$] (green line). The crystal structure of the Ni(II) complex is shown in the upper inset. Traces *b* compare the near-edge spectra of Ni(II) species with square planar-type and tetrahedral-type coordination; the green line shows the spectrum of [(n-butyl)$_4$N]$_2$[Ni(mnt)$_2$] and the orange line that of [(n-butyl)$_4$N]$_2$[Ni(SPh)$_4$], the structure of which is shown in the lower inset (protons are not shown for clarity).

---

5 The DFT calculations given here used the DMol3 code employing the generalized gradient approximation (GGA) with the PBEsol functional, were spin-unrestricted with all-electron core treatments, and the DND basis set.

As we discussed in Chapter 9, the near-edge spectrum can also give valuable clues about both geometry and oxidation state. Our formally Ni(III) example [(*n*-butyl)$_4$N][Ni(mnt)$_2$] clearly shows an approximately +1.0 eV shift in its features, relative to the isostructural formally Ni(II) complex [(*n*-butyl)$_4$N]$_2$[Ni(mnt)$_2$]$_2$ (Figure 10.27). Moreover, the complex shows near-edge spectral features that are characteristic of a square planar-type coordination environment, with a small 1s→3d transition, and an intense feature at 8337.4 eV, attributable to a 1s→4p$_z$ + ligand-to-metal charge-transfer shake-down transition. Figure 10.27 also compares the spectrum of a Ni(II) complex with tetrahedral-type coordination with that of the Ni(II) [(*n*-butyl)$_4$N]$_2$[Ni(mnt)$_2$]$_2$, remembering that Ni(III) tetrathiolate complexes with tetrahedral-type coordination are scarce.

Other spectroscopies can also be highly informative and, together with the metal K-edge XAS, can give a more detailed picture of the metal complex. For our example compound, the obvious extension of Ni K-edge XAS would be ligand K-edge XAS (i.e. sulfur K-edge XAS), which would give information on the covalency of the Ni–S bonding and Ni L-edge XAS, which would probe the 3d manifold with intense dipole-allowed transitions. And there is more: Ni(III) is 3d^7 and should be low spin, with the singly occupied molecular orbital having substantial 3d$_{xy}$ character. In solution[6], the complex will give distinctive $S = {}^1\!/_2$ electron paramagnetic resonance (EPR) spectra, whereas Ni(II) would be EPR-silent. The EPR *g*-values of the Ni(III) complex would also inform upon coordination geometry.[7] The point that we are attempting to make in this last section is that XAS should ideally be used alongside other methods, and comparison with other data is vital, as a combination of techniques can be more powerful than any one method alone.

References

[1] McMaster, W. H.; Del Grande, N. K.; Mallett, J. H.; Hubbell, J. H. Compilation of X-Ray Cross Sections. Lawrence Livermore National Laboratory Report UCRL-50174, Sec II, Rev.1 **1969**.

[2] Victoreen, J. A.; The calculation of X-ray mass absorption coefficients. *J. Appl. Phys.* **1949**, *20*, 1141–1147.

[3] Cramer, S. P.; Wahl, R.; Rajagopalan, K. V. Molybdenum sites of sulfite oxidase and xanthine dehydrogenase. A comparison by EXAFS. *J. Am. Chem. Soc.* **1981**, *103*, 7721–7727.

---

6 EPR should be examined on samples that are magnetically dilute, hence solutions would need to be examined and not the solid crystalline complex that was examined by XAS.

7 Square planar complexes Ni(III) should be low-spin with the paramagnetic singly occupied molecular orbital of predominantly 3d$_{xy}$ character, with 3d$_{xz}$, 3d$_{yz}$ and 3d$_{z^2-r^2}$ filled, and a vacant 3d$_{x^2-y^2}$. As EPR *g*-value shifts from the free-electron value have their origins in spin orbit coupling, and as the 3d$_{xy}$ commutes with filled 3d$_{xz}$ and 3d$_{yz}$ we expect positive *g*-shifts relative to the free-electron value for g_x and g_y. The 3d$_{xy}$ orbital commutes with vacant 3d$_{x^2-y^2}$ giving negative *g*-shifts, but with positive contributions arising from the covalency of sulfur bonding. Hence, we expect $g_x \approx g_y \gg g_z > g_e$.

[4]   Cramer, S. P. Chemical Applications of X-ray absorption spectroscopy – nitrogenase and p450. Ph.D. Thesis, Stanford University, December 1977.

[5]   Stern, E. A. Number of relevant independent points in X-ray-absorption fine-structure spectra. *Phys. Rev. B.* **1993**, *48*, 9825–9827.

[6]   Zhang, L.; Pickering, I. J.; Winge, D. R.; George, G. N. X-ray absorption spectroscopy of cuprous-thiolate clusters in *Saccharomyces cerevisiae* metallothionein. *Chem. Biodivers.* **2008**, *5*, 2042–2049.

[7]   Pushie, M. J.; Zhang, L.; Pickering, I. J.; George, G. N. The fictile coordination chemistry of cuprous-thiolate sites in copper chaparones. *Biochim. Biophys. Acta.* **2012**, *1817*, 938–947.

[8]   Musgrave, K. B.; Angove, H. C.; Burgess, B. K.; Hedman, B.; Hodgson, K. O. All-ferrous titanium(III) citrate reduced Fe protein of nitrogenase: An XAS study of electronic and metrical structure. *J. Am. Chem. Soc.* **1998**, *120*, 5325–5326.

[9]   George, G. N.; Prince, R. C.; Gailer, J.; Buttigieg, G. A.; Denton, M. B.; Harris, H. H.; Pickering, I. J. Mercury binding to the chelation therapy agents DMSA and DMPS and the rational design of custom chelators for mercury. *Chem. Res. Toxicol.* **2004**, 17, 999–1006.

[10] James, A. K.; Dolgova, N. V.; Nehzati, S.; Korbas, M.; Cotelesage, J. J. H.; Sokaras, D.; Kroll, T.; O'Donoghue, J. L.; Watson, G. E.; Myers, G. J.; Pickering, I. J.; George, G. N. Molecular fates of organometallic mercury in human brain. *ACS Chem. Neurosci.* **2022**, *13*, 1756–1768.

[11] Summers, K. L.; Sarabisheh, E. K.; Zimmerling, A.; Cotelesage, J. J. H.; Pickering, I. J.; George, G. N.; Price, E. W. Structural characterization of the solution chemistry of zirconium(IV) desferrioxamine: A coordination sphere completed by hydroxides. *Inorg. Chem.* **2020**, *59*, 17443–17452.

[12] Sarbisheh, E. K.; Summers, K. L.; Salih, A. K.; Cotelesage, J. J. H.; Zimmerling, A.; Pickering, I. J.; George, G. N.; Price, E. W. Radiochemical, computational, and spectroscopic evaluation of high-denticity desferrioxamine derivatives DFO2 and DFO2p toward an ideal Zirconium-89 chelate platform. *Inorg. Chem.* **2023**, *62*, 2637–2651.

[13] Groom, C. R.; Bruno, I. J.; Lightfoot, M. P.; Ward, S. C. The Cambridge Structural Database. *Acta Cryst.* **2016**, *B72*, 171–179.

[14] Cotelesage, J. J. H.; Pushie, M. J.; Grochulski, P.; Pickering, I. J.; George, G. N. Metalloprotein active site structure determination: Synergy between X-ray absorption spectroscopy and X-ray crystallography. *J. Inorg. Biochem.* **2012**, *115*, 127–137.

[15] Pushie, M. J.; Doonan, C. J.; Wilson, H. L.; Rajagopalan, K. V.; George, G. N. Nature of halide binding to the molybdenum site of sulfite oxidase. *Inorg. Chem.* **2011**, *50*, 9406–9413.

[16] Laplaza, C. E.; Johnson, M. J. A.; Peters, J. C.; Odom, A. L.; Kim, E.; Cummins, C. C.; George, G. N.; Pickering, I. J. Dinitrogen cleavage by three-coordinate molybdenum(III) complexes: Mechanistic and structural data. *J. Am. Chem. Soc.* **1996**, *118*, 8623–8638.

[17] Curley, J. J.; Cook, T. R.; Reece, S. Y.; Müller, P.; Cummins, C. C. Shining light on dinitrogen cleavage: Structural features, redox chemistry, and photochemistry of the key intermediate bridging dinitrogen complex. *J. Am. Chem. Soc.* **2008**, *130*, 9394–9405.

11 Analysis II – speciation

In this chapter we discuss the analytical methods that are frequently applied to near-edge spectra to provide what is known as chemical speciation information. These methods have gained significant importance in analysis of complex heterogeneous samples relevant to environmental, geological, and biological sciences. The ability of XAS to penetrate matter and excite a specific element regardless of the sample state, with minimal pre-treatment and largely without destroying the sample, makes it the method of choice in many cases. This type of analysis typically uses a library of spectra of standard compounds, relying on the tendency of a particular chemical type or functional group to exhibit a characteristic and distinctive near-edge spectrum. Here we discuss the various analysis methods that are employed, noting that while the examples we provide focus on near-edge spectra, EXAFS spectra also can be subjected to the same kind of analysis. Our description starts with linear combination analysis of a single data set, then details methods applied to sets of related data.

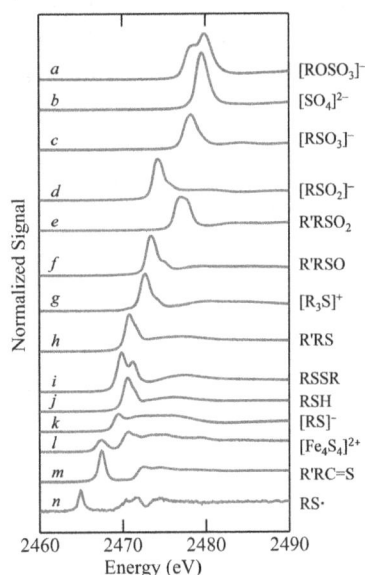

Figure 11.1: Chemical sensitivity of the sulfur K-near-edge X-ray absorption spectra of a range of biologically relevant sulfur compounds, all as 50 mM aqueous solutions at pH 7.4, unless otherwise specified. The amplitudes of some spectra have been scaled down for clarity (). *a* sulfate ester [dextran sulfate] (×½), *b* sulfate (×½), *c* sulfonate [taurine] (×½), *d* sulfinate [hypotaurine] (×⅔), *e* sulfone [methionine sulfone] (×½), *f* sulfoxide [methionine sulfoxide] (×⅔), *g* sulfonium [dimethylsulfonopropionate] (×⅔), *h* organic sulfide [methionine], *i* organic disulfide [cystine], *j* thiol [cysteine], *j* thiolate [cysteinate, pH 11], iron-sulfur cluster [*Pyrococcus furiosus* 4Fe ferredoxin, 5 mM], *m* thione [2,2,4,4-tetramethylcyclobutane-1,3-dithione], and *n* thiyl radical [glutathione-S·].

https://doi.org/10.1515/9783110570441-011

Figure 11.1 shows a series of biologically relevant sulfur K- near-edge spectra, illustrating how the spectra of different chemical types provide a fingerprint that can be used to identify the local environment of sulfur. We start by emphasizing that XAS is sensitive to the local environment; XAS cannot determine long-range structure directly and therefore does not uniquely identify molecules. Instead, standard compounds are used as a proxy for a certain local environment of the absorbing atom. In Figure 11.1, we list both the speciation type represented (e.g. h, organic sulfide, or R'RS) as well as the specific molecule used to collect the spectrum (e.g. [methionine]).

11.1 Linear combination analysis

The basic idea behind linear combination analysis is to fit a linear combination of spectra of standards to the spectrum of an unknown. This approach pivotally depends on the availability of spectra representative of all the compounds that are present in the mixture. The analysis generally uses data that has had the pre-edge and/or background subtracted and is normalized to the edge jump. The typical procedure uses a least squares approach in which the spectrum of the unknown is simulated by summing the spectra of a series of standards, the fractions of which are refined by minimizing the sum of the squares of the differences between experimental and simulated points. The normalization to the edge jump of both standard and unknown spectra means that the fractions can be used to estimate the molar fractions of components. An advantage of this method is that given a suitable library of standards, one can analyse a single spectrum.

All spectra, standards and unknown alike, must be aligned on a common abscissa of energy points. Such alignment typically involves interpolation which, like any manipulation of the data, is undesirable; however in this case it is absolutely necessary to conduct the desired data analysis. In many cases the standard spectra will be of better signal to noise than the unknown, and because interpolating noisy data can be problematic, it is normal practice to interpolate the spectra of the standards onto the abscissa points of the spectrum of the unknown. Once this is done the spectrum of the unknown a, with n X-ray energy points, can be fitted to a linear combination of c standard spectra s by minimizing the sum of the squares of the differences between these, which we will call F, as follows:

$$F = \frac{1}{n-1} \sum_{i=1}^{n} \left(a_i - \sum_{j=1}^{c} f_j s_{ij} \right)^2 \tag{11.1}$$

where f_j are the mole fractions of each component j. In many cases the crude sum of the squares of the differences is used without the denominator $n-1$, while in some cases the so-called reduced F error might be used, which is the sum of the squares of the differences divided by $(n-c)$. The choice of these makes no difference in practice.

While occasionally it may be necessary to allow small relative energy shifts to the standard spectra, in general this is considered an undesirable practice because it allows too much flexibility in the analysis, with a corresponding loss of confidence in the results.

The most relevant questions for linear combination analysis are how many components to include, and how to decide to exclude those standard spectra that contribute to an insufficient degree, and we will address these in Section 11.3. The first question really depends upon the type of spectra being analysed. Figure 11.2 shows an early example [1], which was of horse blood that had been centrifuged to separate red blood cells and plasma. Here we have a biological sample, high in protein, and we can be very confident about some of the sulfur compounds that must be present. It will contain the two sulfur-based amino acids, cysteine (a thiol, RSH) and methionine (an organic sulfide, RSR'), together with cystine, the oxidized form of cysteine containing a disulfide (RSSR). The spectra also clearly show a small but well-resolved peak at just below 2,480 eV, which is characteristic of sulfate ($[SO_4]^{2-}$), which totals at least four components. We can guess that methionine sulfoxides (RSOR', containing an $S = O$ group) may also be present as this is an aerobic sample. Thus, the initial five components in the linear combination analysis are decided based on bio-chemical expectations, and the observation of a sulfate peak. A potential sixth component might also be present, taurine, the most abundant free amino acid in humans, 2-aminoethanesulfonic acid.

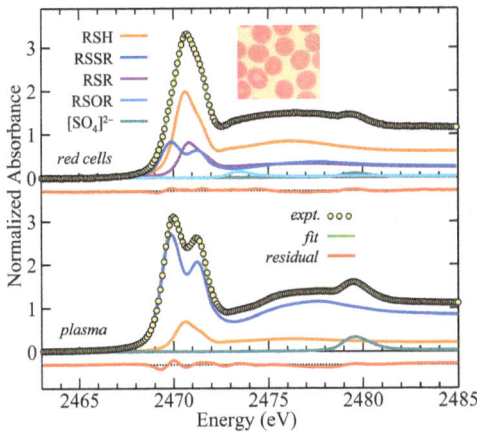

Figure 11.2: Linear combination analysis of horse red blood cells or erythrocytes (upper plot) and horse blood plasma (lower plot). The spectra of the standard compounds used to perform the linear combination analysis are shown with each experimental spectrum, weighted by the refined fractions f_j. The caption of Figure 11.1 specifies the identities of standards. The inset in the upper plot shows the erythrocytes in a micrograph of whole blood.

Figure 11.2 shows the fits using standard spectra representative of the first five forms discussed above, with very good correspondence between data and fit, and about which we will provide some further discussion below.[1] The standard compounds selected for linear combination analyses should be thought of as representative of the functional groups (in this case of sulfur), which are likely to be present. The precise speciation is typically not identified, so that while organic disulfides (RSSR), for example, are likely to be accurately determined, the precise organic form of the disulfide will not be. In many cases this is a strength of this speciation method. For example, in crude oils there will be hundreds of thousands of individual sulfur compounds, and some sulfur-specific speciation methods such as two-dimensional gas-chromatography with sulfur chemiluminescence detection (called GC-GC-SCD) are overwhelmed by a myriad of peaks, some resolved, some not, which are obtained from a crude oil. The analysis of the sulfur K-edge XAS, however, is comparatively simple as one only needs to consider the local speciation or functional group, with a handful of different chemical species as representatives of those functional groups being adequate to provide a speciation analysis. The results of the sulfur speciation by sulfur K-near-edge linear combination analysis are summarized in Table 11.1. Values in parentheses are estimated standard deviations (precisions) in the last digit(s) obtained from the diagonals of the variance-covariance matrix, for example, 21.4 ± 0.1 is written as 21.4(1).

Table 11.1: Linear combination of sulfur K-edge XAS of horse erythrocytes and plasma.

Chemical type %	RSSR	RSH	RSR'	RSOR'	$[SO_4]^{2-}$
Erythrocytes	21.4(1)	54.4(1)	21.3(1)	2.1(1)	0.8(1)
Plasma	76.5(1)	20.6(1)	–	–	2.9(1)

The results of these fits show some trends that are both clear and readily understandable. Erythrocytes show a predominance of reduced sulfur forms (i.e. thiols, RSH) while plasma shows oxidized sulfur forms (i.e. disulfides, RSSR), confirming the well-known idea that the inside of cells (i.e. the erythrocytes) is much more reducing than the outside (the plasma). The small presence of some sulfoxide (RSOR') is not unexpected in the erythrocytes, as oxidative chemistry tends to be a challenge for these cells, and methionine sulfoxide formation is known to occur. Later work from our group on cultures of epithelial cell cultures showed another important difference, that the epithelia contained abundant taurine, the level of which was sensitive to mechanical stimulation and other factors [2]. The lack of abundant taurine detected by XAS in mammalian blood is not unexpected, as taurine is known to not be present at

---

1 As an interesting aside to this analysis, we were initially doubtful about very low estimates for the methionine levels in the plasma because we knew that plasma contains plentiful serum albumin and we had examined sequence data for *human* serum albumin, assuming that horse serum albumin would be similar. In fact, unlike the human protein, horse serum albumin lacks methionine, which matched our analysis very well. This reinforced our confidence in our method.

high levels in mammalian erythrocytes. Interestingly, bird erythrocytes do possess high levels of taurine [3], although they appear to lack the ability to synthesize it [3]. One well-known difference between avian and mammalian erythrocytes is that bird erythrocytes are nucleated, whereas mammalian erythrocytes are not; however, the presence of taurine seems to be unrelated to this, as amphibian erythrocytes are also nucleated and these lack taurine [3]. In other vertebrate cell types taurine plays a role in osmoregulation – cells export taurine in response to hypotonic media, and bird erythrocytes appear to do the same.

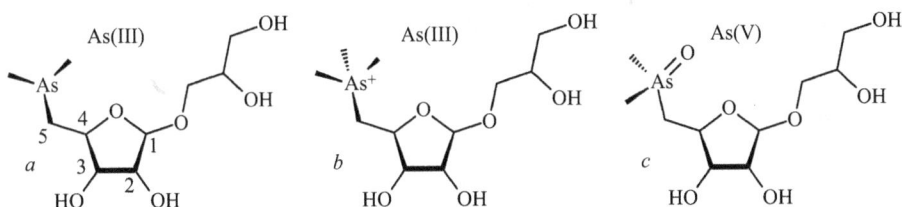

Figure 11.3: Structures of selected arsenosugars. The ring-numbering of the carbons in the ribose ring is shown in (*a*). The arsenic is methylated and attached to the ribose via C5. The pendant group attached at C1 of the ribose is here shown as glycerol, although a range of other forms are also found, including glycerol sulfonates, and glycerol phosphates. Representative arsenic functional forms include formally As(III) methylated arsine (*a*) [R₃As], arsonium cationic species (*b*) [R₄As]⁺, and the formally As(V) arsine oxide (*c*) [R₄AsO]. Thio-arsenosugars also occur, with As = S in place of As = O in (*c*).

Before moving to discuss standard compounds, and how rejection of unnecessary components in the linear combination analysis might work, we give one more example of a linear combination analysis, that of the seaweed *Ulva lactula* [4], also known as sea lettuce. All seaweeds naturally contain appreciable arsenic, and predominantly this occurs in a group of water-soluble ribose-based compounds known as **arsenosugars**. Arsenosugars are present in seaweed at concentrations that range between 0.1 and 1 mM. They occur as As(V) and As(III) compounds, and many lack the high levels of toxicity for which arsenic compounds are known. However, studies have shown that they can certainly elicit a toxic response at higher levels. Representative structures of selected arsenosugars are shown in Figure 11.3.

Here, the fitting of spectra involves standards whose spectral features partially overlap, for which fitting of higher derivatives can allow greater sensitivity to subtle features in the spectrum. Figure 11.4A shows the arsenic K-edge XAS of *U. lactula*, which possesses a broad peak in the near-edge with clearly defined shoulders. The second derivative spectrum, in Figure 11.4B, shows these shoulders as clearly defined downward structures in the second derivative. In principle the minimum obtained by fitting a higher derivative should be exactly the same as that which would be obtained by fitting the absorption spectrum. However, in practice the use of higher derivatives allows the refinement to converge on a minimum more rapidly. If the derivative alone is fitted then

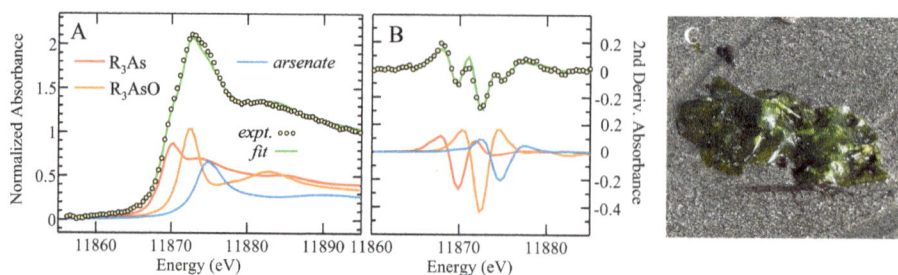

Figure 11.4: Linear combination analysis of the seaweed *Ulva lactula*. Linear combination analysis used a simultaneous fit to both the experimental spectrum (A) and its second derivative (B). The fit indicated fractions f_j 0.41(1) of R_3As, 0.36(1) R_3AsO and 0.23(1) of inorganic arsenate, with values in parentheses being estimated standard deviations. C shows a photograph of *Ulva lactula* (photograph courtesy of Simon George).

the fractions become less reliable as the importance of the edge step is de-emphasized in the derivatives, and the fractions, which ought to total to unity, can sum to a value significantly more or less than unity. The use of a combined absorption-derivative fit circumvents this problem. Here, the function to be minimized includes a sum-of-squares deviation for both the derivative spectrum and the original absorption spectrum, although their relative contributions to the error need careful weighting, with the choice of this value not always straightforward. Alternatively, one can fit to the derivative and constrain the sum of all f_j to be unity. Another issue is that calculation of derivatives invariably increases the apparent noise level, as high frequency noise is relatively amplified, with common practice being to use the coefficients of a least-squares polynomial fitted over some ordinate window to mitigate this,[2] combining a polynomial smooth with the evaluation of the higher derivative. Application of any smoothing function invariably involves some loss or distortion of data, so this may be undesirable. Figure 11.4 shows a successful example of a combined simultaneous fit to the absorption spectrum and its second derivative, calculated using the coefficients of a least-squares quartic polynomial fit over a 1.4 eV energy range.

11.2 Choice of standard compounds

This brings us to another issue, which is the choice of standard compounds. We have already noted that the linear combination analysis method requires a substantial library of spectra of standards. If the sample to be analysed contains a compound for which no standard spectra is available, this can effectively prevent any analysis based on standards. Alternatively, there may be compounds present, which have quite differ-

---

2 The ith coefficient c_i of a polynomial $y = \sum_{i=0}^{n} (x - x_j)^i c_i$ gives the ith derivative of y at x_j.

ent chemical structures but show similar spectra. An example of this, relevant to the *Ulva* arsenic analysis discussed above, is shown in Figure 11.5. The near-edge spectra of the formally As(III) arsonium cation [R$_4$As]$^+$, and the formally As(V) arsine oxide R$_3$As = O show strong similarities, especially in the energy of their dipole-allowed major peak, and if the data being fitted or the standard spectra were of poor signal to noise then distinguishing between these species might be impossible.

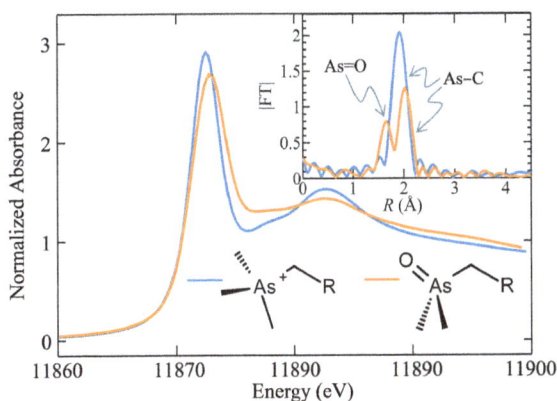

Figure 11.5: Chemically distinct arsenic species with similar near-edge spectra. The As K-edge near-edge spectra for a formally As(V) arsine oxide (orange line) and a formally As(III) arsonium compound (blue line) are compared. The two compounds show similar near-edge spectra, but distinct EXAFS, as shown by the EXAFS Fourier transforms (As–C phase-corrected) in the inset to the figure.

Choice of the relevant physical state for standards is also critical. For example, the spectra of standards prepared as solids may look quite different from the same compound in solution. Figure 11.6 shows an example of such differences for the amino acid L-methionine. The solution spectrum shows a more intense major peak while the solid shows additional structure at higher energies. The differences in Figure 11.6 are by no means atypical with both of these effects often observed when comparing solids and solutions. The question now arises as to why such spectroscopic differences occur. In the case of L-methionine, which crystallizes with space group P21, there are two somewhat different molecular conformations in the unit cell (shown in the inset to Figure 11.6a and b), which have subtly different spectra. Hence, the spectrum of the solid must be a superposition of the spectra for these two discrete conformations. If one compares the density functional theory energy-minimized geometry-optimized structure for L-methionine, using a COSMO field to approximate water, then the conformation shown in Figure 11.6a more resembles that expected for solution; we can estimate that the two differ in energy by 7.05 kJ·mol^{-1}, *a* being lower, and *b* being thermally accessible in solution at room temperature at the 5% level. The solid will have different hydrogen bonding, for example conformation *b* has several close contacts with other molecules with two C–H···S contacts within 3 Å, whereas conformation *a* has no long-range contacts within this distance.

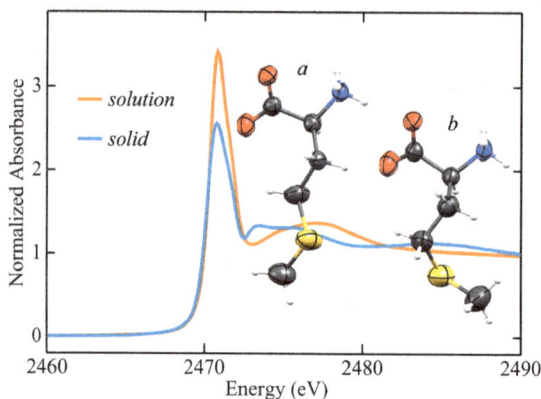

Figure 11.6: Sulfur K-edge X-ray absorption near-edge spectra of L-methionine as a finely powdered crystalline solid, and as an aqueous solution pH 7.4. The inset shows the structures of the two molecular conformations found in the crystal structure (85% thermal ellipsoids).

Irrespective of the reasons behind the differences between solid and solution, for biological samples the best choice of standard spectrum generally is a solution, and not the crystalline solid. The same goes for other factors such as pH, with physiological pH values being the most useful. Figure 11.7 illustrates the systematic variation in the sulfur K-edge near-edge spectrum of cysteine with pH, due to the ionization of the thiol group at high pH values $-SH \leftrightarrow -S^- + H^+$.

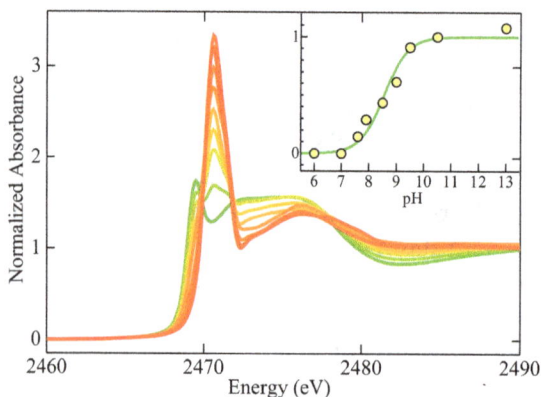

Figure 11.7: pH titration of an aqueous solution of L-cysteine. The near-edge spectra are shown from pH 6 (red, thiol form, RSH) to pH 13 (green, thiolate form, RS⁻). The points in the inset show the fraction of thiolate determined using fits to the titration spectral end members, with the line corresponding to the known cysteine thiolate pK_a of 8.33.

11.3 Determining the number of components in linear combination analysis

One obvious challenge in linear combination analysis is how to determine whether a fitted component is really present or not, especially if the fraction of that component has a low value. If the spectrum is isolated well from those of the other components, then very accurate determination is possible (e.g. sulfate in Figure 11.2). If, on the other hand, the series of standards possess substantially overlapping spectral features, then the situation may be more challenging. In some cases it may be possible to obtain close to the same goodness-of-fit index with different sets of components, with discrimination between these requiring additional information, such as knowledge of the chemistry of the system. The quality of the data is another important factor since if there is appreciable noise, obtaining an unambiguous analysis will be more challenging. One approach that may be helpful is to use a weighted error function, which will include a measure of the noise:

$$F' = \frac{\sum_{i=1}^{n} \left(a_i - \sum_{j=1}^{c} f_j s_{ij} \right)^2}{\sum_{i=1}^{n} a_i^2} \tag{11.2}$$

Statistical information in the form of precisions is available from what is called the variance-covariance matrix that is used during the non-linear optimization procedure. The diagonal of this matrix can be used to give estimated standard deviations for the fraction f_j of each component in the fit. A simple rule that can often give helpful results is that if the fraction f_j of a component is less or equal to its estimated standard deviation then that component can be excluded from the mix without compromising the quality of the fit.

11.4 Principal component analysis

Principal component analysis (PCA) is essentially a method of reducing a large set of data to a smaller one. It employs a transformation of a data matrix made up of a set of spectra using a mathematical technique called singular value decomposition (often abbreviated SVD). In the case of X-ray absorption spectra, this applies to a set of spectra, which are intimately related in some way, for example, in the kinetic time course of a chemical reaction, or micro-XAS spectra of adjacent spatial locations on sample. Similar to linear combination analysis, all spectra must be aligned on a common abscissae, which requires interpolation, which, while undesirable is unavoidable. If we have m spectra, each with n energy points, with individual intensities a_{ij} we can construct an $n \times m$ matrix of spectra **A**, in which a_{ij} is the intensity at the i th energy point of the j th spectrum:

$$A = \begin{pmatrix} a_{11} & a_{12} & \cdots & a_{1m} \\ a_{21} & a_{22} & \cdots & a_{2m} \\ \vdots & \vdots & \vdots & \vdots \\ a_{n1} & a_{n2} & \cdots & a_{nm} \end{pmatrix} \tag{11.3}$$

PCA provides a way of effectively reducing the dimensions of this data matrix while retaining essentially all the information content. We call this reduced set of data the principal components, or alternatively the eigenvectors. We can subject the data matrix to SVD that is based on a theorem in linear algebra, which states that any $n \times m$ matrix with a number of rows n that is greater than or equal the number of columns m can be written as the product of an $n \times m$ column orthogonal matrix, which we will call U, an $m \times m$ diagonal matrix with only positive or zero elements, which we will call W and the transpose of an $m \times m$ orthogonal matrix, which we will call V^T. The decomposition of the data matrix A into U, W and V^T is central to PCA, and to much of data science:

$$A = U \cdot W \cdot V^T \tag{11.4}$$

$$A = \begin{pmatrix} u_{11} & u_{12} & \cdots & u_{1m} \\ u_{21} & u_{22} & \cdots & u_{2m} \\ \vdots & \vdots & \vdots & \vdots \\ u_{n1} & u_{n2} & \cdots & u_{nm} \end{pmatrix} \cdot \begin{pmatrix} w_{11} & & & \\ & w_{22} & & \\ & & \ddots & \\ & & & w_{mm} \end{pmatrix} \cdot \begin{pmatrix} V_{11} & V_{21} & \cdots & V_{m1} \\ V_{12} & V_{22} & \cdots & V_{m2} \\ \vdots & \vdots & \vdots & \vdots \\ V_{1m} & V_{2m} & \cdots & V_{mm} \end{pmatrix}$$

$$\tag{11.5}$$

The matrix U is the same shape as our original matrix of spectra A, and the matrices U and V are each orthogonal so that $U^T \cdot U = V^T \cdot V = I$, where I is the identity matrix. The procedure of SVD is, in effect, a rotation, but the details of this are beyond the scope of this chapter and so we will not delve into them here. The matrix U, called the **eigenvector matrix**, gives the principal components of the data matrix A with each of its columns $U_{(j)}$ being the j th principal component. Unfortunately, the use of the word "component" can be confusing when this nomenclature is used in the context of mixtures of spectra; the principal components are just mathematical constructs and do not directly represent the spectra of physical components of mixtures. The $m \times m$ square diagonal matrix W is called the **eigenvalue matrix** or the **singular matrix** and is comprised of what are called singular values w_j. We can refer to these using just a single subscript as the matrix is diagonal. The singular values w_j are a measure of the relative contribution of each eigenvector $U_{(j)}$ to the data matrix A, and their squares are the eigenvalues of $A^T \cdot A$. The matrix V is essentially a values table indicating how much of each principal component is used to make up each column of the data matrix A. Typically we would plot the principal components as the products $U \cdot W$ versus incident energy.

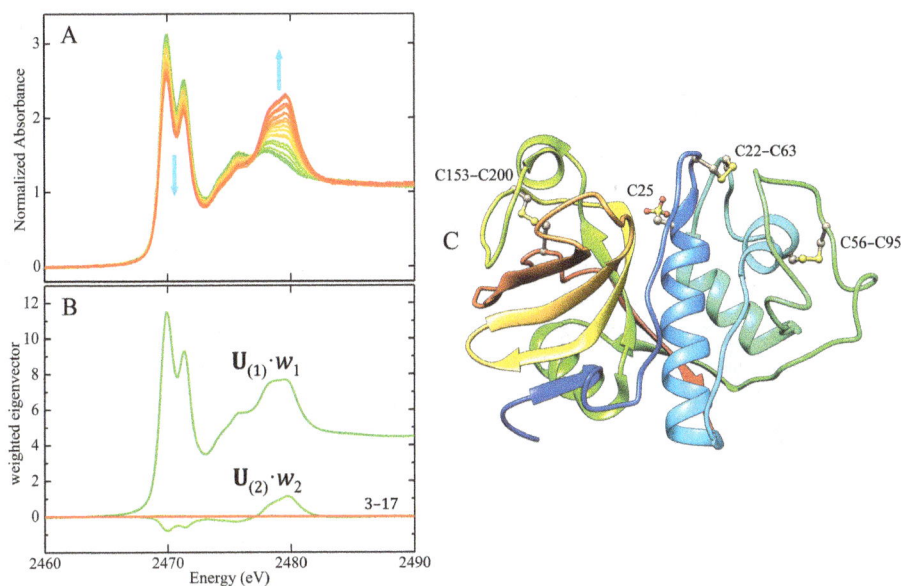

Figure 11.8: Example of principal component analysis. Kinetic time-course data (A) together with the results of principal component analysis (B) showing all eigenvectors (principal components), eigenvalue-weighted as $\mathbf{U} \cdot \mathbf{W}$. The time-course data (A) is the sulfur K-edge XAS of an aqueous solution of the thiol protease papain from *Carica papaya* (crystal structure [5] shown in (C)), with spectra collected over 4 h of X-ray beam exposure. The progressive changes in the spectra are due to X-ray beam-induced photo-oxidation of the sulfur sites in the sample.

The first principal component (first eigenvector) $\mathbf{U}_{(1)}$ is typically the numerical average of all the spectra in the data matrix $\mathbf{A}$. With near-edge data the first eigenvector therefore always looks just like an ordinary spectrum. Like the first, the second principal component will also be a linear combination of the data, but this time it will be one that captures the remaining variance in the data, and is uncorrelated with the first. This will typically have both positive and negative values, and will be orthogonal to the first. If there were only two components making up all the spectra in the data matrix $\mathbf{A}$ then the first eigenvector would be the average of the two, and the second would be the difference between the two. Another important point is that PCA is sensitive only to variations among the spectra in our data matrix $\mathbf{A}$. Thus, if all contain the same fraction of some species, then PCA will not pick up that constant physical component as a separate principal component. In practice, the first principal component might be "upside down", in that it resembles an inverted spectral average. This is not important mathematically, as this will be reflected in the $\mathbf{V}$ matrix and will not change the reconstruction of the spectra, but in such cases it is common practice to flip the plots vertically so that the first eigenvector is shown the "right way up". In most cases the high order components of $\mathbf{W}$ are sufficiently small that these may be set to zero without changing anything other than the noise, effectively reducing the data dimensionality.

To illustrate some of the features of PCA we will consider a kinetic time course, due to X-ray beam-induced photo-oxidation [6] of an aqueous solution of the enzyme papain from *Carica papaya*, which we have chosen both because of its simplicity and because it illustrates both strengths and weaknesses of the method. The experimental data have all been edge-normalized and background-subtracted, and are shown in Figure 11.8A. We will note at this point that here we depart from some other workers who prefer to use raw data, with no edge-jump normalization or other pre-treatment. Papain contains six structural cysteine residues, which are all disulfides (Figure 11.8C) and one active site cysteine thiol (C25), which in the crystal structure was oxidized to a sulfonate (Figure 11.8C). As expected, the initial spectrum strongly resembles that of a disulfide (e.g. Figure 11.1*i*), which should be 6 out of 7 total sulfurs. The 7th active site sulfur might be a cysteine thiolate or perhaps oxidized to sulfonate. In the time sequence we see an increase in intensity at ~2,480 eV concomitant with a decrease in the disulfide double peak centred at ~2,471 eV. The decrease in dimensionality in the time sequential total data set that is afforded by PCA is immediately apparent upon examining the principal components in Figure 11.8B. As expected, the first principal component represents an average of the input spectra, with major variances being represented by the second principal component, showing a positive structure at energies expected for oxidized sulfur at ca. 2,480 eV, and a negative structure at the double peak region of the disulfide centred at about 2,471 eV.

The number of principal components that provide a significant contribution to the data set can be estimated using several different methods, with two discussed here. The first is a so-called scree plot[3] and the second is by computing what is called the Malinowski indicator function. Most PCA software orders the output so that the eigenvalues w_j are arranged from largest to smallest; a scree plot simply shows the eigenvalues plotted in this order. A scree plot will tend to turn a corner, often referred to as an elbow, when the eigenvectors become unimportant, contributing to the noise. Figure 11.9A shows a scree plot from the data of Figure 11.8. The elbow here is found at the number of components c of 2, or possibly 3.

Scree plots have been criticized as unreliable because they can have multiple "elbows", which can make it challenging to estimate the best number of principal components to retain. Indeed, as we have noted, the scree plot of Figure 11.9A is slightly ambiguous, as one can imagine an elbow at $c = 3$, rather than $c = 2$. Possibly a better method is given by what is called Malinowski's indicator function, $I(c)$. $I(c)$ can be

3 A scree plot is so-called because it was thought to look a bit like a scree slope, a steep and unstable mountain rock formation made up of a jumble of rock fragments. Scree running used to be a popular mountain pastime in the UK. This allows for a quick decent as the scree slides with you as you run down the mountainside, although one that is hard on the boots. In the UK, many scree slopes have been used up by runners with all the scree now at the bottom.

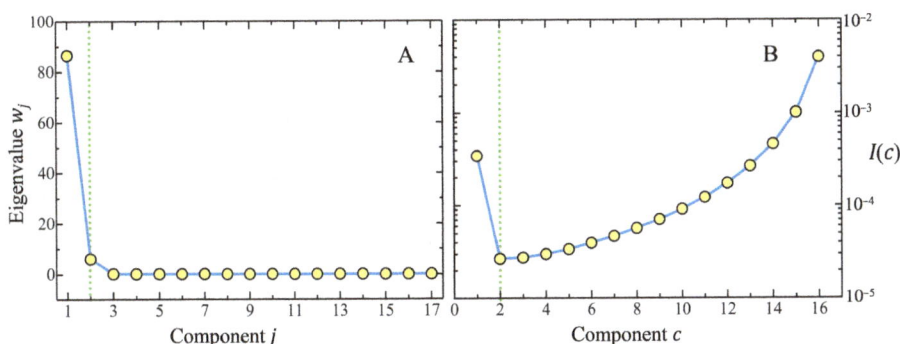

Figure 11.9: (A) Scree plot from the principal component analysis of the data in Figure 11.8. (B) Plot of the Malinowski indicator function $I(c)$ for the same data. In both (A) and (B), the green vertical broken lines indicate c, the number of non-noise principal components.

computed using eq. (11.6) by running through the number of components included, which again we will call c. A plot of $I(c)$ versus c will show a minimum at the required number of components:

$$I(c) = \frac{1}{(m-c)^2} \left(\frac{1}{n(m-c)} \sum_{j=c+1}^{m} w_j^2 \right)^{\frac{1}{2}} \tag{11.6}$$

The term raised to the power of ½ in $I(c)$ is an estimate of the inherent error space R_E computed by summation of all the eigenvalues that are higher than c, which for the correct c, would contain only noise contributions to the data. The function $I(c)$ has been developed empirically, but in practice it does work reasonably well. A plot of $I(c)$ versus c, again using the data from Figure 11.8, is shown in Figure 11.9B. This shows a clearer view of the most appropriate number of significant principal components, with a clear minimum at 2. A final test is to reconstruct the entire data matrix using $\mathbf{A} = \mathbf{U} \cdot \mathbf{W} \cdot \mathbf{V}^T$ employing the minimum number of components c, and then compare the original experimental data and the reconstructions. As data sets can be extensive, instead of going through each and every reconstruction, one might choose to examine the residual as a function of component or spectrum number j, as shown in Figure 11.10A, and pick the worst reconstructions to examine for adequacy, as in Figure 11.10B.

There is a widely held assumption, which in practice seems to be justified but for which we know no proof, that the number of principal components c needed to reconstruct a set of data is the same as the number of physical components (i.e. different chemical species) represented in the original data matrix. There are some obvious restrictions to this, which occur when the ratios between physical components do not vary, or when one or more physical components is constant throughout. In the former case, if the samples all contain two different physical components which are always

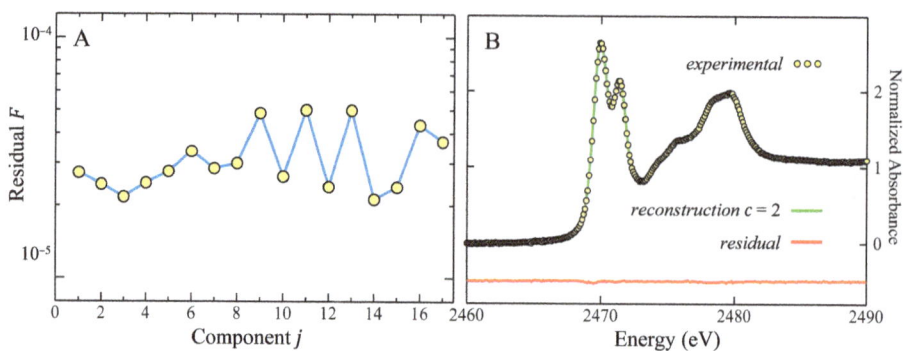

Figure 11.10: A: Plot of residuals computed using eq. (11.1) of the individual spectra in the data matrix a versus the spectrum number j. B: Comparison of the experimental data for the reconstruction with the worst residual ($j = 11$), showing excellent reproducibility. Reconstructions used the minimum number of components, $c = 2$, determined from the plots in Figure 11.9.

present in the same relative amounts, then PCA of the spectra will fail to detect that there are two components. In the latter case the presence of an invariant component would be missed altogether.

With our example data set of Figure 11.8 the observant reader may note that there must be at least three different sulfur forms present. These are (1) the active site sulfur, (2) the structural cystine disulfides, which are the presumed target of photo-oxidation and (3) the products of photo-oxidation. How can this be reconciled with our determination of $c = 2$? The most probable answer has just been discussed. A likely target for the sulfur photo-oxidation is the cystine disulfides, as it is well known that S–S bonds in proteins are susceptible to radiolytic damage [7, 8]. We note that the measurements on papain discussed in this chapter were conducted in air,[4] over a long time period (4 h) and at room temperature, so that in addition to the products of water radiolysis[5] there were plenty of other oxygen species present. If the active site sulfur were not as susceptible to photo-oxidation it would have been an invariant component. As such, it would appear only as part of the first principal component and would not have been detected by PCA, hence the value of $c = 2$. We will return to discuss the chemistry of Figure 11.8 later, but before this we will turn to the task of

---

4 The reader may remember that it does not take very much air to wipe out the beam at 2.4 keV. What is called a cut-off box was used that employed a helium flight path terminated with a very thin polymer window, with the sample in air, again enclosed by a very thin polymer window, separated from the helium flight path by 1 mm of air through which X-ray transmittance was more than adequate.

5 Water radiolysis primarily produces hydrated electrons and hydroxyl radicals, with the latter being potent oxidizers; the hydrated electrons could react with any O_2 present to form reactive oxygen species such as $O_2^{\cdot -}$.

determining what physical components might be present in the mix, and to this end the method of choice can be target factor analysis.

11.5 Target factor analysis

Target factor analysis, involving what are often called target transforms, goes hand in hand with PCA. The goal here is to determine whether a spectrum that was not included in the original data matrix **A**, for which we will use a column matrix **T**, can be adequately represented by the principal components of **A**. The spectrum in question would typically be that of a standard compound, which is suspected to be representative of the species present in the original data. Thus, target transformation effectively tests to see whether a physical component is present in our original samples. As before, the data contained in **T** must be aligned with the energy points of the original data matrix and have the same number of points n, so once again interpolation is required. The target transform **T*** can be computed using eq. (11.7):

$$\mathbf{T}^* = \mathbf{U} \cdot \mathbf{U}^T \cdot \mathbf{T} \tag{11.7}$$

$$
\begin{pmatrix} T_1^* \\ T_2^* \\ \vdots \\ T_n^* \end{pmatrix} =
\begin{pmatrix} u_{11} & u_{12} & \cdots & u_{1m} \\ u_{21} & u_{22} & \cdots & u_{2m} \\ \vdots & \vdots & \vdots & \vdots \\ u_{n1} & u_{n2} & \cdots & u_{nm} \end{pmatrix} \cdot
\begin{pmatrix} u_{11} & u_{21} & \cdots & u_{n1} \\ u_{12} & u_{22} & \cdots & u_{n2} \\ \vdots & \vdots & \vdots & \vdots \\ u_{1n} & u_{2n} & \cdots & u_{mn} \end{pmatrix} \cdot
\begin{pmatrix} T_1 \\ T_2 \\ \vdots \\ T_n \end{pmatrix} \tag{11.8}
$$

When computing the target transform **T***, **U** can be restricted to include only the significant principal components, c, determined in the previous section. If the target transform **T*** is essentially the same as the original spectrum **T** of the candidate standard being tested, then we conclude that the candidate is a part of the mixture. While a by-eye estimation of similarity between **T*** and **T** can be hard to beat, an abstract way of determining this is useful, and this is provided by the spoil function, which we will call Spoil(c), and which can be calculated using eq. (11.11). The spoil is so called because it estimates the degree to which the reconstruction of **T*** is **spoiled**, and is often written in all capitals as SPOIL. The equation for Spoil includes (amongst other terms) a reduced residual for the target transform F_T (eq. 11.9) (sometimes called the apparent error in the test vector, or AET) and an estimate of the inherent error space R_E (eq. (11.10)) from the summation of all the eigenvalues that are higher than c, which lack non-noise contributions to the data:

$$F_T = \left(\frac{1}{(n-c)} \sum_{i=1}^{n} \left(T_i^* - T_i \right)^2 \right)^{\frac{1}{2}} \tag{11.9}$$

$$R_E = \left(\frac{1}{n(m-c)} \left| \sum_{j=c+1}^{m} w_j^2 \right| \right)^{\frac{1}{2}}$$ (11.10)

$$\text{Spoil}(c) = \sqrt{\frac{F_T}{R_E \sum_{k=1}^{c} \left(\frac{\sum_{i=1}^{n} u_{ik} T_i}{w_k} \right)^2} - 1}$$ (11.11)

The spoil function gives a dimensionless positive number, which arbitrarily gives a goodness of fit for the target transform $\mathbf{T}^*$, in which values of < 3 are considered acceptable, 3–6 moderately acceptable and > 6 unacceptable. With near-edge spectra a certain amount of caution must be exercised when examining Spoil because in many cases simply adjusting the energy range of the data included in the PCA calculation can change Spoil from being unacceptable with a narrow range to acceptable with a wider range. This is because with the wider energy ranges the edge-step tends to dominate the calculation of Spoil, which has the unfortunate consequence of making just about everything acceptable. Conversely, when dealing with well-resolved and spectroscopically detailed spectra, if exactly appropriate standard compounds are not available then Spoil may indicate that $\mathbf{T}^*$ is unacceptable even if there is a match in speciation. When PCA is used with EXAFS [9], on the other hand, no such difficulties exist, and here Spoil may be very useful indeed.

There are some obvious complications with target factor analysis. Firstly, if the original data matrix from which our eigenvectors are derived contains an invariant component then this component will not be identified by computing its target transform. Moreover, in such a case the Spoil of the other components, even if perfectly identified, may be "spoiled" to give values that suggest that these are not present in the mix either. Thus, if we use the PCA of our *C. papaya* papain photo-oxidation example and test various alternatives using target transforms, none of these give Spoil values in the acceptable range, although linear combination analysis of individual scans from 11.8 gives excellent three-component fits employing the standard spectra of Figure 11.1 *a*, *c* and *i*, suggesting that the end product of the photo-oxidation is an oxygen-linked sulfate ester (i.e. as in Figure 11.1a). The other obvious complication would be if two or more components are always present in invariant ratios. In this case the target transforms for all the other components in the mixtures would all give acceptable Spoil, but the target transforms of the compounds present with the same ratio would give an unacceptable Spoil, however a combined standard spectrum consisting of a linear combination of the individual compounds with the correct ratio would give an acceptable Spoil.

Because of these challenges, rather than attempting to demonstrate target transforms using the data of Figure 11.8 we have instead chosen to use the cysteine pH titration data of Figure 11.7. Figure 11.11 compares the target and target transforms of two standards that should represent the species expected to be present, and two

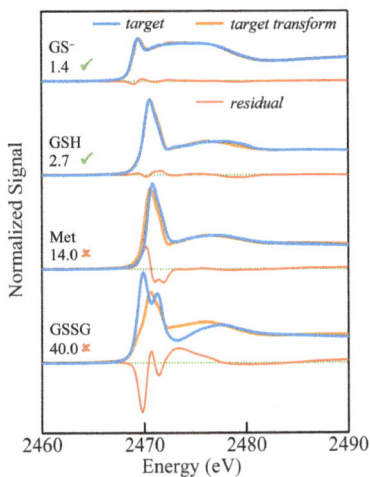

Figure 11.11: Target factor analysis of the data of Figure 11.7 using reduced high pH glutathione (GS⁻, pH 11.5), reduced low-pH glutathione (GSH, pH 6.0), methionine (Met, pH 7.4) and oxidized glutathione (GSSG, pH 7.4) and $c = 2$. The Spoil value is given to the left of each plot, showing a clear pass for GS⁻ and GSH and a fail for the other two compounds.

standards that should correspond to species that are not expected to be present. Here, we have chosen to use glutathione (γ-glutamylcysteinylglycine) in place of cysteine, which shows subtly different spectra, because using spectra for cysteine itself would be too circular to be valid. The target transforms of both the thiol and thiolate forms of glutathione clearly pass the Spoil test, while both methionine and oxidized glutathione fail it.

Once a series of standard spectra have been identified as acceptable components using the methods of target factor analysis, then linear combination analyses can be carried out to determine the fractions of each standard spectrum in the original data **A**, giving a quantitative speciation analysis. This can be done either component by component, as described in Section 11.1, or by using the methods that we will describe below, in Section 11.6. Before moving on, we should note one practice which we consider undesirable, which is to use the target transforms of the standard spectra in place of standard spectra themselves for the quantitative speciation analysis; doing this may be a little too circular as it is effectively fitting the data with itself. Irrespective of the choice of standards, linear combination fitting of all of the component spectra in the data matrix can be a laborious task. Fortunately, there is a numerical method that can quickly determine the solution, which we will now discuss.

11.6 Linear combination fits from singular value decomposition

The least squares solution of a linear combination fit is available directly from single value decomposition. If we consider our $n \times m$ data matrix **A**, containing the spectra m of a series of related unknowns each with n data points, then the vector $\mathbf{A}_{(j)}$ can be an individual spectrum. In addition we can consider an $n \times c$ matrix of standard spectra that we can call **s**, and a vector of length c consisting of the fractions **f**. We seek the **f** which will provide the best approximation to $A_{(j)}$, minimizing the sum of the squares of the differences, and satisfying

$$\left| \mathbf{s} \cdot \mathbf{f} - \mathbf{A}_{(j)} \right|^2 \rightarrow 0 \tag{11.12}$$

If we subject the matrix **s** to SVD to obtain its eigenvectors **U** with singular values **W** and $\mathbf{V}^T$ (noting that of course these are discrete from our previous application of SVD to our data matrix **A**), as follows:

$$\mathbf{s} = \mathbf{U} \cdot \mathbf{W} \cdot \mathbf{V}^T \tag{11.13}$$

here, the eigenvectors are contained in the $n \times c$ matrix **U**, and the eigenvalues are in the diagonal $c \times c$ matrix **W**. The vector **f** that best satisfies eq. (11.12) can be computed using the following equation:

$$\mathbf{f} = \sum_{j=1}^{c} \left(\frac{\mathbf{U}_{(j)}^T \cdot \mathbf{A}_{(j)}}{w_j} \right) \mathbf{V}_{(j)} \tag{11.14}$$

Thus, in one stroke, given a suitable matrix of standard spectra **s**, this method provides the least-squares fit to the data $A_{(j)}$, and it can be quite easily applied to each of the spectra making up the m columns in **A**. The estimated standard deviations σ^2 for each fraction f_j are also available from the values within the **V** matrix, according to

$$\sigma^2_{(f_j)} = \sum_{j=1}^{c} \left(\frac{V_{ij}}{w_j} \right)^2 \tag{11.15}$$

As with all of the procedures discussed in this chapter, this method absolutely requires that all of the data and all of the standard spectra be on the same energy grid, which typically means that both data and standard spectra must be subjected to interpolation. We note that the results of this computation will not be any different from the non-linear optimization methods typically used for fitting a single spectrum as described in Section 11.1, and that no small relative energy shifts of reference spectra can be included, which can be treated as refinable parameters using non-linear optimization methods. Finally, the combined fitting to absorption spectrum and higher derivative, as shown in Figure 11.4 can be more difficult to manage. Nonetheless, the use of SVD to identify least-squares solutions is certainly a powerful addition to our analytical toolbox.

11.7 Principal component analysis of the EXAFS

While PCA of near-edge spectra are quite frequently used, especially in environmental applications of XAS, PCA applications to the EXAFS are not nearly as common [10]. In part, this is because collecting the EXAFS is more challenging in terms of signal to noise, and because many environmental samples tend to be dilute, making the near-edge the more tractable of the two. PCA can be used with EXAFS together with a library of standards, in exactly the same way as we have described for near-edge spectra, with target transforms and linear combination fitting. But there is slightly more to the PCA of EXAFS than this. Ideally, the discussion that will follow would have belonged in Chapter 10, but would not have been as useful there without our discussion here of SVD. As we have noted above, with near-edge spectra, only the first principal component actually looks like a near-edge spectrum, and all the rest of the principal components are comprised of structure that goes both above and below the baseline. If one calculates PCA of a set of EXAFS spectra, on the other hand, all of the non-noise principal components look like EXAFS, and moreover, they can be analysed using the standard fitting methods that we described in Chapter 10. This difference arises from the fact that each principal component is orthogonal to all of the others, just as the individual (resolved) Fourier components of an EXAFS spectrum are orthogonal to each other. As a result, the components can be Fourier-transformed to give radial structure functions and fitted to extract structural details. Of course, there is a good chance that the eigenvectors will be upside-down relative to conventional EXAFS, with the sign compensation embedded within the V^T matrix. Just as the eigenvectors from PCA of near-edge spectra may need to be flipped vertically so as to be right-side-up, so might the eigenvectors from PCA of EXAFS, or alternatively one could just use negative coordination numbers in the refinement. Like the near-edge case the eigenvectors arising from PCA of EXAFS are still just mathematical constructs, and do not directly relate to physical components of mixtures. Figure 11.12 shows an example, which is an analysis of the first two eigenvectors of the EXAFS spectra derived from mixtures of a linear digonal cuprous-oxygen complex with Cu–O bond lengths of 1.802 Å, and a cuprous thiolate cluster with a $[Cu_4SPh_6]^{2-}$ core possessing three-coordinate Cu–S with bond lengths of 2.272 Å, with each copper having two Cu⋯Cu contacts at 2.696 Å and one at 2.847 Å.

The chief advantages in doing PCA of EXAFS in this manner, if any, may relate to the reduction in data dimensionality. Hence, if there are a substantial number of EXAFS spectra in the set of data, instead of completing curve-fitting analyses of each individual spectrum making up the total data, one would only need to fit the c non-noise principal components (eigenvectors).

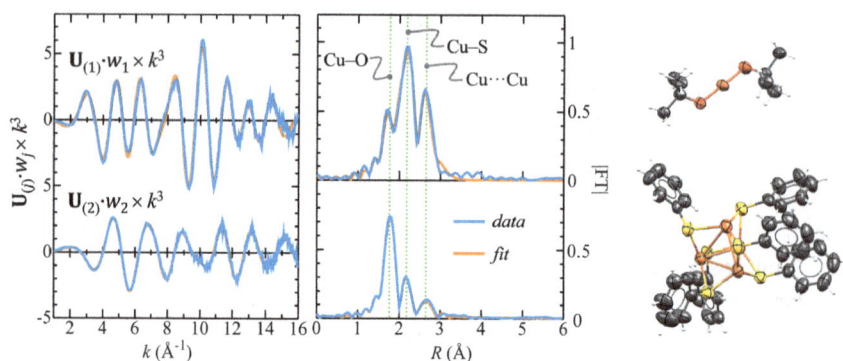

Figure 11.12: Principal component analysis of EXAFS. EXAFS curve-fitting analysis of the first two principal components of the EXAFS spectra of a series of mixtures with different proportions of a digonal cuprous oxygen species, and the cuprous thiolate cluster $[Cu_4SPh_6]^{2-}$, the crystal structures of which are shown to the right of the plot. The EXAFS principal components were analysed using standard EXAFS curve-fitting approaches. The EXAFS Fourier transforms (right) were computed using Cu–S phase correction, with the vertical green broken lines shown to guide the eye to the peak correspondence for the different interatomic distances in the mixtures.

11.8 Data analytics

In the early days of XAS researchers were often challenged to get just a single spectrum,[6] but in modern times, and as we have discussed above, data sets can be extensive. The availability of such extensive experimental data opens the possibility of new approaches to analysis. One such approach is hierarchical data clustering. Here, each individual spectrum in an extensive set of related spectra would be compared with all the others, and a metric of some sort describing the differences would be employed. This might be as simple as the square root of the sum of the squares of the differences between two spectra (called the **Euclidean distance** in the language of data analytics) or some other metric, to produce a set of what are called **distances** representing how different each spectrum is from all the others. This is expressed in eq. (11.16), where a are the spectra, and d_{ij} is the Euclidean distance between spectrum i and spectrum j. The summation is over all data points n included within the calculation:

---

6 One of us (George) remembers his first XAS experiments during which he chatted with famed muscle researcher Hugh E. Huxley who was conducting his acclaimed muscle X-ray diffraction measurements on an adjacent beamline. Huxley remarked that he was envious of George, because all he (George) needed was a single spectrum. Since those early days George estimates that he has collected close to 250,000 spectra.

$$d_{ij} = \left[\frac{1}{n-1} \sum_{k=1}^{n} \left(a_{ki} - a_{kj} \right)^2 \right]^{\frac{1}{2}} \qquad (11.16)$$

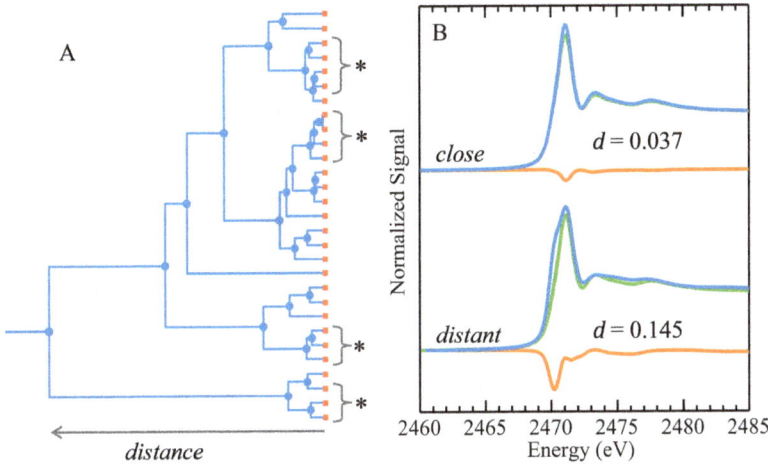

Figure 11.13: Hierarchical data clustering of sulfur K near-edge spectra of fossil fuel samples; the clustering analysis automatically groups many of the spectra (those marked using an asterisk*) by geographical region; the dendrogram is shown in (A), while (B) shows two pairs of spectra corresponding to small and large values of *d* (top and bottom, respectively). In (B), the orange lines show the difference spectra (green–blue). Data and plots are shown with courtesy of Dr Julien Cotelesage.

A disadvantage of data analytics shared with many of the methods that we have discussed in this chapter is that the spectra *a* must possess common abscissae, and hence require some kind of interpolation onto a common grid. Also, the choice of energy range can substantially modify the values of d_{ij}; if a very wide energy range is used, this will emphasize the edge-step and diminish the importance of the near-edge details. Moreover, wide ranges also emphasize issues with background removal that may affect individual spectra, and this may introduce differences in the distances that are not due to variations in speciation. Notwithstanding such issues, once the distances d_{ij} are computed, these can then be subjected to analysis to show clusters of related spectra, the results of which can be presented in the form of a dendrogram, shown in Figure 11.13A.

A more sophisticated analysis of the data in Figure 11.13A might take into account the geographical locations from which the samples were derived and a range of geological conditions related to formation and maturity of the samples. In such cases other methods for visualizing the data might be employed, for example, *t*-distributed stochastic neighbor embedding or *t*-SNE (often pronounced "tee-snee"), which is an accepted non-linear method for dimensionality reduction (PCA being a linear dimension reduction method), typically to a two-dimensional plane. Here, pairs of spectra are assigned what is called a probability value, modelled as a Gaussian distribution, which would have higher values for similar spectra, and lower values for spectra that are less similar. The values are then clustered in a lower dimensional map by minimizing the divergence between the distributions with respect to the locations on the map, using a Student's t-distribution[7] (the t in t-SNE). The more pronounced tails in the t-distribution help keep the values from overlapping in the reduced dimensional result. t-SNE contains an adjustable parameter called the **perplexity**, and some optimization parameters (i.e. number of iterations, learning rate and momentum), which must be chosen by the experimenter; there are a number of approximations that are inherent within the t-SNE method, which might artefactually suggest relationships where none exist.

To date, data analytics such as discussed in this section have been very little used in XAS. For example, we know of only one application of t-SNE [11], and while interesting, this study employed only computed spectra, which unfortunately did not resemble previously reported experimental data. Given the simplicity of many of the methods, and the range of readily accessible tools that are now available, we anticipate an increase in applications of these methods in the years to come.

References

[1] Pickering, I. J.; Prince, R. C.; Divers, T. C.; George, G. N. Sulfur K-edge X-ray absorption spectroscopy for determining the chemical speciation of sulfur in biological systems. *FEBS Lett.* **1998**, *441*, 11–14.

[2] Gnida, M.; Sneeden, E. Y.; Whitin, J. C.; Prince, R. C.; Pickering, I. J.; Korbas, M.; George, G. N. Sulfur X-ray absorption spectroscopy of living mammalian cells: An enabling tool for sulfur metabolomics. *In situ* observation of taurine uptake into MDCK cells. *Biochemistry.* **2007**, *46*, 14735–14741.

[3] Shihabi, Z. K.; Goodman, H. O.; Holmes, R. P.; O'Connor, M. L. The taurine content of avian erythrocytes and its role in osmoregulation. *Comp. Biochem. Physiol. A.* **1989**, *92*, 545–549.

[4] George, G. N.; Prince, R. C.; Singh, S. P.; Pickering, I. J. Arsenic K-edge X-ray absorption spectroscopy of arsenic in seafood. *Mol. Nutr. Food Res.* **2009**, *53*, 552–557.

[5] Kamphuis, I. G.; Kalk, K. H.; Swarte, M. B.; Drenth, J. Structure of papain refined at 1.65 Å resolution. *J. Mol. Biol.* **1984**, *179*, 233–256.

[6] George, G. N.; Pickering, I. J.; Pushie, M. J.; Nienaber, K.; Hackett, M. J.; Ascone, I.; Hedman, B.; Hodgson, K. O.; Aitken, J. B.; Levina, A.; Glover, C.; Lay, P. A. X-ray induced photo-chemistry and X-ray absorption spectroscopy of biological samples. *J. Synchrotron Radiat.* **2012**, *19*, 875–886.

[7] Bhattacharyya, R.; Dhar, J.; Ghosh Dastidar, S.; Chakrabarti, P.; Weiss, M. S. The susceptibility of disulfide bonds towards radiation damage may be explained by S⋯O interactions. *IUCrJ.* **2020**, *7*, 825–834.

---

7 Student's t-distribution was reported by William Gosset in 1908. His employer was the well-known brewery Guinness, who required him to use a pseudonym and Gosset chose "Student" because he was a student of Mathematics. The t-distribution with one degree of freedom is equivalent to a Lorentzian or Cauchy distribution.

[8]   Frank, P.; Sarangi, R.; Hedman, B.; Hodgson, K. O. Synchrotron X-radiolysis of L-cysteine at the sulfur K-edge: Sulfurous products, experimental surprises, and dioxygen as an oxidoreductant. *J. Chem. Phys.* **2019**, *150*, 105101/1–18.

[9]   Manceau, A.; Marcus, M. A.; Tamura, N. Quantitative speciation of heavy metals in soils and sediments by synchrotron X-ray techniques. *Rev. Minerol. Geochem.* **2002**, *49*, 341–428.

[10]   Wasserman, S. R.; Allen, P. G.; Shuh, D. K.; Bucher, J. J.; Edlestein, N. M. EXAFS and principal component analysis: A new shell game. *J. Synchrotron Radiat.* **1999**, *6*, 284–286.

[11]   Tetef, S.; Govind, N.; Seidler, G. T. Unsupervised machine learning for unbiased chemical classification in X-ray absorption spectroscopy and X-ray emission spectroscopy. *Phys. Chem. Chem. Phys.* **2021**, *23*, 23586–23601.

12 Experimental artefacts

12.1 Introduction

Many of the methods discussed in this book are prone to experimental artefacts, which are usually associated with choices made by the experimenter. Providing an exhaustive description of all the things that might go wrong is not reasonable, and we admit here that there will probably be some artefacts not discussed in this chapter. Notwithstanding this caveat, we will attempt to cover what we consider the most common artefacts, garnered from extensive experience with our experiments and assisting other users, both in how to conduct and occasionally in how not to conduct the best possible experiment. The purpose of this chapter, therefore, is to review such experimental artefacts to help future experimenters avoid them.

12.2 Transmittance leakage effects

Among the most common artefacts are transmittance pinhole or leakage effects, which have already been noted in the context of sample preparation in Chapter 7 (see Section 7.2.2). There are two main sources of transmittance leakage effects: the preparation of samples that are overall too absorbing or thick, with slight imperfections allowing leakage, and the preparation of samples that have cracks or holes in them, such as pressed disks that have fragmented. For the former, we have discussed the preparation of good transmittance samples in Chapter 7 and noted that many XAS software packages have tools to calculate how much to dilute a transmittance sample, which can help the user avoid leakage artefacts. Figure 12.1 shows a schematic diagram of an irregular sample with beam leakage.

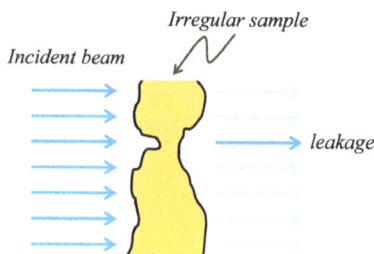

Figure 12.1: Transmittance pinhole or leakage effects. Disproportionately more photons transmit through the thin part of the sample, anomalously increasing the signal.

We define leakage $a(E)$ as a fraction of the transmitted beam that passes through sample pinholes or thin areas. We write $a(E)$ as a function of energy because if there are thin spots, the leakage will depend upon the energy. It is simple to show that this will result in a distorted absorbance $A'(E)$ according to the following equation:

https://doi.org/10.1515/9783110570441-012

$$A'(E) = \log \frac{1 + a(E)}{\exp[-A(E)] + a(E)} \tag{12.1}$$

This predicts that a specific fractional leakage will affect a strongly absorbing sample relatively more than a weakly absorbing sample. This is shown in Figure 12.2a, together with an example in Figure 12.2b, which is the Mn K-edge near-edge spectrum of $MnCl_2$ recorded in transmittance with two samples: one deliberately made too thick, showing clear distortions from the leakage effect, and the other correctly made with no significant distortions. Like many of the artefacts discussed here, leakage artefacts manifest as a squashing of intense features in the spectrum, such as the peak on top of the absorption edge at ~6,551 eV, while enhancing any pre-edge features, such as the peak at ~6,540 eV.

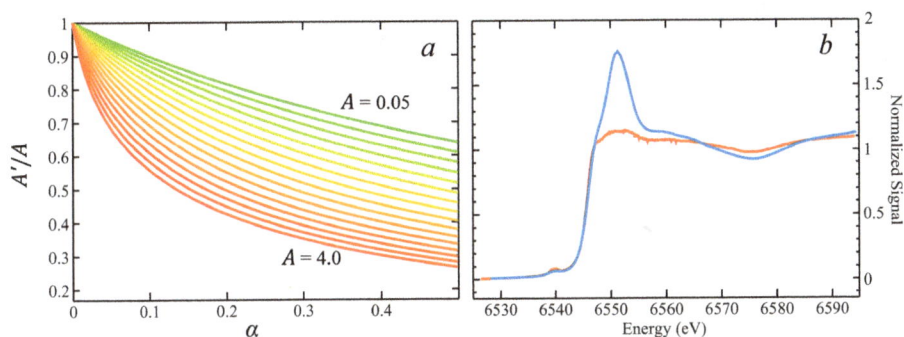

Figure 12.2: Panel (*a*) shows the ratio A'/A versus a for a range of values A, going from 0.05 (green) to 4.0 (red) in 0.025 increments. The leakage distortions are much more pronounced with higher absorption coefficients (red). Panel (*b*) compares two experimental spectra of $MnCl_2$, both of which have been pre-edge subtracted and normalized to the edge jump, one having been deliberately made too thick (red line) and the other correctly made (blue line).

As discussed by George et al. [1], in the case of spheres of elemental α-sulfur, which are ideally arranged in a close-packed single layer to minimize leakage, substantial leakage distortions of the spectra will be present for spherical particle sizes greater than about 1 μm. The results of these calculations are reproduced in Figure 12.3. In practice, depending on the material, powders with particle sizes below about 20 μm are very difficult to produce by simple grinding, even for quite soft materials, no matter how energetically they are ground.

12.3 Fluorescence self-absorption

Fluorescence self-absorption occurs when a sample is either too concentrated, too thick or both. It is a very common artefact in X-ray absorption spectroscopy since the fluorescence signal from such a sample usually will be seductively low in noise yet

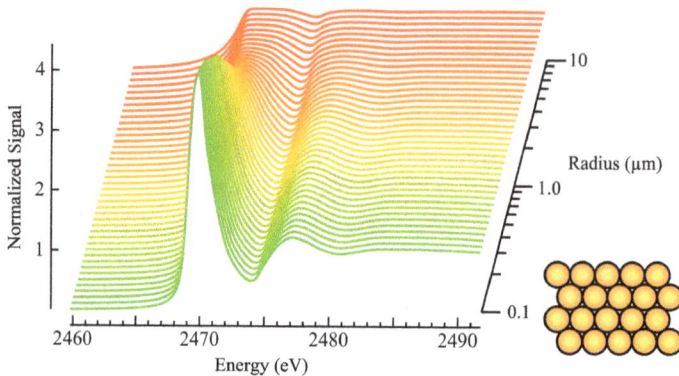

Figure 12.3: Simulated spectral distortions of the sulfur K-edge XAS arising from transmittance leakage for close-packed spheres (see inset, bottom right) of α-sulfur with different radii.

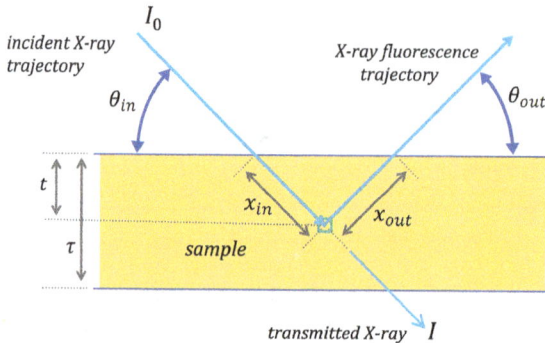

Figure 12.4: Coordinate system for describing fluorescence self-absorption.

will contain distortions affecting the results of both near-edge and EXAFS analyses. Fluorescence self-absorption arises when there is a significant variation through the absorption edge in the level of absorption of the fluorescent photons by the sample. We recall from Chapter 4 that the X-ray absorption $A(E)$ is specified by the following equation (also see eq. (4.14)):

$$A(E) = \mu(E)x = \log\frac{I_0(E)}{I(E)} \tag{12.2}$$

Here E is the incident X-ray energy, $\mu(E)$ is the X-ray absorption coefficient, x is the X-ray path length, with $I_0(E)$ and $I(E)$ being the incident and transmitted X-ray intensities, respectively. Figure 12.4 shows the geometry relevant to a typical X-ray fluorescence measurement with the incident X-ray (x_{in}) and fluorescent X-ray path lengths (x_{out}) to a sample volume element with angles of X-ray incidence and fluorescence exit of θ_{in} and θ_{out}, respectively. We consider a perpendicular depth t of the volume element within the sample, and the fluorescence F will be described as follows:

$$dF(E) \propto \mu'(E) \exp(-\mu(E)x_{in} - \mu_F x_{out})dt \qquad (12.3)$$

or

$$dF(E) \propto \mu'(E) \exp\left(-\left[\frac{\mu(E)}{\sin\theta_{in}} + \frac{\mu_F}{\sin\theta_{out}}\right]t\right)dt \qquad (12.4)$$

In eqs. (12.3) and (12.4) $\mu'(E)$ is the **partial** X-ray absorption coefficient as a function of incident energy E for the specific absorption edge being investigated (e.g. the S K-edge), which will be zero below the absorption edge, $\mu(E)$ is the corresponding **total** sample X-ray absorption coefficient function (of which $\mu'(E)$ is a part) and μ_F is the X-ray absorption coefficient specific to the energy of the fluorescence X-rays; each includes contributions from all elements in the sample. By integrating eq. (12.4) over the total perpendicular thickness τ (Figure 12.4), we can obtain:

$$F(E) = \kappa \frac{\mu'(E)}{\left(\frac{\mu(E)}{\sin\theta_{in}} + \frac{\mu_F}{\sin\theta_{out}}\right)}\left[1 - \exp\left(-\left[\frac{\mu(E)}{\sin\theta_{in}} + \frac{\mu_F}{\sin\theta_{out}}\right]\tau\right)\right] \qquad (12.5)$$

The constant κ can include both the fluorescence yield and the efficiency of the fluorescence detector. When a sample is dilute, $\mu'(E) \ll \mu(E)$ and $F(E) \propto \mu'(E)$, which is the requirement for samples to be suitable for fluorescence detection. However, in practice, fluorescence may be used for samples that are too thick or too concentrated, and indeed for some systems, there may be no alternative method of detection. In these cases, fluorescence self-absorption will occur. A plot of $F(E)$ for a hypothetical thin film of $FeSO_4$ is shown in Figure 12.5 using tabulated X-ray cross-sections and assuming a density of 2.84 and $\theta_{in} = \theta_{out} = 45°$, with incident X-ray energies just above the sulfur and iron K-edges. In the figure, $F(E)$ is normalized by dividing by $F_{max} = \lim_{\tau \to \infty}[F(E)]$, and the abscissa is plotted as a function of absorbance, which is directly proportional to the sample (hypothetical film) thickness τ.

For distortion-free detection F/F_{max} versus absorbance should be a straight line (see broken lines in Figure 12.5). However, as Figure 12.5 shows, as the absorbance increases, fluorescence self-absorption results in an increasingly nonlinear fluorescence signal. Consequently, self-absorption will cause a squashing of any sharp features in the spectrum with higher absorbance values when the XAS is detected by fluorescence.

With soft and tender X-ray XAS, it is common for the sample thickness to be substantially greater than the penetration depth of the incident and fluorescent X-rays. In such cases, the exponential term in eq. (12.5) becomes vanishingly small, so that

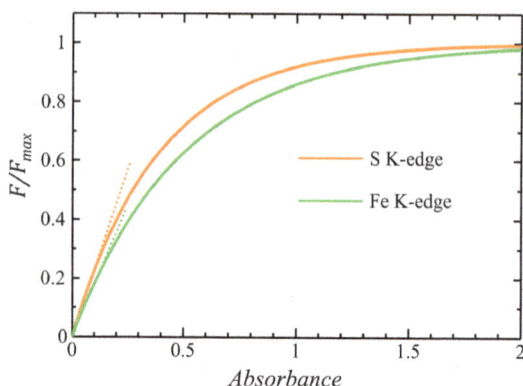

Figure 12.5: Fluorescence self-absorption calculated for a sample of FeSO$_4$ at incident X-ray energies just above the sulfur (orange line) and iron K-edge (green line). Fluorescence is expressed as normalized to the maximum fluorescence signal. The broken lines show how the self-absorption curves depart from near linearity at relatively low values of absorbance.

$$F(E) = \kappa \frac{\mu'(E)}{\left(\frac{\mu(E)}{\sin \theta_{\text{in}}} + \frac{\mu_F}{\sin \theta_{\text{out}}}\right)} \tag{12.6}$$

Frequently, the sample will be planar and inclined at 45° to the incident X-ray beam, with the fluorescence detector positioned at 90° to the incident X-ray beam in the horizontal plane to minimize the scattered radiation. In this case, $\theta_{\text{in}} = \theta_{\text{out}} = 45°$, with eq. (12.6) further simplified to the following equation:

$$F(E) \propto \frac{\mu'(E)}{(\mu(E) + \mu_F)} \tag{12.7}$$

Figure 12.6 shows a typical example of a fluorescence-detected XAS spectrum with distortions arising from self-absorption. The sulfur K-edge XAS of elemental sulfur is compared with an essentially undistorted spectrum of the same sample (measured simultaneously) using electron yield detection.

A number of different fluorescence self-absorption correction algorithms are built into various XAS analysis codes, all working in more or less the same way. Here, we will describe our code, which was first implemented as part of the exafspak analysis suite in 1987. Of the terms in eq. (12.5) (or 12.7 for thick samples), both $\mu(E)$ and $\mu'(E)$ show spectroscopic structure. To reflect this, we can write the total absorption coefficient as a sum of the structured $\mu'(E)$ (sulfur K-edge-only absorption coefficient for the sulfur K-edge) and a smoothly varying background $\mu_b(E)$ from everything else contributing to the absorption coefficient, such as lower absorption edges and all the other elements contained in the sample:

$$\mu(E) = \mu'(E) + \mu_b(E) \tag{12.8}$$

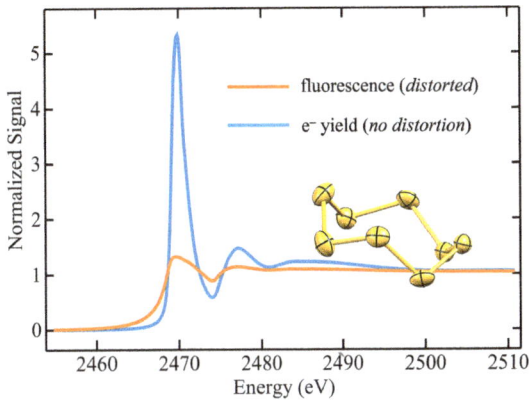

Figure 12.6: Fluorescence self-absorption at the sulfur K-edge. The orange curve shows the X-ray fluorescence detected spectrum of a thick sample of elemental α-sulfur, while the blue curve shows the relatively undistorted total electron yield spectrum (measured simultaneously with the fluorescence). The inset shows the structure of α-sulfur.

If we know the composition of the sample, we can calculate $\mu_b(E)$ from tabulated X-ray absorption coefficients. We can also compute a relatively unstructured version of $\mu'(E)$, which we will call $\mu''(E)$, from the same tabulated absorption coefficients. This is a simple edge-step function with zero below the edge and a smooth fall-off in absorption above it. Equations (12.5) or (12.7) (for thick samples) can then be used to compute $F_u(E)$, an unstructured version of $F(E)$ over the same energy points as the distorted experimental spectrum. Now, excluding the structured energy region of the experimental spectrum, which might be between 2,460 and 2,495 eV for the data in Figure 12.6, the distorted experimental spectrum can be scaled to match the computed structure-free $F_u(E)$, and employing the various components discussed above, we can solve for $\mu'(E)$. With thick samples, this can be done directly using eq. (12.7), but in the more general case of eq. (12.5) nonlinear methods must be used to obtain the solution. Such code can effectively recover undistorted spectra even from severely distorted fluorescence data, such as that shown in Figure 12.6. The use of these algorithms is not without challenges. For example, subtle variations in sample thickness can make a large difference and can be very difficult to account for, and so self-absorption correction should be treated with a generous degree of caution.

Self-absorption can give rise to some quite unusual spectra, which can be surprising to the novice experimenter. One such example, that of ammonium thiomolybdate, is shown in Figure 12.7, again with the fluorescence compared with total electron yield. Here, the sulfur K-edge measured in fluorescence looks like a normal spectrum, albeit a distorted one, while the Mo L_{III} edge appears inverted, and the Mo L_{II} has almost vanished. This strange-seeming experimental data is relatively simple to explain. The sulfur K-edge fluorescence yield is approximately 0.078, more than double the fluorescent yield of the Mo L_{III} and Mo L_{II} edges, 0.037 (Lα12) and 0.034 (Lβ1), re-

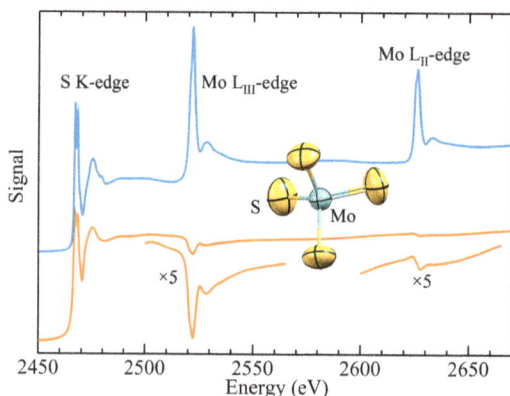

Figure 12.7: Self-absorption distortions of the fluorescence XAS of solid ammonium thiomolybdate $(NH_4)_2MoS_4$ (orange lines), showing inverted Mo L_{III} and L_{II} edges. The crystal structure is shown in the inset (85% thermal ellipsoids). The blue line shows the undistorted total electron yield XAS spectrum. Spectra have been vertically offset for clarity.

spectively [2]. When the energy of the incident X-ray beam is sufficient to excite the Mo $2p_{3/2}$, giving rise to the Mo L_{III} absorption, the total fluorescence yield signal drops because the molybdenum excitation effectively steals photons from the sulfur, as the sample is self-absorbing. At the L_{II} edge, a similar phenomenon occurs, but here because the incident and fluorescent X-ray beams are more penetrating, the net change in total fluorescence is almost zero. If the sample were physically thin, then this distortion would not be observed. We note in passing that eq. (12.5) could, in principle, be used to correct even these data, but a summation over all absorption edges and fluorescence lines contributing would need to be included, together with their respective fluorescence yields. Another curious fluorescence self-absorption artefact can be observed at the L_{III} and L_{II}-edges of rhodium chloride. Here, while both L_{III} and L_{II} are much diminished by self-absorption, one can observe a normal-looking L_{III} edge and an inverted L_{II} edge. This is because the primary fluorescence lines for the Rh L_{III} edge (the $L\alpha 12$) fall below the chlorine K-edge, while those for the L_{II} edge ($L\beta 1$) fall just above it. Hence, in this case, the self-absorption is more pronounced for the L_{II} fluorescence, despite the more penetrating beams at higher energies.

With all these problems related to fluorescence detection and with every case we have discussed using electron yield to give undistorted spectra, the reader might wonder why experimenters would ever bother with fluorescence detection and not just use electron yield. There are two answers to this: of the two methods, fluorescence detection is much more sensitive, facilitating the examination of all kinds of dilute samples, while, as we will see, electron yield detection is certainly not without its own pitfalls. Meanwhile, it is worth noting that fluorescence self-absorption has been exploited to provide additional information – an independent estimate of the size of sulfur storage globules in sulfur bacteria, together with their characteristic sulfur K-edge

XAS [3]. This spectroscopic study effectively resolved a long-standing chemical conundrum related to the chemical nature of the sulfur in the storage globules [1, 3]. Sulfur bacteria are photosynthetic organisms that grow by oxidizing sulfide to sulfate, and in the presence of plentiful sulfide store sulfur as elemental sulfur, which they can convert to sulfate when sulfides are in short supply. When viewed under the optical microscope, these sulfur storage globules can seem to shimmer and appear liquid [4]. In part because of this, there were various proposals of exotic forms such as micelles consisting of multilayers of long-chain polythionates. XAS showed that the sulfur storage globules, in fact, contained a form resembling the most common allotrope, α-sulfur, rather than any exotic forms. The fact that the XAS-derived size of the globules (0.7 μm) matched that observed microscopically meant that the *in vivo* density of the sulfur globules was close to that of α-sulfur ($2.1 \text{ g} \cdot \text{cm}^{-3}$), rather than lower values suggested previously (e.g., $1.3 \text{ g} \cdot \text{cm}^{-3}$). Pickering and co-workers proposed that the globules consist of a core of fragments with local structures resembling α-sulfur, with a modified globule surface conferring hydrophilic properties. This might be due to the proteins that are known to be associated with the globules or modification of surface sulfur by incorporation of polar groups such as thionates [5].

12.4 Electron yield sample charging effects

Chapter 5 detailed several ways in which electron yield can be measured. When data collection is carried out under vacuum, one can either use an electron yield detector, such as a channeltron, or measure the current flow from an electrically isolated sample to ground, often called the 'drain current'. In either case, sample charging may be problematic and will occur when the sample is insufficiently conductive. As charge builds up on the sample, it will progressively inhibit electron loss, resulting in a net decrease in signal amplitude over time, with the signal observed to diminish with repeated sweeps. The situation differs with gas-amplified electron yield detection, and while it is often stated that sample charging is not a problem with such experiments, this is not completely true. While charge build-up over time does not occur in the same way as with the sample in a vacuum, when the sample is insufficiently conductive or the sweeping voltage is too low, sample charging effects may dampen the signal. Changes from sweep to sweep are usually not observed, and in helium, an equilibrium appears to be established quite quickly. Figure 12.8 shows an example of this, in which data acquired using a sample of finely powdered red phosphorus dusted onto Mylar adhesive tape is compared with the data resulting from an essentially identical sample but prepared on conducting carbon tape.

Other charging-related phenomena that might be observed in gas-amplified electron yield include spikes in the data, signal steps and discontinuities, and noise that appears and disappears. If gas-amplified electron yield is misbehaving, a quick means of diagnosis can be to arbitrarily cut the incident intensity by partly closing an up-

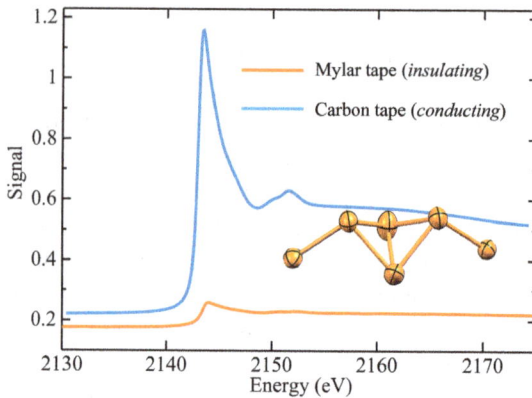

Figure 12.8: Electron yield charging effects in an XAS measurement of the phosphorus K-edge of red phosphorus. The blue line shows the raw data from a sample made by dusting finely powdered material on to conducting carbon tape, and the orange line shows a nearly identical sample made using insulating Mylar tape. Spectra are plotted on the same scale. The inset shows a partial structure of red phosphorus (the central P_4 unit repeats).

stream slit. If such measures mitigate the problems, it is very likely that sample charging effects are occurring, with a possible easy remedy being to increase the sweeping voltage for the electron yield. However, we note that such an increase should be limited by safety considerations.

12.5 Solid-state detector dead time effects and related phenomena

The challenges of saturating count rates when using solid-state detectors for fluorescence XAS have already been discussed in Chapter 5. In general, the type of dead time most frequently encountered is known as paralysing dead time, where the detector readout drops at very high count rates. As also discussed in Chapter 5, we note that in almost all cases, these problems could be avoided by using a more appropriate choice of detector system. An extreme example of a dead-time-saturated detector system is shown in Figure 12.9. This data was given to us by an initially perplexed SSRL user on condition of anonymity, after requesting help because their data did not appear as expected. Here, the detector was so badly saturated that at the peak of the signal, the spectrum is inverted as a consequence of the detector spending nearly all of its time in a dead-time condition.

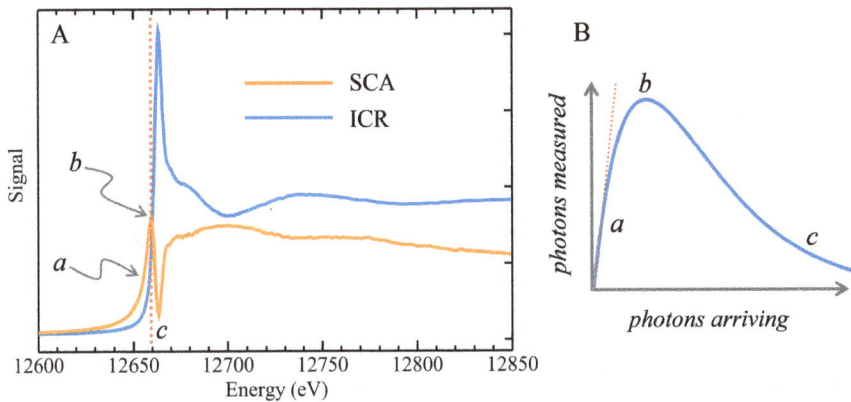

Figure 12.9: Extreme effects of paralysing detector deadtimes. Panel (A) compares the total incoming count rate (ICR, blue) and windowed count rate (SCA, orange) of a selenium-containing sample. Panel (B) shows a schematic diagram of detector dead time, with relevant points indicated. The red broken line in (B) shows the linear response for low count rates, which are applicable to region *a* in both panels. At point *b* in panel (A) (also indicated by the broken red line in A), the detector output turns over, meaning that additional photons arriving result in fewer measured photons. Thus, the SCA data in (A) can be seen to be inverted at the portions corresponding to the maximum rate of photon arrival at the detector (point *c*), resulting in a negative rather than a positive peak.

12.5.1 Ice diffraction and detector dead time

We debated whether ice diffraction effects should be given their own section, but on balance, we decided that they belonged here because the root of the problem relates to detector non-linearities. Notwithstanding this condition, we note that nearly all the artefacts we discuss reflect some interplay between the choice of sample preparation, choice of detectors and the like. For aqueous solutions of biological or environmental samples, low temperatures are frequently used to protect the sample from radiation damage or to enhance the EXAFS by freezing out vibrational modes. In such cases, crystalline ice in the sample can cause challenges. As the X-ray energy is scanned during the XAS experiment, the angles of the ice diffraction change so that diffraction spots can translate across the face of the solid-state detector array, illuminating different array elements. When such illumination occurs, the count rates can be substantially elevated, causing the detector count rates to exceed the pseudo-linear regime, leading to a decrease in observed windowed counts. Figure 12.10 shows an example in which a dilute copper-containing metalloprotein sample was found to contain significant crystalline ice. Figure 12.10 also shows the effects of thawing the sample and adding a glassing agent, which served to eliminate crystalline ice.

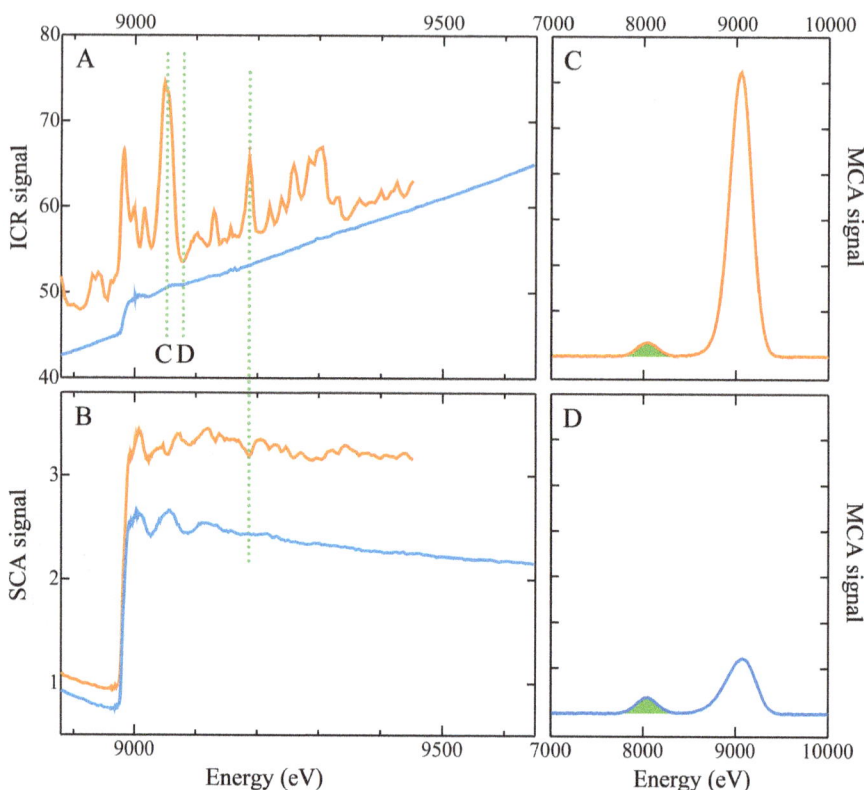

Figure 12.10: Ice diffraction from a frozen solution of a dilute copper-containing metalloprotein sample. Panels (A) and (B) compare the initial sample (orange lines) showing the total incoming counts (ICR) and Cu Kα windowed counts (SCA). Panels (C) and (D) show the multi-channel analyser (MCA) spectra, the response of the detector element as a function of emitted energy, respectively, on and off a diffraction feature in the ICR plot (panel A, indicated by green broken lines labelled C and D). The blue lines in panels (A) and (B) show the spectra after the addition of 25% v/v glycerol as a glassing agent. The third green broken line projecting into panel (B) shows a dip in windowed fluorescence due to detector nonlinearities when ice diffraction is being registered. The green shaded areas in panels (C) and (D) show the windowed counts, corresponding to the Cu Kα.

Two different methods can be used to help eliminate crystalline ice from frozen aqueous samples. The first is to add a glassing agent such as glycerol or polyethylene glycol (PEG), although caution is warranted to ensure that this does not increase photoreduction or interact structurally with the element of interest. The second is to freeze the sample rapidly, which gives well-defined ice crystals less opportunity to properly form, as discussed in Chapter 7 (Section 7.4.1).

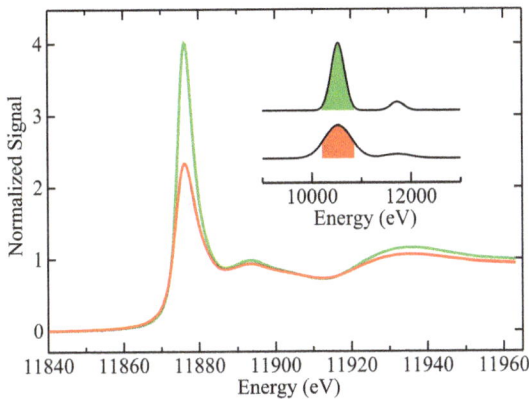

Figure 12.11: Artefacts from automatic dynamic peaking time adjustment. The green curve shows an undistorted spectrum for a sample of As(V)-*tris*-glycerolate [6], and the red curve shows the distortions caused by dynamic peaking time adjustment at high count rates. The inset shows the MCA spectra with the electronic windows on the As Kα fluorescence line shown as solid; the upper curve (filled green) shows low count rates and long peaking time, and the lower curve (filled red) shows high count rate and short peaking time, illustrating the 'missing' wings of the peak beyond the red fill.

12.5.2 Variations in peaking time of digital signal processing hardware

Some modern digital signal processing hardware employs dynamic setting of the peaking time (see Chapter 5, Section 5.3.5). This is a deliberate design feature: for low count-rate operations, long peaking times will give superior energy resolution, whereas for high count-rates, shorter peaking times mean that the detector is not saturated, with the caveat that the energy resolution of the output becomes worse. The challenge here occurs when the counts under a fluorescence line are summed between pre-defined limits (the 'window' and the most common method of measuring the fluorescence) and when the count rates change substantially across the spectrum. In this case, the shape of the measured fluorescence line changes substantially, becoming much broader in the high count-rate regions of the spectrum, with the consequence that the wings of the line may be absent from the summed (windowed) fluorescence signal. Figure 12.11 shows an example of artefacts arising from dynamically adjusting peaking times. This effect will always be present with hardware that uses dynamic setting of peaking times, but only becomes a problem when the sample is concentrated and the bulk of the total signal comes from the fluorescence. Like almost all the artefacts that we have discussed thus far, peaking time variations cause a damping of intense features of the spectrum, with distorted EXAFS amplitudes. This could easily be avoided by choosing the correct detector for the experiment or by setting the electronic window sufficiently wide so that the deterioration in solid-state detector resolution due to dynamic adjustment of peaking time will be minimal. Moreover, if dynamic peaking time artefacts are not only present but substantial, it is likely that other artefacts, such as self-absorption, may be present too.

12.6 Energy calibration standard fluorescence leakage

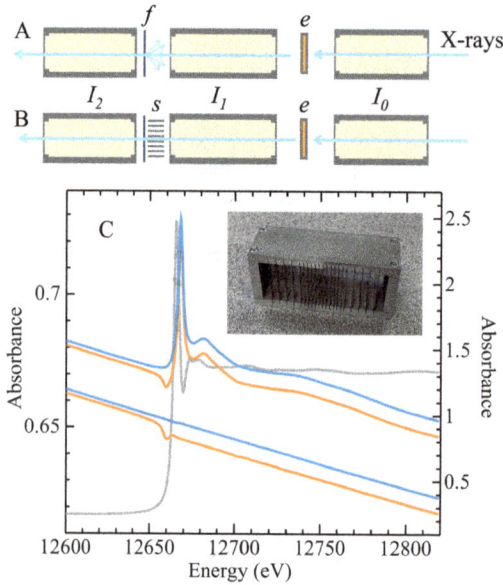

Figure 12.12: Fluorescence leakage from calibration standard. (A) A schematic of a three-ion chamber transmittance experiment of sample (e) with calibration standard fluorescence entering the I_1 ion chamber. (B) The use of Soller slits (s) before the calibration standard or foil (f). (C) Measure XAS $\log(I_0/I_1)$ for frozen aqueous selenate (upper curves, 2 mM, pH 7, 10 K) with (blue line) and without (orange line) Soller slits (shown in the inset). The lower pair of lines display spectra from a blank sample containing no selenium, showing a small, inverted, self-absorption distorted signal from the calibration foil without Soller slits (orange line) that is absent when Soller slits are used (blue line). The XAS of the calibration standard $\log(I_1/I_2)$, a sample of grey hexagonal elemental selenium, is shown by the grey line (right absorbance scale). The inset in (C) shows a photograph of the Soller slits described by Tse et al. [7].

There are some categories of samples that are too concentrated for fluorescence measurements but are close to being too dilute for transmittance. As discussed in Chapter 6, in a standard hard X-ray experiment, one might employ three gas ionization chambers: the first to measure the incident X-ray beam intensity (I_0), the second positioned after the sample to measure the transmitted intensity (I_1) and the third positioned after a calibration standard (I_2). The absorbance of the sample would then be given by $\log(I_0/I_1)$ and that of the calibration standard by $\log(I_1/I_2)$ (shown in Figure 12.12).

The calibration standard, when illuminated by the incident X-ray beam at energies higher than its absorption edge, will produce X-ray fluorescence distributed over 4π steradians (Chapter 4). Since the standard is relatively concentrated, often an elemental foil, this calibration fluorescence will be strong. Unless precautions are taken, some of this fluorescence will enter the I_1 ion chamber and will contribute significantly to the measured I_1, producing a larger signal in I_1 and distorting (decreasing) the transmittance data of the sample (e in Figure 12.12). Very often, the calibration standard, being a metal foil or an elemental preparation of some sort, might have a lower energy of XAS onset than the sample. In such cases, the artefact will manifest as a dip just below the sample absorption, as shown in Figure 12.12. The use of collimating Soller slits [7] can essentially eliminate such calibration standard back-fluorescence artefacts.

12.7 Radiation damage

Methods that employ ionizing radiation, such as X-rays, run the risk of altering the sample; XAS is not immune to this [8]. Energetic photons have the potential to break bonds, causing the production of free radicals, some of which will be mobile. We will begin by discussing some of the radiation chemistry of water and then consider other sample types.

The passage of ionizing radiation, such as an X-ray photon, through water creates what is known as a **sparse track** of ionization and excitation events [9]. This initiates a complex series of chemical reactions, many of which begin with the water radical cation $[H_2O]^{\bullet+}$:

$$H_2O + (\text{ionizing radiation}) \rightarrow [H_2O]^{\bullet+} + e_{aq}^- \qquad (12.9)$$

In which e_{aq}^- is a hydrated electron. For a single H_2O molecule, the excited electron can originate from any of the five occupied molecular orbitals, each giving rise to a different excited state. The decay of these electrons can produce photons or secondary electrons, which in turn can cause further excitations, creating **ionization spurs** emanating from the original sparse track. The water radical cation can further react with water to produce a hydroxyl radical and a hydroxonium cation, which are transiently associated via an oxygen $\cdots$oxygen intermolecular bond [10]:

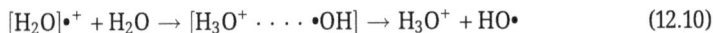

$$[H_2O]^{\bullet+} + H_2O \rightarrow [H_3O^+ \cdots \cdots \bullet OH] \rightarrow H_3O^+ + HO\bullet \qquad (12.10)$$

The primary products of water are thus hydrated electrons and hydroxyl radicals, which are respectively powerful reducing and oxidizing agents. In the absence of other chemical reactions, the majority of these products will react to regenerate water and heat, as follows:

$$HO\bullet + H_3O^+ + e_{aq}^- \rightarrow 2H_2O + \Delta \qquad (12.11)$$

A number of additional reactions are known. Two HO• can react to form H_2O_2 (eq. (12.12)), which can then react with hydrated electrons to generate a hydroxyl radical and a hydroxide anion (eq. (12.13)). In addition, two hydrated electrons can combine with water to produce two hydroxide anions and molecular hydrogen (eq. (12.14)), which explains the bubbles that can be observed when irradiating water at room temperature with X-rays from a modern XAS beamline:

$$HO• + HO• \rightarrow H_2O_2 \tag{12.12}$$

$$e_{aq}^- + H_2O_2 \rightarrow HO• + OH^- \tag{12.13}$$

$$e_{aq}^- + e_{aq}^- + 2H_2O \rightarrow H_2 + 2OH^- \tag{12.14}$$

The effects of ionizing radiation can cause either photo-oxidation or photo-reduction. For example, irradiation of an aqueous solution of As(III) arsenite $[As(OH)_3]$ at physiological pH causes rapid photo-oxidation to As(V) arsenate, while a Se(VI) selenate $[SeO_4]^{2-}$ solution at a similar pH will show ready photo-reduction to Se(IV) selenite. In Chapter 11, we used the time course for the photo-oxidation of sulfur in papain as an example to illustrate some of the strengths and weaknesses of principal component analysis. Of the two, photo-reduction is by far the most common; Figure 12.13 shows an example: a frozen solution of the chelate complex between Cu(II) ethylenediaminetetraacetic acid (EDTA).

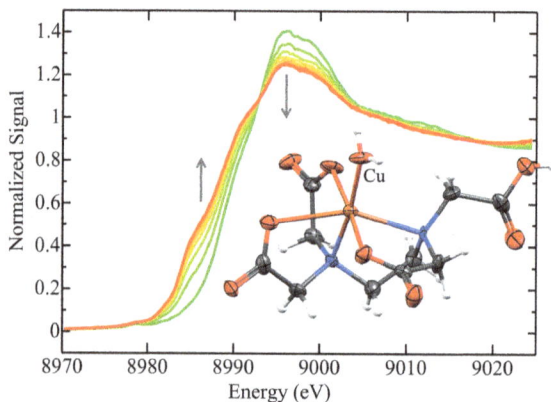

Figure 12.13: Time course of photo-reduction of an aqueous solution sample of Cu(II) EDTA. The reaction occurred in the presence of 25% v/v glycerol, using rapid sequential sweeps of the Cu K near-edge region. The green trace shows the initial state, which is essentially all Cu(II). Subsequent sweeps show progressive changes with photo-reduction towards the Cu(I) form, indicated by the red trace. The inset shows the crystal structure of the solid. The arrows indicate changes with exposure.

In this case, the presence of glycerol, normally added as a glassing agent, considerably increases the rate of photo-reduction [11]. This is thought to work by dint of the fact that glycerol, and other commonly added glassing agents such as other alcohols or ethers (e.g., polyethylene glycols), are highly effective scavengers of hydroxyl radicals. They achieve this through the abstraction of a proton, forming water and creating a relatively stable carbon-centred radical. Hydroxyl radical scavenging may consume HO• that would normally recombine with e_{aq}^- forming form water and heat (eq. (12.11)), leaving an excess of e_{aq}^- potentiating the reduction. Figure 12.14 shows this potentiation of photo-reduction for a frozen solution of $CuCl_2$ in water [11], where the presence of 25% v/v glycerol causes a more than 10-fold increase in the rate of photo-reduction.

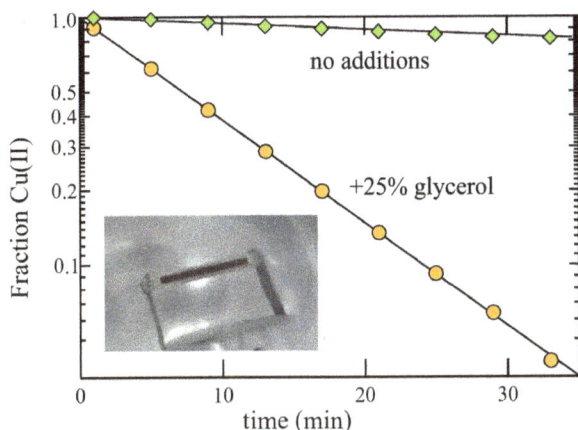

Figure 12.14: Exacerbation of the rate of photo-reduction at 10 K of aqueous Cu(II) by the addition of 25% v/v glycerol. Green: no addition; yellow: with 25% v/v glycerol. The inset shows the X-ray beam mark on an aqueous sample following irradiation – the purple colour is due to the presence of hydrated electrons trapped in the ice matrix. (Data replotted from Neinaber et al. [11].).

In many cases, the products of photo-reduction can have different structures than those of chemical reduction. The first observation of this phenomenon that we know of is with Cu K-edge XAS studies of copper bound to a fragment of the prion protein [12]. The prion protein is perhaps best known for its role in neurodegenerative diseases such as Creutzfeldt–Jakob disease (CJD) and bovine spongiform encephalopathy, which is also known as 'Mad Cow disease'. These diseases are associated with an accumulation of a misfolded form of the protein, which can be infectious, with ingestion of the misfolded protein potentially transmitting the disease from one host to another. The native protein plays many roles, one of which is related to binding extracellular copper [13] by repeating residues that form part of its N-terminus. The oxidized Cu(II)-containing peptide has four close ligands to copper in an approximately square planar geometry. The chemically reduced protein has two ligands to the Cu(I) with dig-onal coordination geometry, whereas the X-ray photo-reduced peptide, formed by ex-posure at a temperature of 10 K, has a three-coordinate structure [12]. Presumably, this difference is due to the low temperatures limiting the motion of the coordinating li-gands to the copper. A similar situation has been observed with metmyoglobin, where low-flux density beamlines allow measurement of the fully oxidized (formally ferric) protein, but measurements on high-flux density beamlines resulted in quantitative photo-reduction to a form distinct from the conventional chemically reduced de-oxy myoglobin, most likely six coordinate with the sixth ligand locked in place by the low temperatures used during data acquisition (again 10 K) [8].

The use of non-aqueous solvents can engender some quite dramatic radiation chemistry. An example is Zn(II)-*bis*-diethyldithiocarbamate, which as a solid is rela-tively stable to X-ray irradiation but, as a solution in acetone, is rapidly degraded through photo-oxidation of the ligand, as shown in Figure 12.15.

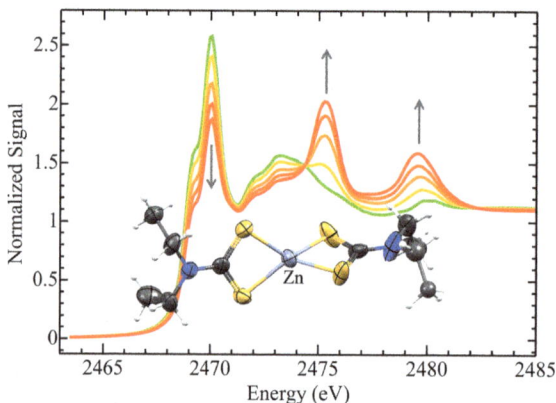

Figure 12.15: X-ray beam-induced photo-oxidation of a solution of Zn(II)-*bis*-diethyldithiocarbamate in acetone with degradation of the diethyldithiocarbamate ligand. The green trace shows the initial sweep, and the red the final sweep; the arrows indicate the direction of changes with exposure. The inset shows the crystal structure of the solid.

Measures that can be taken to mitigate the extent of photo-damage to samples have been reviewed previously [8]. These include the use of photon shutters to restrict the sample exposure to times when data is being actively acquired and the use of rapid scanning with translation of the sample so that the beam interrogates a fresh spot with each scan. Other methods include lowering the flux density by omitting specular optics (X-ray mirrors) from the beamline. One example of this strategy is SSRL's beamline 7-3, which has only an upstream vertically collimating mirror[1] and no horizontal focusing. This gives a relatively large beam, which allows measurements of samples that are sensitive to photo-damage. Other measures that may be used to prevent photo-damage include the removal of solvent (e.g., water) from the samples and the use of free-radical scavengers such as nitrate and sulfate, which can scavenge hydrated electrons, at least at room temperature, to form nitrite and sulfite, respectively.

Finally, for this section, we note that X-ray-induced radiation chemistry can be used to deliberately generate some chemically interesting species. An example of this is aqueous Hg(0), as a model of solvated elemental mercury [14], which plays vital roles in the mercury biogeochemical cycle. This work shows that photochemically generated solvated Hg(0) has no short-range coordination, confirming notions about how this species might be present in solution. Other such novel radiation chemistry applications include the serendipitous preparation of chemically important but reactive and hence elusive species such as the thiyl (R–S•) [15] and selenyl (R–Se•) [16] free radicals. Figure 12.16 shows the generation of thiyl by X-ray irradiation of a stable sulfenic acid in which the sulfur functionality is protected by a bowl-shaped cyclophane.

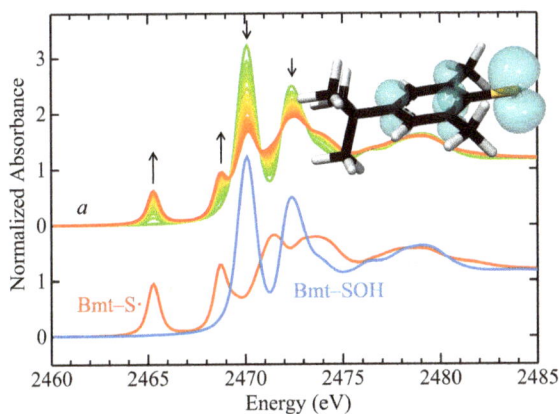

Figure 12.16: X-ray beam-induced changes in the sulfur K-edge X-ray absorption spectrum of a stable sulfenic acid. The sulfur functionality is protected by a bowl-shaped cyclophane framework, 1,3-bis[[2,6-bis (2,6-dimethylphenyl)phenyl]methyl]-5-tert-butyl-benzene, which is abbreviated Bmt. The series of spectra *a* show a time-series of sulfur K-edge X-ray absorption spectra in which each spectrum is separated by 21 min of X-ray irradiation, with the green spectrum showing the starting material and the red spectrum showing the result following 7 h of irradiation. The arrows indicate the direction of spectral change over time. The lower curves show the spectra for the starting material (Bmt–SOH; blue) and the thiyl radical (Bmt–S•; red), with the inset showing the calculated spin density (0.04 spins per a.u.^{3}) for the thiyl radical, using 2,6-dibenzyl-4-tert-butylbenzene in place of the Bmt cyclophane to speed up computations.

---

1 At the time of writing the upstream collimating mirror on SSRL 7-3 has experienced photon-induced damage giving significant slope errors so that it is typically not used.

Moving away from the deliberate exploitation of experimental artefacts, and as a final segment in this chapter, we will catalogue some common artefacts that relate to mistakes or issues with data acquisition.

12.8 Problems with dark currents (offsets)

Many combinations of detectors and electronics will show a signal even when the beam is off. Gas ionization chambers using current-to-voltage amplifiers are a frequently encountered case of this. When using proper experimental practice, dark currents (or offsets) should be measured with the X-ray beam off, and these readings should then be subtracted from all subsequent experimental readings with the beam on. The neglect of

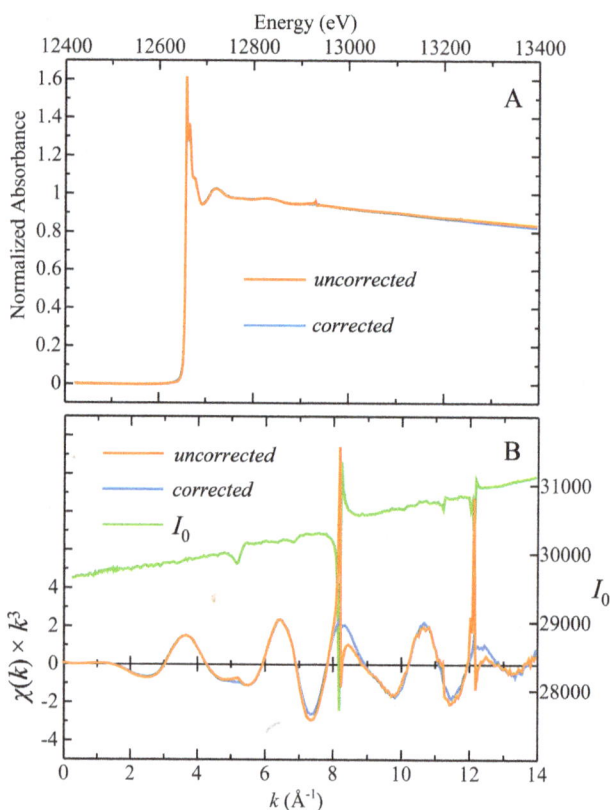

Figure 12.17: Comparison of transmittance data $\log(I_0/I_1)$ with and without dark current subtraction for I_0. (A) Comparison of the pre-edge subtracted data, showing negligible difference. (B) Comparison of the k^3 weighted EXAFS data, showing that monochromator crystal glitches ratio adequately when dark currents are subtracted (corrected, blue) but not without dark current subtraction (uncorrected, orange). The sample is 100 mM selenophene (C_4H_4Se) dissolved in toluene, measured at 10 K.

this simple procedure is surprisingly commonplace, perhaps because the distortions of the data appear superficially quite subtle. However, when dark current subtraction is omitted, distortion of the measured spectrum can result. Thus, proper, quantitative XAS must use dark current subtraction, especially if the EXAFS oscillations are to be analysed. Figure 12.17 shows an example, comparing transmittance data with and without subtraction of the dark current reading from I_0. While the normalized pre-edge subtracted data show effects that seem subtle, the EXAFS oscillations with k^3 weighting clearly show the advantages of dark current subtraction.

12.9 Inadequate ion chamber sweeping voltage

While many beamlines now use photodiodes in place of gas ionization chambers, the latter are still commonplace, meriting a brief mention of an important potential artefact. In Chapter 5, we discussed at some length the requirement for a correct sweeping voltage in gas ionization chambers. If too low a sweeping voltage is used, X-ray fluorescence data may appear noisy with imperfect ratioing with I_0, frequently with a series of scans failing to overlay properly. Similarly, transmittance measurements may not overlay correctly. That the sweeping voltage is in the plateau region can be easily checked by simply increasing the voltage and observing whether the ion chamber readouts increase. If they do, then increasing the sweeping voltage might be useful, always assuming that the voltage remains insufficient to cause arcing between the ion chamber plates.

12.10 Preferred orientation of powders

Preferred orientation of powder samples is a well-known problem in powder X-ray diffraction. The crystal habit manifests in the powdered sample even in very finely ground powders, and if the crystallites are plates or needles, then they tend to physically pack in the powder with some preferred orientation. Preparation of powdered solid samples for XAS was discussed in Chapter 7; a very commonly used diluent is boron nitride (BN), which itself has a plate-like morphology. When mixed with a compound that tends to form plate-like crystallites, it can promote preferred orientation – partial alignment of sample crystallites. The fact that the X-ray beam from synchrotron sources is polarized means that polarization effects can be present in the XAS from such cases. One can test for this by rotating the sample relative to the X-ray **e**-vector and looking for changes in the XAS with rotation. A solution is to use a different diluent, such as lithium carbonate or sucrose, remembering our caution from Section 7.2 to avoid substances that can react chemically with the sample.

We reiterate that there may be other artefacts that are not considered here. During our years of experience, we have come across many, some of which may be outside the experimenter's control, including, for example, issues with data acquisition software. In closing, we note that despite anyone's best efforts, there is probably no experimental measurement that is perfect in all respects. There is a risk to productivity in striving for perfection – one must make reasonable progress towards one's goals. The trick is to do so without making serious mistakes. We hope though that describing such pitfalls and artefacts might help others diagnose them or, better yet, avoid them altogether.

References

[1] George, G. N.; Gnida, M.; Bazylinski, D. A.; Prince, R. C.; Pickering, I. J. X-ray absorption spectroscopy as a probe of microbial sulfur biochemistry: The nature of bacterial sulfur globules revisited. *J. Bacteriol.* **2008**, *190*, 6379–6383.

[2] Krause, M. O. Atomic radiative and radiationless yields for K and L shells. *J. Phys. Chem. Ref. Data.* **1979**, *8*, 307–327.

[3] Pickering, I. J.; George, G. N.; Yu, E. Y.; Brune, D. C.; Tuschak, C.; Overmann, J.; Beatty, J. T.; Prince, R. C. Analysis of sulfur biochemistry of sulfur bacteria using X-ray absorption spectroscopy. *Biochemistry.* **2001**, *40*, 8138–8145.

[4] Winogradsky, S. Über schwefelbakterien. *Botanische Zeitung.* **1887**, *31*, 490–507.

[5] Steudel, R. On the nature of the "elemental sulfur" (S⁰) produced by sulfur-oxidizing bacteria—a model for S⁰ globules. In: *Autotrophic Bacteria* (Eds. Schlegel, H. G.; Bothwien, B.) Springer-Verlag: Berlin, 1989, pp. 289–303.

[6] Andrahennadi, R.; Fu, J.; Pushie, M. J.; Wiramanaden, C. I. E.; George, G. N.; Pickering, I. J. Insect excretes unusual six-coordinate pentavalent arsenic species. *Environ. Chem.* **2009**, *6*, 298–304.

[7] Tse, J. J.; George, G. N.; Pickering, I. J. Use of Soller slits to remove reference foil fluorescence from transmission data. *J. Synchrotron Radiat.* **2011**, *18*, 527–529.

[8] George, G. N.; Pickering, I. J.; Pushie, M. J.; Nienaber, K.; Hackett, M. J.; Ascone, I.; Hedman, B.; Hodgson, K. O.; Aitken, J. B.; Levina, A.; Glover, C.; Lay, P. A. X-ray-induced photo-chemistry and X-ray absorption spectroscopy of biological samples. *J. Synchrotron Radiat.* **2012**, *19*, 875–886.

[9] Garrett, B. C.; Dixon, D. A., *et al.* Role of Water in electron-initiated processes and radical chemistry: Issues and scientific advances. *Chem. Rev.* **2005**, *105*, 355–390.

[10] Lin, M.-F.; Singh, N.; Liang, S.; Mo, M.; Nunes, J. P. F.; Ledbetter, K.; Yang, J.; Kozina, M.; Weathersby, S.; Shen, X.; Cordones, A. A.; Wolf, T. J. A.; Pemmaraju, C. D.; Ihme, M.; Wang, X. J. Imaging the short-lived hydroxyl-hydronium pair in ionized liquid water. *Science.* **2021**, *374*, 92–95.

[11] Nienaber, K. H.; Pushie, M. J.; Cotelesage, J. J. H.; Pickering, I. J.; George, G. N. Cryoprotectants severely exacerbate X-ray-induced photoreduction. *J. Phys. Chem. Lett.* **2018**, *9*, 540–544.

[12] McDonald, A.; Pushie, M. J.; Millhauser, G. L.; George, G. N. New insights into metal interactions with the prion protein: EXAFS analysis and structure calculations of copper binding to a single octarepeat from the prion protein. *J. Phys. Chem. B.* **2013**, *117*, 13822–13841.

[13] Pushie, M. J.; Pickering, I. J.; Martin, G. R.; Tsutsui, S.; Jirik, F. R.; George, G. N. Prion protein Expression level alters regional copper, iron and zinc content in the mouse brain. *Metallomics.* **2011**, *3*, 206–214.

[14] Nienaber, K. H.; Nehzati, S.; Cotelesage, J. J. H.; Pickering, I. J.; George, G. N. X-ray induced photoreduction of Hg(II) in aqueous frozen solution yields nearly monatomic Hg(0). *Inorg. Chem.* **2018**, *57*, 8205–8210.

[15]   Sneeden, E. Y.; Hackett, M. J.; Cotelesage, J. J. H.; Prince, R. C.; Barney, M.; Goto, K.; Block, E.; Pickering, I. J.; George, G. N. Photochemically-generated thiyl free radicals observed by X-ray absorption spectroscopy. *J. Am. Chem. Soc.* **2017**, *139*, 11519–11526.

[16]   Nehzati, S.; Dolgova, N. V.; Sokaras, D.; Kroll, T.; Cotelesage, J. J. H.; Pickering, I. J.; George, G. N. A photochemically-generated selenyl free radical observed by high-energy resolution fluorescence detected X-ray absorption spectroscopy. *Inorg. Chem.* **2018**, *57*, 10867–10872.

13 XAS imaging

13.1 Introduction

The availability of micro- and nano-focused X-ray beams has enabled a range of different applications. There are several excellent texts and reviews of these methods [1, 2], and an outstanding book entitled *X-ray Microscopy* provides more detail than here [3]. Because these texts cover many aspects of the field quite adequately, and because this book is focused on X-ray absorption spectroscopy, our primary goal in this chapter will be to examine the insights that XAS can give when micro-beam methods are used. Before discussing applications of XAS, we will first review the various methods for the production of X-ray micro-beams, but because this is well covered elsewhere, we will keep our treatment brief.

13.2 Micro-focus optics

There are essentially four different ways in which small X-ray beams can be made. The first, a trivial solution that does not involve any focusing at all, is to use an aperture to exclude all but a small part of the beam. This is useful only when a relatively large beam is required and has the advantages of being inexpensive and quick to implement. Early applications of this type used a double set of aligned precision mechanical slits [4], calibrated with a laser using the spacing of exterior Fraunhofer diffraction bands. For most modern applications, an aperture usually consists of a laser-drilled pinhole 20–100 μm in size, typically made from a high atomic number material such as tungsten or tantalum. Actual focusing optics can be divided into three types based on X-ray refraction, reflection and diffraction.

13.2.1 Refractive optics

Refractive optics are based on the same physical principles as ordinary refractive lenses used for visible light and found in cameras, telescopes, etc. With visible light, convex lenses will focus light rays, causing a parallel beam to converge, while concave lenses will cause light rays to diverge. With X-rays, however, as we discussed in Chapter 3 (Section 3.4), because the refractive index for X-rays is less than 1, the situation is reversed, and it is concave lenses that can be used to focus a beam of X-rays. For a material of refractive index η in a medium of refractive index η_0, the focal length f is given by what is called the "lens-maker's equation":

https://doi.org/10.1515/9783110570441-013

$$\frac{1}{f} = \frac{\eta - \eta_0}{\eta_0}\left(\frac{1}{r_1} - \frac{1}{r_2} + \frac{\eta - \eta_0}{\eta_0}\frac{d}{\eta r_1 r_2}\right) \qquad (13.1)$$

In which r_1 and r_2 are the radii of curvature of the two faces of the lens, and d is the thickness of the lens. If we assume that our lens is in a vacuum, so that $\eta_0 = 1$, that d is very small compared to r_1 and r_2, and remembering that for X-ray energies far from any absorption edges $\eta = 1 - \delta$, then we can write as follows:

$$\frac{1}{f} = -\delta\left(\frac{1}{r_1} - \frac{1}{r_2}\right) \qquad (13.2)$$

For a biconcave lens with equal radii of curvature on the two surfaces ($r_1 = -r_2 = r$), eq. (13.2) gives the following equation:

$$f = \frac{r}{2\delta} \qquad (13.3)$$

As we have previously discussed, δ will be very small (ca. 10^{-6} for hard X-rays), and thus the focal length f of a biconcave lens of practical dimensions will be very large. Thus, a single lens will not be realistically useful as a focusing optic. However, if one stacks n such lenses to form a compound refractive lens (Figure 13.1), then the total focal length is given as follows:

$$f = \frac{r}{2\delta}\frac{1}{n} \qquad (13.4)$$

Typically, such devices are made of low-Z material, such as beryllium to absorb as little of the incident X-ray beam as possible and are considered useful between 5 and 40 keV. To help minimize X-ray absorption, other fabrication methods, including saw-tooth and kinoform lenses, have been considered [5]. Remembering from Chapter 3 (eq. (3.8)) that $\delta \propto \lambda^2 \propto 1/E^2$, we note that the focal length f will increase as the square of the X-ray energy.

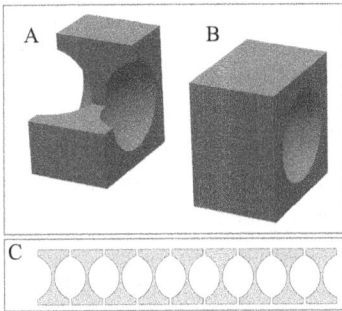

Figure 13.1: Schematic diagram of compound refractive lens X-ray focusing optics: (A) a cut-away view of a single lens, (B) a view of an intact lens and (C) a cross-section of a compound lens consisting of nine individual lenses.

Our major interest in this book is spectroscopy; hence, the ease of scanning the incident X-ray energy is an important consideration. For a 350 eV scan over the selenium K-edge, the focal length of a compound refractive lens will increase by ~5%, while near the iron K-edge, it will increase by ~10%. While we know of no X-ray spectroscopic studies to date using compound refractive lenses, the dependency of focal length on energy might complicate such experiments, particularly at lower X-ray energies.

13.2.2 Reflective optics – X-ray capillaries and mirrors

If apertures are the simplest and easiest way to obtain a small X-ray beam, then capillaries are probably the next easiest to use. Capillary optics are often made of glass, but other materials including metals have been explored. All X-ray capillary micro-focus optics work using X-ray incidence angles with the interior surface of the capillary that are less than the critical angle $\theta_c = \sqrt{2\delta}$ (see Chapter 3, Section 3.4). The number of bounces on the inner surface of the capillary further divides them into single-bounce and multiple-bounce variants. Capillaries can be divided into mono-capillaries and poly-capillaries, with mono-capillaries also falling into three different types: tapered, ellipsoidal and parabolic. Tapered capillaries are typically multiple-bounce devices and are easier to manufacture but have a lower efficiency than single-bounce devices because there is some loss of beam with each bounce. Some commercial glass capillaries have a thin interior coating, which limits their use to hard X-ray applications because the coating absorbs too much of the incident beam in the soft and tender X-ray regimes. Single-bounce capillaries can have an interior shaped into a specific optical figure, such as a parabolic or an ellipsoidal shape. For example, a parabolic figure would focus a parallel beam of X-rays to a point. Figure 13.2 shows schematic diagrams of both a parabolic capillary and a tapered capillary, together with characterization of a tapered metal capillary [6–8]. Poly-capillaries are tapered capillary bundles, typically made of glass, which have a relatively large incident X-ray aperture and a focus as small as 10–20 μm.

Possibly the most extensively used micro-focus optics are Kirkpatrick–Baez mirrors, frequently abbreviated as K-B mirrors or K-B pairs. These are pairs of mirrors fabricated with an elliptical figure, focusing in the vertical and horizontal directions, with vertical and horizontal working directions in front of and behind the optical axis. Figure 13.3 shows a schematic of a K-B mirror pair, together with some relevant dimensions. Typically, the source might be a so-called virtual source, created by using a set of beam-defining slits within the beamline.

If the X-ray angles of incidence on the vertical and horizontal mirrors, both with curvature r, are respectively θ_v and θ_h, then the working distance q is related to the mirror separation k and source distance p (see Figure 13.3) by the following equations:

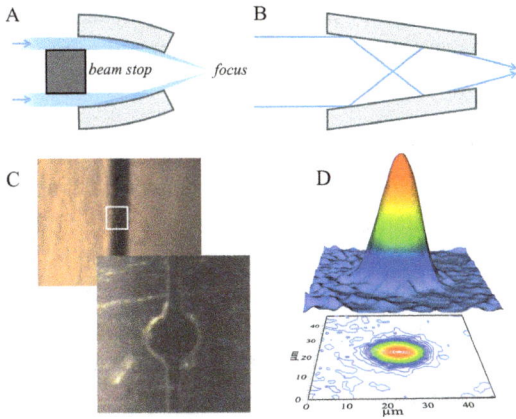

Figure 13.2: Capillary micro-focus optics: (A) a schematic (cross-section) of a parabolic single-bounce capillary, (B) a tapered multiple-bounce capillary, (C) micrographs of the exit aperture of a tapered copper metal capillary and (D) the output of the capillary at incident energy of 2,470 eV measured with a CdWO$_4$ scintillator at the focal point of capillary C.

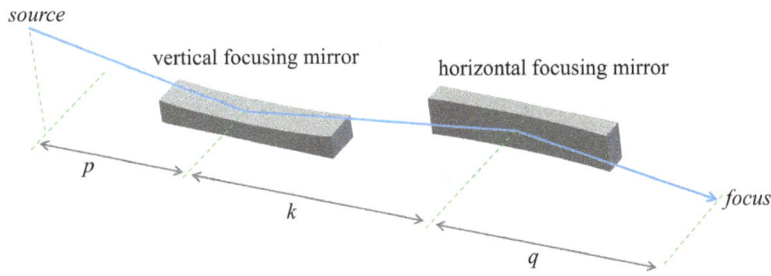

Figure 13.3: Schematic of a Kirkpatrick–Baez mirror pair. The source-to-first-mirror distance p, inter-mirror distance k and second-mirror-to-focus distance q are discussed in the text.

$$\frac{1}{p} + \frac{1}{k+q} = \frac{2}{r \sin \theta_v} \tag{13.5}$$

$$\frac{1}{p+k} + \frac{1}{q} = \frac{2}{r \sin \theta_h} \tag{13.6}$$

The mirrors themselves can often be silicon structures with metallic coatings, as described in Chapter 3, but we note that systems using multilayer optics can result in significantly higher performance.

There are many different X-ray mirror-based micro-focus systems available. Describing all these different systems in detail is outside the scope of this work, so we will restrict ourselves to a brief mention of two. **Montel mirrors** consist of two identical elliptical mirrors, usually fabricated independently and then assembled to form a single unit, with the reflecting surfaces at 90° to each other on the inside of the device – in effect, a side-by-side K-B mirror pair. They provide a more compact device capable of higher demagnification than a conventional K-B pair. **Wolter mirrors** consist of ellipsoidal and hyperboloidal surfaces combined on the same surface and come in a range of configurations.

For our purposes, the major advantage of reflection-based optics is that the energy of the incident beam can easily be changed without significantly altering the focus of the beam on the sample, thereby facilitating spectroscopy measurements. For capillary optics, the working distance from the end of the optic to the sample can often be small. For example, in our previous studies using tapered metal capillaries (e.g. Figure 13.2), the working distance was only 100 µm. For mirrors, the working distance tends to be much larger. While this depends on the mirror system, it will typically be several centimetres, which is more convenient.

13.2.3 Diffractive optics – Fresnel zone plates

The Fresnel zone plate, often called a 'zone plate' for short, is in effect a diffraction-based lens for monochromatic X-rays. It consists of a series of concentric rings that become narrower at larger radii until the last and narrowest ring is reached. Zone plates are probably the most frequently used X-ray micro-focusing optics for high-resolution and very high-resolution applications. In effect, zone plates are circular diffraction gratings, usually fabricated lithographically, in which every other zone either blocks the X-ray light or phase-shifts it by $\lambda/2$. The radii of the zone plate edges are given as follows:

$$r_n^2 = nf\lambda + \frac{n^2\lambda^2}{4} \tag{13.7}$$

where n is the zone number, and f is the focal length for first-order diffraction. A schematic of a zone plate with eight zones and four zone pairs is shown in Figure 13.4. The resolution of the device is controlled by the X-ray wavelength λ, and by the outermost zone width $\Delta r_N = r_N - r_{N-1}$, where N is the total number of zones, and the coherence of the incident X-ray beam. Zone plate optics, when used with fourth-generation synchrotron radiation sources, can generate very small hard X-ray beams, with resolutions close to 10 nm now achievable [9]. This is much more challenging with hard X-rays than with soft X-rays since the degree of coherence is lower in the hard X-ray regime, especially for third-generation sources. Conversely, the working distance tends to be more convenient for hard X-rays at several centimetres, while with soft X-rays, the working distance might be ½ mm or even less with the highest resolution devices. In practice, commercial Fresnel zone plates are available with hundreds or even thousands of zones, with diameters D ranging between approximately 0.1 and 1 mm. Experimental arrangements may include an order-sorting aperture (an X-ray opaque plate with a central hole) placed after the zone plate to restrict the light falling on the sample to the first-order diffraction, and a central beam-stop placed before the zone plate.

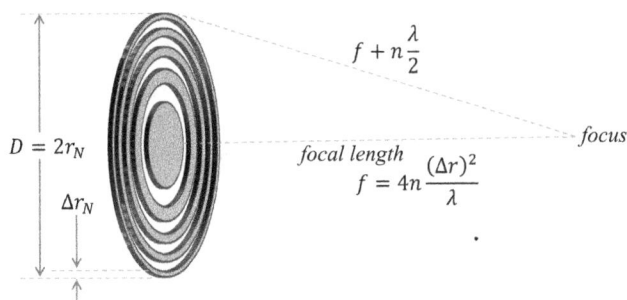

$$f + n\frac{\lambda}{2}$$

$$D = 2r_N$$

$$\Delta r_N$$

$$\text{focal length} \quad f = 4n\frac{(\Delta r)^2}{\lambda}$$

$$focus$$

Figure 13.4: Schematic of a Fresnel zone plate focusing optic.

Zone plates are frequently fabricated from gold, providing opaque zones, on a relatively X-ray transparent support, usually consisting of silicon nitride. For more details on focusing optics in general, and Fresnel zone plate optics in particular, the reader is referred to Chris Jacobsen's excellent book [3].

For spectroscopic applications, the focal length of a zone plate increases linearly with the incident X-ray energy. For soft X-ray measurements, the energy scan range represents a larger fraction of incident energy than with hard X-rays. Moreover, the shorter Fresnel zone plate working distances for soft X-rays mean that the focus degrades more rapidly with small positional errors inside or outside the focal point. Hence, exact positioning of the sample is much more important with soft X-rays than with hard X-rays. Consequently, while the change in focal length in practice may not matter very much for hard X-ray near-edge studies, with soft and tender X-rays, the sample position must be adjusted to accommodate the changes in focal length. We note that with hard X-ray studies if EXAFS is required, then the change in Fresnel zone plate focus may provide additional challenges, just as it does with soft X-ray near-edge studies. Table 13.1 summarizes the different micro-focus optics discussed in this section, with some of their advantages and limitations.

Table 13.1: Summary of X-ray micro-focus optics.

Optic	Spot size	Advantages and limitations
Aperture	20–100 μm	Trivial alignment, low flux
Compound refractive lens	5–50 μm	>5 keV, the focal length changes with X-ray energy.
K-B mirrors	1–10 μm	Good general-purpose optics
Polycapillary	15–35 μm	Easy alignment
Capillary	1–15 μm	Easy alignment
Fresnel zone plate	10–500 nm	Focal length changes with X-ray energy.

13.3 Raster scanning the sample

Before discussing micro-beam applications relevant to XAS, we briefly consider different methods of raster scanning the sample. Typically, the micro-focused X-ray beam is allowed to interrogate different parts of a spatially heterogeneous sample, which is moved relative to the beam. The scanning can be done using a mechanical stage, which might be based on stepper motors, DC-servos plus encoders or piezoelectric transducer-based devices for the highest resolution applications. Stepper motor-driven stages allow scanning of large samples with a precision that is typically about 1 μm. DC-servo stages combine direct current motors with gear reducers and a position-sensing encoder in a closed feedback system. They have the advantage of rapid motion and are approximately as widely used as stepper-motor alternatives. The ultimate mechanical stage resolution is provided by the piezoelectric transducer-based stage; various types are available, but all depend upon the deformation of a piezoelectric crystal in an applied electric field. They have a small range of motion, which can be mitigated by a combined stage configuration, stacking a piezoelectric stage on top of a DC-servo or stepper motor-based stage, with the latter providing coarse motion over a wide range and the piezoelectric stage being available for fine adjustments.

There are two different ways in which the sample might be scanned. The first is a move–stop–measure strategy. Here, the sample is moved to a location, a measurement is made, possibly after a mechanical settling time, the data is read out and the sample is then moved to the next location for another measurement. This process is repeated until an image or map of the sample is built up. This point-to-point strategy often results in very long data acquisition times. An alternative to this acquisition method is the so-called fly-scan, in which the sample is continuously moved, with data acquired on the fly. In both point-to-point and fly-scan strategies, the sample might be raster scanned, so that the beam records stripes across the sample (e.g. in a horizontal direction) before the sample is moved incrementally in the other direction (e.g. vertically) so that the next raster can be acquired. Data can be acquired in what is called unidirectional scanning, in which the sample is scanned in one direction with data being collected on the fly, then rapidly returned to the start of the next raster, or in a bidirectional mode where left-to-right and right-to-left scans both collect data and are interlaced. Using fly-scan acquisition, the effective dwell times per pixel can be reduced to milliseconds when signal-to-noise ratio allows, and overhead is significantly reduced, although there will inevitably be some motion blurring of details.

In spectroscopic applications, which are our primary focus here, the additional complexity of scanning the incident X-ray energy must be considered. For soft X-ray scanning transmission X-ray microscopy (STXM), complete two-dimensional raster scans of the sample are collected for every energy point in a complete data set. In contrast, hard X-ray full-spectrum near-edge imaging experiments, in which a spectrum is collected for every pixel, can use continuous scanning of the monochromator on a per-pixel basis. Another strategy is to collect the same horizontal raster for each incident energy, moving vertically to the next raster only when all energy points for the first vertical position have been collected. This method minimizes the time difference between energies and allows the accumulation of a complete (in terms of energies) partial map, which can be partially analysed during the experiment.

13.4 Spectroscopic imaging – beyond elemental mapping

Spectroscopic imaging seeks to use the power of XAS to study speciation in spatially heterogeneous samples to extract location-sensitive chemical information. We will consider three different approaches: first, XAS measurements of single points of special interest on a sample, which we will call μ-XAS; secondly, the measurement of full maps using a small number of carefully selected incident X-ray energies, chosen based on spectroscopic features in the XAS and finally, full-spectrum imaging. Because many, but not all, of the analysis methods that we will discuss involve X-ray fluorescence, as a preliminary to our main discussion, we will review some specifics of this.

13.4.1 X-ray fluorescence imaging

As a preliminary to this section, we need to offer a brief discussion of nomenclature. There are several different nomenclatures and acronyms used for what is essentially the same technique. These include X-ray fluorescence (XRF), synchrotron X-ray fluorescence (SXRF), X-ray fluorescence imaging (XFI), synchrotron X-ray fluorescence imaging (SXFI), synchrotron radiation-induced X-ray emission (SRIXE), rapid scan X-ray fluorescence (RSXRF), as well as X-ray fluorescence microscopy (XFM) and X-ray fluorescence mapping (XFM). The use of the word imaging in some of these nomenclatures is a hot-button topic for some, who may insist that what we do with XFI is mapping and not imaging. With this nomenclature, mapping would be when the data is collected pixel by pixel, and imaging would be when the data is collected all at once, such as in an old-fashioned medical radiograph using a photographic plate. In fact, modern radiology uses digital X-ray detectors (charge-coupled devices or similar technology) that must be read out pixel by pixel, and so in the modern context this semantic distinction makes no sense, and we are content with our use of X-ray fluorescence imaging.

In almost all cases, XFI depends upon solid-state detectors of the types reviewed in Chapter 5. Such detectors can simultaneously read out the fluorescence from any element in a sample with absorption edge energy below that of the incident X-ray beam. As we have discussed, modern analysis of X-ray fluorescence emission spectra from solid-state detectors is best done using a peak-fitting approach. In this method, fluorescence lines are usually approximated by a Gaussian function, possibly including a low-energy tail due to incomplete charge collection (Section 5.4.6), with the elastic and inelastic scatter approximated by two or more asymmetric peaks. One challenge of the peak-fitting method relates to estimating the areas of small peaks in the presence of a background signal, with the whole emission spectrum spanning a very large dynamic range. Visualization of both small fluorescence lines and the shape of the background often uses a common, but purely arbitrary, method, which is to use log-log-square root (LLS) scaling of the data, as in eq. (13.8), in which E specifies the energy of the emitted X-ray sensed by the detector and $F(E)$ are the detector counts. LLS is usually applied to the raw detector counts, and the 3 +1s in the equation are included to ensure that zeros or negative values never occur. Alternatively, the raw data can be simply displayed on a logarithmic scale, in which values that are close to zero fall off the lower edge of the plot:

$$F'(E) = \log\left(\log\left(\sqrt{F(E)+1}+1\right)+1\right) \tag{13.8}$$

The background is variously accounted for either by subtraction of measured backgrounds from blank regions of the sample [10] or by a per-pixel baseline removal process, which can involve various types of smoothing applied to the data. Prominent among these smoothing methods is the so-called SNIP method [11], which is frequently applied to LLS data directly. Both methods have advantages and limitations (e.g. a blank region of the sample might not be available), but it has been demonstrated that per-pixel baselines can give serious errors in quantification [10]. Figure 13.5 shows a peak-fitting analysis of an MCA spectrum[1] of a sample of human brain tissue. In agreement with normal practice, this is shown with a $\log_{10}$ ordinate scale; consequently, the bottoms of all the peaks are missing, and a few smaller peaks are off the bottom of the plot. In such analyses, the positions of the peaks corresponding to emission lines are typically fixed using tabulated values, as are the relative intensities of associated peaks. For example, the positions of the Zn Kα2, Zn Kα1 and Zn Kβ1,3 would be fixed at 8,615.8, 8,638.9 and 9,572.0 eV, respectively, with respective relative intensities of 0.51, 1.0 and 0.17 (we note that Kα and Lα individual peaks typically are not resolved using conventional solid-state detectors).

In hard X-ray XFI of thin samples, quantification is usually determined relative to some standard, which needs to be measured independent of the sample. This is typically quoted as an areal density, often using the somewhat unorthodox units of μg cm^{-2} or ng cm^{-2}, representing the estimate of the total mass of an element of interest in

---

1 The term MCA spectrum refers to the output spectrum of an energy dispersive solid-state detector (discussed in Chapter 5). Historically, MCA stands for multi-channel analyser, a stand-alone instrument with histogramming memory that would display detector counts as a function of incident photon energy. Nowadays, the function of the multi-channel analyser is fulfilled by the data acquisition computer, although the term MCA spectrum persists despite the fact that it would difficult to find a multi-channel analyser at a modern synchrotron facility.

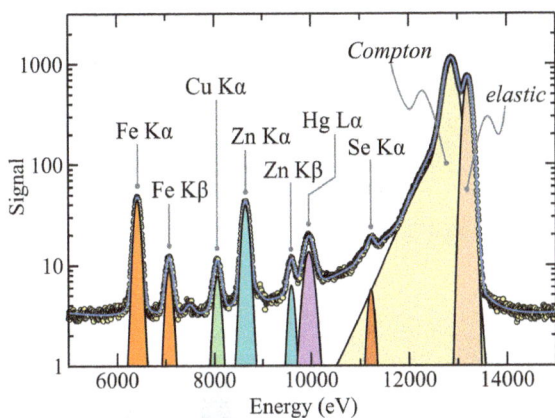

Figure 13.5: MCA peak-fitting analysis. The experimental data are represented by the points, and the fit is shown by the blue line. Prominent contributions to the spectrum are shown by the filled curves.

micro- or nanograms within the sample, expressed as a function of surface area, neglecting sample thickness.

Multimodal approaches, where different techniques can be applied to the same or adjacent sections, are also important. For biological tissues, Fourier transform infrared spectroscopic imaging can complement the methods discussed to provide information on biological molecules co-locating with metal species. Optical microscopy, sometimes in combination with optical staining of adjacent sections, is also an essential complement to XFI for identifying structural features and for aligning in the beam. For both FTIR and XFI, care must be taken in sample mounting to ensure that the signal from any supports does not interfere with that from the sample.

13.4.2 Experimental geometries

For many hard and tender X-ray experiments, while a variety of geometrical arrangements of the sample are possible, the most common are with the sample positioned at either 45° or 90° to the incident X-ray beam, as shown in Figure 13.6. The advantage of the 45° geometry is that it allows the detector to be at 90° to the incident X-ray beam, at which angle X-ray scattering from the sample will be minimal. The disadvantages of this geometry are that the footprint of the micro-focused X-ray beam on the sample will be increased in the horizontal manner by $\sqrt{2}$. Moreover, if there are any surface imperfections on the sample, then there may be shadowing artefacts in the data. The sample will also be effectively thicker, again by $\sqrt{2}$, which may mean that some samples are too thick for good sample transmittance data to be measured, especially with tender X-ray experiments. For samples of a columnar nature, such as a cross-section of a plant stem, the use of 45° geometry will also give parallax distortions of the images. This geometry may also pose challenges in aligning with optical microscope images measured at 90° on the same sample. The 90° geometry, which is also shown in Figure 13.7, will have a smaller beam footprint and fewer shadowing artefacts, but a considerably larger fraction of the total signal will be due to X-ray scattering. This may mean that the detector will become saturated and may degrade the signal-to-noise ratio of fluorescence lines close in energy to that of excitation. With soft X-ray experiments, the 90° geometry is more standard, especially for STXM measurements, which we will discuss later.

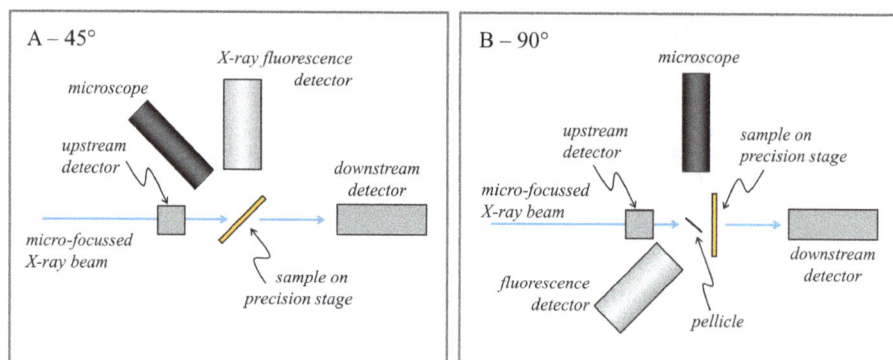

Figure 13.6: Hard X-ray fluorescence imaging experimental arrangements. Schematic plan views of the experiments are shown for 45° (A) and 90° (B) geometries of the sample with respect to the incident beam. In practice, the microscope might be positioned vertically above the sample to avoid crowding. In (B), the pellicle is an X-ray transparent mirror reflecting an image of the sample for observation by the microscope.

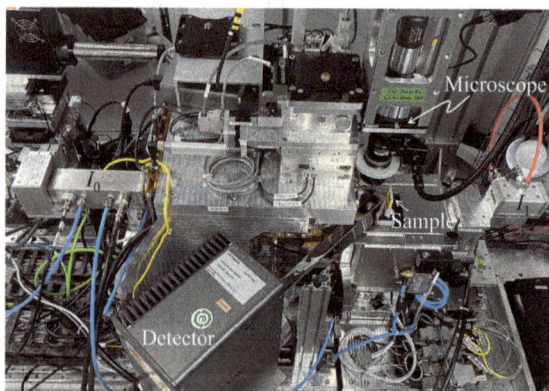

Figure 13.7: Photograph of a 90° geometry hard X-ray fluorescence imaging experiment on the BioXAS imaging beamline at the Canadian Light Source. The incident X-ray beam enters from the left side of the picture, passes through the I_0 ion chamber, is focused by K-B mirrors (which are hidden from view by the central metal support structure) and impinges on the sample. Photograph courtesy of Gosia Korbas, Canadian Light Source.

Before proceeding to discuss spectroscopic applications, we note that simple X-ray fluorescence maps can often be quite beautiful, irrespective of any scientific value. Figure 13.8 shows an early example: an intact freshwater minnow.

Figure 13.8: X-ray fluorescence imaging of an intact freshwater minnow. All the signals, especially the calcium signal, are attenuated by the thickness of the sample (maximum ~3 mm) and come predominantly from the surface of the sample.

13.4.3 μ-XAS

Perhaps the simplest X-ray spectroscopic method that can be applied to structured samples is μ-XAS. Here, a conventional XFI map of a sample is collected, and then used to determine regions of interest based on the measured levels of the different elements or other structural differentials. These regions of interest might be locations of either low or high concentration, or perhaps deriving from observed correlations with other elements. An example of this is shown in Figure 13.9, in which a thin section of larval stage zebrafish shows clear correlations between Hg and both Se and S [12]. MacDonald et al. observed co-localizations with Hg:Se molar stoichiometries varying between 7.5:1 and 1.6:1. Mercury L_{III} near-edge spectra show quite subtle variation with chemical form; nonetheless the Hg L_{III} μ-XAS data of MacDonald et al., together with the observed elemental correlations, strongly suggested that the chemical form of the localized regions of correlation was nanoparticulate mixed chalcogenide $HgS_xSe_{(1-x)}$ [12], in which x varied from 0.44 to 0.87. The chalcogenides HgSe and β-HgS are iso-structural, both with the zincblende structure, and geological formations of HgSe (called tiemannite) are in most cases the mixed chalcogenide $HgS_xSe_{(1-x)}$ with quite variable x.

Here may be as good a place as any to briefly discuss correlations and correlation plots. It is easy to find examples in which linear regression has been used to examine correlations between two elements with XFI data. This is an incorrect practice that we discourage because linear regression assumes zero errors for the abscissae, whereas with XFI data for two elemental signals, errors will be present in both abscissae and ordinate data. As discussed by MacDonald et al., a more satisfactory method for determining the best linear fits to XFI correlation plots is to minimize the sum of the squares of the perpendicular (closest) distance from each point to the line, the slope of which estimates the ratio of the two elements for the data set. If adequate analysis software is not available, then a reasonable approximation that would be valid if there are similar errors for the elements concerned might be to simply calculate two different linear regression results by swapping abscissae and ordinate data and then averaging the two different lines.

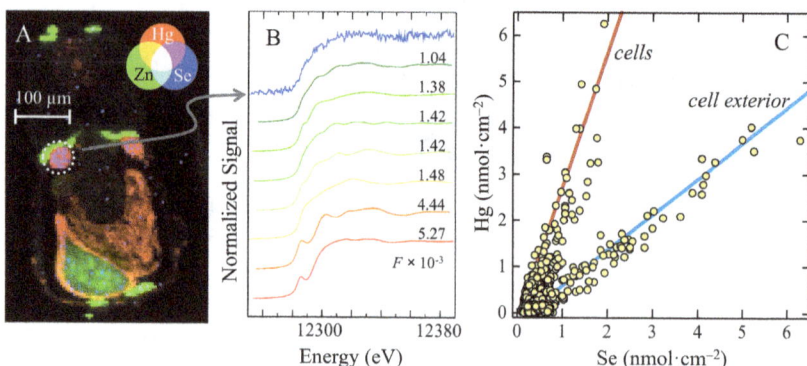

Figure 13.9: Colocalization of mercury and selenium in larval stage zebrafish kidney. (A) The X-ray fluorescence image with mercury in red, zinc in green and selenium in blue. The developing kidneys (pronephric ducts) can be seen near the middle of the image. (B) The Hg L_{III} μ-XAS of the left kidney, together with a series of standard compounds and least-squares errors. The most similar of the standard compounds examined was black mercuric sulphide (top) or β-HgS. Panel C shows a correlation plot from a high-resolution scan (not illustrated), indicating that intra- and extra-cellular mercury and selenium occur in different molar ratios. The data are consistent with the presence of nanoparticles of the mixed chalcogenide $HgS_xSe_{(1-x)}$.

13.4.4 Spectroscopic imaging using a small number of incident energies

The use of spectroscopic imaging with hard X-rays was first introduced using a small number of incident X-ray energies. The first attempts along these lines by Kinney and co-workers used transmittance measurements and included both two-dimensional measurements [13] and three-dimensional reconstructions [14], but only standards were reported, and new insights into systems were obtained more than a decade later when the first chemically specific fluorescence XFI was reported [4]. The idea behind this method is to record a series of maps with the incident X-ray energy exciting maxima in the XAS of constituent compounds across the absorption edge of the element of interest. The method exploits the previously discussed sensitivity of near-edge spectra to chemical speciation and requires prior knowledge of the likely chemical constituents of the sample to choose appropriate energies. A good starting point is to conduct bulk XAS first to establish an average speciation. When using an energy-dispersive detector and with reasonable background rejection, the maximum number of incident energies required is equal to the number of chemical species to be mapped [4]. We can express the fluorescence using the following equation:

$$F(E) = k_s \sum_i m_i I_i(E) \tag{13.9}$$

Here, E is the incident X-ray energy (note that this is different from eq. (13.8)), $F(E)$ is the measured fluorescence at incident energy E, k_s is a constant derived from measurements of samples of known concentration, m_i is the molar fraction of component i and $I_i(E)$ is the normalized intensity of component i at incident energy E, values for which would be derived from XAS of standard compounds. The summation is over all components i. The equation can be solved for m_i giving speciation maps of each component. If there is poor separation between background and the fluorescence of the element of interest, then additional energy points may be required, and this would also be the case if a non-dispersive fluorescence detector is used. If we assume that the total measured signal $T(E)$ is the result of a background signal $B(E)$ and the fluorescence signal $F(E)$, where E is the incident X-ray energy, and ignoring any change in penetration depth with energy, then we can write:

$$T(E) = B(E) + F(E) \tag{13.10}$$

Here, the background might include contributions from elastic and inelastic scattering and X-ray fluorescence from other elements in the sample. In experiments at the sulfur K-edge, Pickering et al. used a non-dispersive detector because the separation between elastic scattering and S Kα fluorescence is only about 160 eV, which is sufficiently small that resolving the fluorescence and scattering is challenging using a conventional solid-state detector [7]. These researchers examined tissues of onion using four chemical components, employing six incident energies: one below the sulfur absorption edge and one above, and four for each of the components. Equation (13.9) can be solved by matrix inversion, incorporating additional terms for the background when eq. (13.10) is more appropriate [7]. Figure 13.10 shows the process of separation of species for the arsenic hyperaccumulating brake fern *Pteris vittata* [15].

This fern will take up relatively non-toxic, formally As(V) arsenate (a mixture of $[H_2AsO_4]^-$ and $[HAsO_4]^{2-}$ at neutral pH) and convert it to the more toxic, formally As(III) oxyanion arsenite ($As(OH)_3$ at neutral pH), which is stored in the tissues of the plant. To selectively image arsenate and arsenite, Pickering et al. used two incident X-ray energies: one at the most intense XAS peak of the arsenite spectrum and the other at the most intense XAS peak of the arsenate spectrum (Figure 13.10).

In the same study, Pickering et al. examined a range of tissues from both the sporophyte and gametophyte generations of the fern. A selection of these results is shown in Figure 13.11, which clearly shows the presence of arsenate in the transport vessels and that the reproductive tissues lack any substantial presence of toxic arse-

Figure 13.10: Chemically specific X-ray fluorescence imaging of arsenic in the brake fern *Pteris vittata* [15]. (*a*) Comparison of the As K-edge XAS for arsenite and arsenate at pH 7.4, with the X-ray energies selected for imaging (E_1 and E_2) indicated by the vertical broken green lines. (*b*) A photograph of the intact *Pteris vittata* plant, with the red box indicating the area selected for imaging, shown to scale in panel (*c*), the tip of one pinnule (frond or leaf). (*d*)–(*f*) The X-ray absorbance $A = \log(I_0/I_1)$ and the As Kα fluorescence normalized with I_0 collected data at energies E_1 and E_2, respectively. (*g*) and (*h*) The results of solving eq. (13.9) giving chemically selective imaging of arsenite and arsenate, respectively. The sample thickness was calculated from the absorbance, assuming that the sample had an X-ray cross-section corresponding to that of water. Data were measured at a resolution of 5 μm using a tapered metal capillary [6] similar to the device shown in Figure 13.2. Adapted with permission from Ref. [15]. Copyright 2006 American Chemical Society.

nite. Since these early studies on *P. vittata*, others have used similar methods to extend this work to investigate *P. vittata* in more detail and to investigate the arsenic biochemistry of the related arsenic hyperaccumulating species *Pityrogramma calomelanos* [16–19].

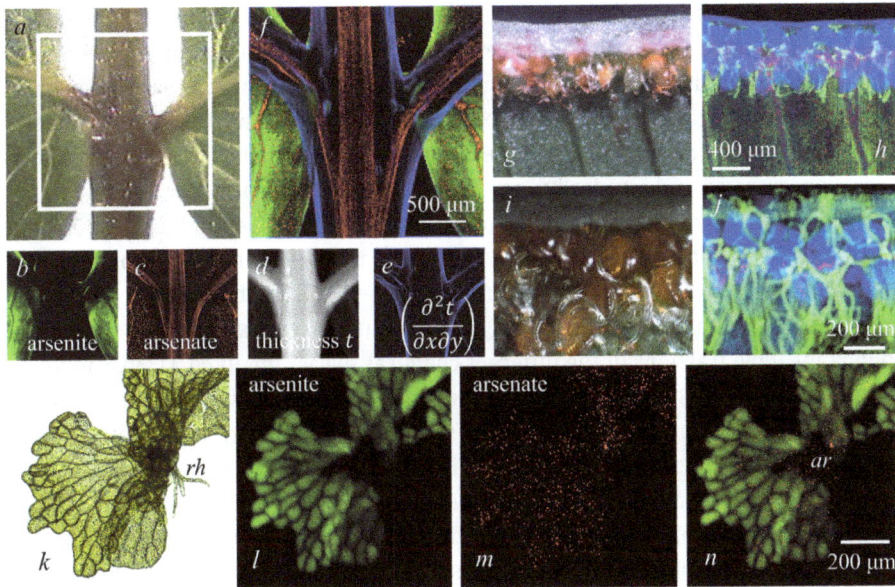

Figure 13.11: Chemically selective XFI for various *P. vittata* tissues. (*a*) A micro-graph of a rachis (central stem) from the same plant shown in Figure 13.10, with two pinnules emerging. The white box shows the area examined by XFI. (*b*)–(*d*) The images of arsenite, arsenate and the thickness, determined from the total absorbance. (*e*) The second derivative of the thickness, computed to show the edges of the tissue. (*f*) A tri-colour composite clearly indicating the location of arsenate (red) in the central transport vessels and arsenite (green) in the tissues. (*g*)–(*j*) The reproductive tissues of the sporophyte (arsenite: green, arsenate: red, absorbance: blue), with the sori shown in (*g*) and (*h*), and the sporangia with internal spores visible in (*i*) and (*j*). The filamentous structures, called paraphyses (thought to fulfil a protective function), contain toxic arsenite, but the reproductive tissues do not. (*k*)–(*n*) The one-cell-thick gametophyte, with rhizoids (*rh*) and the future site of archegonial development indicated (*ar*). Here, arsenite (green) can be seen to be localized in the interior of the cells, most likely in the large central vacuole, and arsenate is present in hot spots that are ~5 µm or less across. This suggests that arsenate (red) may be located within cellular Golgi bodies. Again, toxic arsenite is notably absent from cells near the site of reproductive tissue development (*ar*). Adapted with permission from ref. [15]. Copyright 2006 American Chemical Society.

13.4.5 Full-spectrum spectroscopic imaging

Here, a complete XAS spectrum of an element of interest is collected on a per-pixel basis, yielding what is called an 'image stack'. Arguably, the state of the art in full-spectrum spectroscopic imaging is achieved in the soft X-ray regime on specialized beamlines using what is called 'scanning transmission X-ray microscopy' (STXM). With STXM, Fresnel zone plate optics are used on very thin, relatively concentrated samples, with rapid scanning of the sample position at selected X-ray energies across an absorption edge of interest. This collection of images is usually measured in transmittance and is usually called an 'image stack', and with STXM studies the method of spectroscopic imaging is sometimes referred to as 'spectromicroscopy'. Most STXM studies have used the carbon K-edge, but other soft X-ray absorption edges have also been studied, including the K-edges of other light elements as well as transition metal L-edges. The major disadvantages of the STXM method are that samples must be relatively concentrated and physically very thin. Hard X-ray full-spectrum XAS imaging is also sometimes carried out, and in this case the method is to rapidly scan the XAS on a per-pixel basis with the sample being scanned in point-to-point mode. To date, hard X-ray full-spectrum XAS imaging has only been used in a limited number of studies. When dealing with full-spectrum imaging, the simple matrix inversion method of solving for mole fractions of the different chemical species, m_i in eq. (13.9), cannot be used because the system is overdetermined, with substantially more energy points than components. To obtain quantitative speciation (m_i values), the individual XAS per-pixel spectra must either be least-squares fitted to a sum of standard spectra, as described in Section 11.1, or the least-squares problem must be solved using single value decomposition, as described in Section 11.6. Full-spectrum imaging provides very rich spectroscopic imaging data and is ideal for application of many of the methods discussed in Chapter 11.

13.4.6 Full-field and wide-field spectroscopic imaging

Full-field imaging exploits transmittance measurements using a collimated monochromatic X-ray beam with a position-sensitive detector, such as a scintillator (e.g. $CdWO_4$), and a camera to capture projection images of a sample. Such measurements can allow the rapid collection of a full-spectrum XAS image stack by the simple expedient of scanning the monochromator. Figure 13.12 shows an example that was collected using a simple set-up on a beamline normally used for spectroscopy of bulk samples.

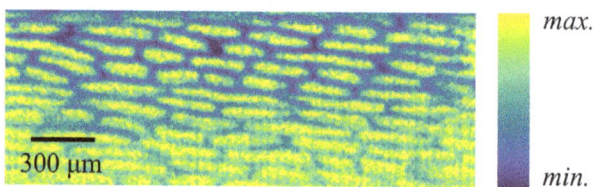

Figure 13.12: Full-field projection image of an epidermal peel of onion. The measurement shown was conducted just above the K-edge of sulfur (2,485 eV) using a portable set-up consisting of a CMOS camera triggered after a monochromator move, $CdWO_4$ scintillator, with samples supported on a Si_3N_4 cover slip. The high sulfur content of the cell interiors is clearly visible. Experiments were conducted using SSRL BL4-3 with an upstream collimating mirror and a Si(111) double crystal monochromator [20].

A novel wide-field selenium K-edge spectroscopic imaging method using bent crystal Laue diffraction has been reported by Qi et al. [21]. In this method, a polychromatic X-ray beam passes through a Laue geometry monochromator that is bent about the horizontal axis, with the diffraction resulting in an initial vertical convergence of the X-ray beam to a line focus at which the sample is positioned. After passing through the sample, the beam diverges with separation of wavelengths until it falls on an area detector, which registers the transmittance of the sample with energy separation vertically. The bent Laue geometry crystal meets what Qi et al. call the 'magic condition' [21], under which the geometrical and polychromatic foci of the crystal overlap. Consequently, the resulting spectra show moderately good energy resolution, which is sufficient for the speciation of a range of selenium chemical forms. The sample can then be translated to construct a two-dimensional image or rotated to allow tomographic reconstruction of a slice (see Section 13.4.7). Combining tomography followed by vertical translation allows a full three-dimensional reconstruction, with an XAS spectrum of every voxel.

13.4.7 Three-dimensional methods – tomography

In tomography, a sample is typically rotated, with data collected across its length. For example, with XAS, one might rotate about the vertical axis, which we can call φ and translate the sample in the horizontal direction, which we can call x. A plot of measured X-ray fluorescence or transmittance against φ and x is called a 'sinogram', and tomography is the process of reconstructing a cross-section in the horizontal plane from the sinogram data. Conventional X-ray tomography typically uses a full-field or partial-field arrangement, exploiting either sample X-ray absorption or phase-contrast. The latter is sometimes called 'propagation-based imaging' because it requires large distances between the sample and the detector. It exploits interference between refracted and transmitted X-rays to effectively visualize the changes in the real part of the refractive index within a sample. X-ray fluorescence tomography uses basically the same principles for reconstruction. In this case, a small number of excitation energies can be used to construct speciation images, as described in Section 13.4.4. Similarly, for concentrated samples, the transmittance could be used to reconstruct speciation images, a method which is routinely employed at soft and tender X-ray energies, at what are called transmission X-ray microscopy beamlines (abbreviated TXM). An early example of a chemically specific tomographic reconstruction is shown in Figure 13.13. In this case, the sample absorbance was measured alongside the fluorescence and used to correct for attenuation of the X-ray fluorescence by the sample. If the transmission of the sample is not measured, then correction for absorption can be challenging, and for this reason, it is often neglected. The main disadvantage of tomographic methods is that the X-ray dose tends to be relatively high. Moreover, when employing X-ray fluorescence, extraction of a single μ-XAS spectrum would involve a tomographic reconstruction at a very large number of incident X-ray energies and is not practical. Neither of these difficulties is present with confocal X-ray fluorescence imaging.

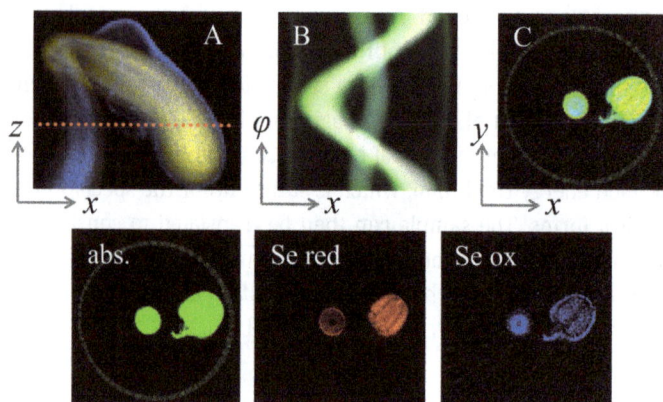

Figure 13.13: Chemically specific X-ray fluorescence tomography of an *Astragalus bisulcatus* seedling. *A. bisulcatus* is a selenium hyperaccumulator that accumulates Se-methyl-L-selenocysteine in its tissues [1, 4, 8], and not, as has been claimed by others, in the trichomes. The seedling was measured live, having been germinated on an aqueous solution of sodium selenate, using two incident energies (12,661.1 and 12,667.3 eV) corresponding to the maxima of Se-methyl-L-selenocysteine (Se red) and selenate (Se ox), respectively [4, 8]. (A) A tri-colour conventional chemically specific X-ray fluorescence image of the sample (red: Se red, green: absorption, blue: Se ox). (B) The sinogram corresponding to the red dotted line in (A). (C) The tomographic reconstruction as a tri-colour plot with the individual components shown below. Adapted with permission from ref. [1]. Copyright 2014 American Chemical Society.

13.4.8 Three-dimensional methods – confocal X-ray fluorescence imaging

With confocal X-ray fluorescence imaging, two focusing optics are used: one to generate the incident micro-focused X-ray beam and a second optic, often a poly-capillary device, located on the X-ray fluorescence detector, as shown in Figure 3.14.

The confocal experimental set-up must be very carefully aligned so that the focal point of the incident X-ray beam coincides with the focal point of the optic feeding X-rays from the sample to the detector. This arrangement allows the experimenter to effectively look below the surface of the sample and collect an X-ray fluorescence image of a section within it. Points of interest can be located, and μ-XAS spectra can easily be collected at the coordinates of any voxel. For example, Choudhury et al. used confocal XFI to examine archaeological bone samples from individuals who had been exposed to lead [22]. The archaeological samples in question were weathered and of a fragile consistency, so conventional sectioning was not possible. These researchers also used a novel spoked-channel array on the detector together with a K-B mirror

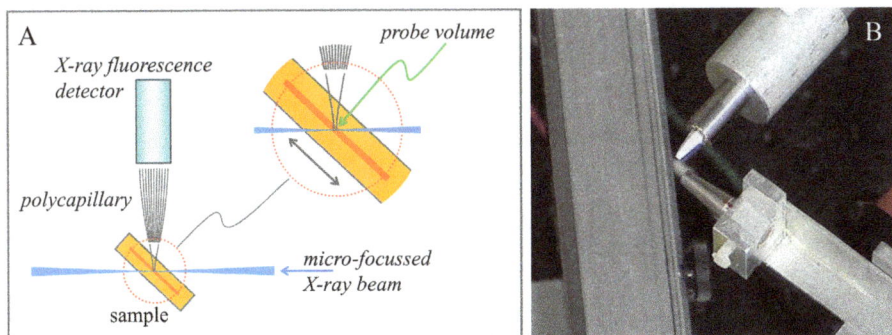

Figure 13.14: Confocal X-ray fluorescence imaging: (A) a schematic of the experimental set-up and (B) a photograph of the experimental arrangement. In (B), the sample is hidden by the exterior of its holder, and the incident beam is admitted from the bottom right, with the detector and focusing optic (white) at the top right.

pair for the incident beam to achieve 2 μm spatial resolution – approximately an order of magnitude improvement over that available from conventional polycapillary optics. Moreover, Pb L_{III} μ-XAS was used to speciate the lead in the samples with a voxel dimension of $2.0 \times 2.5 \times 1.5$ μm^3 [22]. Like tomography, confocal slices can be used to construct a three-dimensional view of a sample, although, also like tomography, the X-ray dose of such three-dimensional reconstructions might be prohibitively large. Figure 13.15 shows such a three-dimensional reconstruction using a polycapillary optic.

As a final remark related to the X-ray technique for this chapter, we mention that a combination of what will be discussed in Chapter 14, high-energy resolution fluorescence-detected XAS, and the XAS imaging methods discussed here may be a particularly powerful combination, called 'HERFD-XAS imaging'. Apparatus for this has already been developed [23].

13.5 Sample environment

There are substantial variations in sample environments for XAS imaging, depending on the scientific discipline. For example, with environmental sediments or solid mine tailings, sample preparation might be as simple as placing the material on an appropriate adhesive tape. In contrast, biological tissues require substantially greater attention to detail. Preparation of biological samples can be one of the most important aspects of performing XAS imaging or related experiments [1, 2]. Fortunately, there are reviews [1, 2] that provide significant detail, allowing us to be brief in our discussion here. For tissues, sample mounting and preservation methods tend to be distinct from those used for bulk sample preparation (Section 7.5.2) because the sample envi-

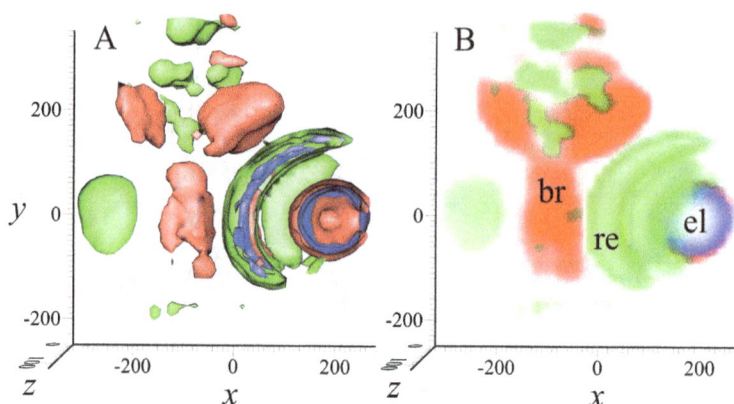

Figure 13.15: Three-dimensional confocal X-ray fluorescence imaging of the head of a larval stage zebrafish exposed to methylmercury chloride. The z-range runs from approximately midway between the dorsal and ventral surfaces down to the ventral surface. Both iso-surface-rendered (A) and volume-rendered (B) images are shown with zinc (green), selenium (blue) and mercury (red). Axis scales are in μm with an arbitrary zero point. The eye-lens (el), brain (br) and retina (re) are marked in (B).

ronment for XAS imaging needs to be less enclosed. With biological samples, tissue sections are frequently studied, and early work benefited from fixing and sectioning methods employed in conventional microscopy, which were designed to preserve structure and not composition. Fixation of tissue samples is now widely acknowledged to change both chemical speciation and to wash out some elements of interest. Cryo-sectioning of frozen samples is now becoming standard practice. Keeping tissue samples frozen following sectioning tends to be impractical, and lyophilization or air drying are alternatives, but they have the disadvantage that the chemical speciation within the sample might be altered. Sample sections can be supported on a thin polymer film or on thin plastic cover slips, provided that these contain no metals or elements of interest. The polymer support should be as thin as possible because it will be a source of unwanted scattered radiation, possibly resulting in saturating count rates predominantly from the scatter. With high resolutions, silicon nitride (Si_3N_4) windows can be used to support samples. These are typically hundreds of nanometres in thickness and hence produce much less scatter. These are particularly useful for cell cultures, which can be studied with high X-ray spatial resolutions, and here too fixing should be avoided. Cell cultures should also be studied close to the time of sample preparation because they tend to change with time, with micro-crystals of NaCl growing on the exterior of the dried cells. Plant tissues provide a special case, as these can often be studied live [1, 4, 15], (Figures 13.10–13.13) especially at low resolutions when the damaging X-ray flux densities that accompany high spatial resolutions are not needed. For samples that require this, cryogenically cooled stages have been used (e.g. this was done with the experiments shown in Figure 13.15) [24], but such devices are typically restricted to above liquid nitrogen temperatures. Moreover, tender and soft X-ray experiments may need to be carried out either in helium (for tender X-rays) [25] or in a vacuum (for soft X-rays). In addition, for multimodal methods (Section 13.4.1), various sample registration methods may be very useful.

References

[1] Pushie, M. J.; Pickering, I. J.; Korbas, M.; Hackett, M. J.; George, G. N. Elemental and chemically specific X-ray fluorescence imaging of biological systems. *Chem. Rev.* **2014**, *114*, 8499–8541.

[2] Pushie, M. J.; Sylvain, N. J.; Hou, H.; Hackett, M. J.; Kelly, M. E.; Webb, S. M. X-ray fluorescence microscopy methods for biological tissues. *Metallomics*. **2022**, *14*, mfac032/1–31.

[3] Jacobsen, C.; *X-ray Microscopy*. Cambridge University Press: Cambridge, UK, **2020**.

[4] Pickering, I. J.; Prince, R. C.; Salt, D. E.; George, G. N. Quantitative, chemically-specific imaging of selenium transformation in plants. *Proc. Natl. Acad. Sci. U.S.A.* **2000**, *97*, 10717–10722.

[5] Alianelli, L.; Sawhney, K. J. S.; Barrett, R.; Pape, I.; Malik, A.; Wilson, M. C. High efficiency nano-focusing kinoform optics for synchrotron radiation. *Opt. Express*. **2011**, *19*, 11120–11127.

[6] Hirsch, G.; Metal capillary optics: Novel fabrication methods and characterization. *X-Ray Spectrom*. **2003**, *32*, 229–238.

[7] Pickering, I. J.; Sneeden, E. Y.; Prince, R. C.; Block, E.; Harris, H. H.; Hirsch, G.; George, G. N. Localizing the chemical forms of sulfur *in vivo* using X-ray fluorescence spectroscopic imaging: Application to onion (*Allium cepa*) tissues. *Biochemistry*. **2009**, *48*, 6846–6853.

[8] Pickering, I. J.; Hirsch, G.; Prince, R. C.; Sneeden, E. Y.; Salt, D. E.; George, G. N. Imaging of selenium in plants using tapered metal monocapillary optics. *J. Synchrotron. Rad*. **2003**, *10*, 289–290.

[9] Da Silva, J. C.; Pacureanu, A.; Yang, Y.; Bohic, S.; Morawe, C.; Barrett, R.; Cloetens, P. Efficient concentration of high-energy X-rays for diffraction-limited imaging resolution. *Optica*. **2017**, *4*, 492–495.

[10] Crawford, A. M.; Deb, A.; Penner-Hahn, J. E. M-BLANK: a program for the fitting of X-ray fluorescence spectra. *J. Synchrotron Rad*. **2019**, *26*, 497–503.

[11] Ryan, C. G.; Clayton, E.; Griffin, W. L.; Sie, S. H.; Cousens, D. R. SNIP, a statistics-sensitive background removal treatment for the analysis of PIXIE spectra in geoscience applications. *Nucl. Instrum. Meth. Phys. Res.* **1988**, *934*, 396–402.

[12] MacDonald, T. C.; Korbas, M.; James, A. K.; Sylvain, N. J.; Hackett, M. J.; Nehzati, S.; Krone, P. H.; George, G. N.; Pickering, I. J. Interaction of mercury and selenium in the larval stage zebrafish vertebrate model. *Metallomics*. **2015**, *7*, 1247–1255.

[13] Kinney, J.; Johnson, Q.; Nichols, M. C.; Bonse, U.; Nusshardt, R. Elemental and chemical-state imaging using synchrotron radiation. *Appl. Opt*. **1986**, *25*, 4583–4585.

[14] Kinney, J. H.; Johnson, Q. C.; Saroyan, R. A.; Nichols, M. C.; Bonse, U.; Nusshardt, R.; Påhl, R. Energy modulated X-ray microtomography. *Rev. Sci. Instrum*. **1988**, *59*, 196–197.

[15] Pickering, I. J.; Gumaelius, L.; Harris, H. H.; Prince, R. C.; Hirsch, G.; Banks, J. A.; Salt, D. E. G.; George, G. N. Localizing the biochemical transformations of arsenate in a hyperaccumulating fern. *Environ. Sci. Technol*. **2006**, *40*, 5010–5014.

[16] Kachenko, A. G.; Gräfe, M.; Singh, B.; Heald, S. M. Arsenic speciation in tissues of the hyperaccumulator *P. calomelanos* var. *austroamericana* using X-ray Absorption Spectroscopy. *Environ. Sci. Technol*. **2010**, *44*, 4735–4740.

[17] van der Ent, A.; de Jonge, M. D.; Spiers, K. M.; Brueckner, D.; Montargès-Pelletier, E.; Echevarria, G.; Wan, X.-M.; Lei, M.; Mak, R.; Lovett, J. H.; Harris, H. H. Confocal volumetric μXRF and fluorescence

computed μ-tomography reveals arsenic three-dimensional distribution within intact *Pteris vittata* fronds. *Environ. Sci. Technol.* **2020**, *54*, 745–757.

[18]   Kashiwabara, T.; Kitajima, N.; Onuma, R.; Fukuda, N.; Endo, S.; Terada, Y.; Abe, T.; Hokura, A.; Nakai, I. Synchrotron micro-X-ray fluorescence imaging of arsenic in frozen-hydrated sections of a root of *Pteris vittata*. *Metallomics*. **2021**, 13, mfab009/1–8.

[19]   Remigio, A. C.; Harris, H. H.; Paterson, D. J.; Edraki, M.; van der Ent, A. Chemical transformations of arsenic in the rhizosphere–root interface of *Pityrogramma calomelanos* and *Pteris vittata*. *Metallomics*. **2023**, *15*, mfad047/1–18.

[20]   Hackett, M. J.; Caine, S.; George, G. N. *Unpublished observations*.

[21]   Qi, P.; Samadi, N.; Martinson, M.; Ponomarenko, O.; Bassey, B.; Gomez, A.; George, G. N.; Pickering, I. J.; Chapman, L. D. Wide field imaging energy dispersive X-ray absorption spectroscopy. *Sci. Rep.* **2019**, *9*, 17734/1–14.

[22]   Choudhury, S.; Agyeman-Budu, D. N.; Woll, A. R.; Swanston, T.; Varney, T. L.; Cooper, D. M. L.; Hallin, E.; George, G. N.; Pickering, I. J.; Coulthard, I. Superior spatial resolution in confocal X-ray techniques using collimating channel array optics: elemental mapping and speciation in archaeological human bone. *J. Anal. At. Spectrom.* **2017**, *32*, 527–537.

[23]   Edwards, N. P.; Bargar, J. R.; van Campen, D.; van Veelen, A.; Sokaras, D.; Bergmann, U.; Webb, S. M. A new μ-high energy resolution fluorescence detection microprobe imaging spectrometer at the Stanford Synchrotron Radiation Lightsource beamline 6–2. *Rev. Sci. Instrum.* **2022**, *93*, 083101/1–10.

[24]   Choudhury, S.; Thomas, J. K.; Sylvain, N. J.; Ponomarenko, O.; Gordon, R. A.; Heald, S. M.; Janz, D. M.; Krone, P. H.; Coulthard, I.; George, G. N.; Pickering, I. J. Selenium preferentially accumulates in the eye lens following embryonic exposure: a confocal X-ray fluorescence imaging study. *Environ. Sci. Technol.* **2015**, *49*, 2255–2261.

[25]   Hackett, M. J.; George, G. N.; Pickering, I. J.; Eames, B. F. Chemical biology in the embryo: in situ imaging of sulfur biochemistry in normal and proteoglycan-deficient cartilage matrix. *Biochemistry*. **2016**, *55*, 2441–2451.

14 High-energy resolution methods

14.1 Introduction

In Chapter 5, we discussed X-ray detector technologies, including crystal analysers that facilitate measurement of X-ray fluorescence or X-ray scatter at resolutions that are better than the natural linewidth. This ability to measure emission with high-energy resolution is in addition to the ability to scan the incident X-ray energy with high resolution, as discussed in Chapter 3. The combination of these two energy axes is important, as we discuss in this chapter, as it allows specialized measurements with substantial advantages over conventional X-ray absorption spectroscopy. This chapter focuses predominantly on X-ray photon-in/photon-out experiments that for the most part employ crystal analysers to measure emission with high-energy resolution, as discussed in Sections 5.7.4–5.7.6. In much of this chapter, the detector system of choice will be the Johann spectrometer, and before discussing details of the spectroscopy, we will first review this spectrometer in somewhat more detail than we did in Chapter 5.

14.2 The Johann geometry array spectrometer

The use of an array of analyser crystals was pioneered by Cramer and co-workers [1], with arrays now frequently used to increase the solid angle, and therefore the total signal, over that from a single analyser crystal. The Johann geometry array is arranged with the Rowland circles for all analyser crystals intersecting at both the sample and the detector, as shown schematically in Figure 14.1.

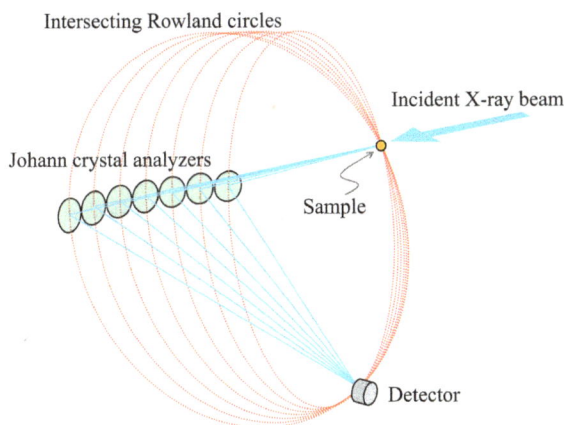

Figure 14.1: Schematic of a Johann analyser array. The detector is positioned directly below the sample with intersecting Rowland circles aligned with each analyser crystal.

https://doi.org/10.1515/9783110570441-014

Figure 14.2 shows photographs of the seven-crystal Johann array at SSRL with 1 m Rowland circles [2]. The experiments that we will describe are photon-limited and require a small beam for optimal resolution, therefore work best with high flux density beamlines. Since sample radiation damage, therefore it provides additional challenges. Use of a helium cryostat is important to mitigate sample damage. Normally, the entire crystal array is enclosed in a helium flight path that allows the beam to pass from the sample (inside a helium cryostat) to the analysers and back to the detector – a close to 2 m round trip – all in helium to minimize atmospheric X-ray absorption.

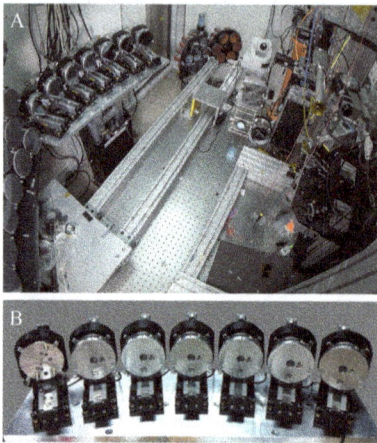

Figure 14.2: Photographs of the seven-crystal Johann array at SSRL. (A) The experimental hutch (SSRL BL 6-2) during setup before installation of the helium flight path. The seven-crystal Johann array is shown on the top left side of the picture. The cryostat and the detector will be positioned on the left. (B) The seven-crystal array prior to installation of the analyser crystals (photographs courtesy of Dr Dimosthenis Sokaras of SSRL).

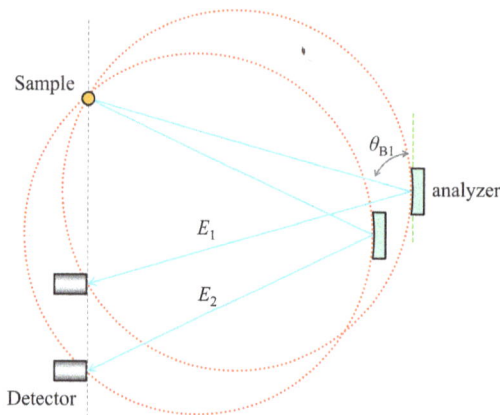

Figure 14.3: Schematic of relative positioning of analyser, detector and sample with a Johann spectrometer for two different analyser energies, where $E_1 > E_2$.

With the Johann spectrometer, the X-ray emission energy can be scanned by moving the relative positions of the analyser, sample and detector. Figure 14.3 shows a schematic of the position change for two energies and a single analyser crystal. In Chapter 5, we discussed the requirement of a near-90° Bragg geometry for good spectrometer energy resolution $|\Delta E/E|$, which is articulated by the following equation (reproduced from eq. (5.7)):

$$\left|\frac{\Delta E}{E}\right| = \omega_D + \Delta\theta_B\cot\theta_B \tag{14.1}$$

Here, ω_D is the Darwin width or the angular width of a Bragg reflection, and $\Delta\theta_B$ is the angular divergence arising from the effective source size, which is the size of the incident X-ray spot illuminating the sample and the acceptance of the analyser crystal. This spot-size contribution to the energy resolution becomes smallest as the Bragg angle θ_B approaches 90°, the condition that is the best possible resolution for a given crystal analyser. Obviously, geometric considerations mean that θ_B cannot be too close to 90° or the sample and detector would be on top of one another (e.g. see Figure 14.1), therefore a practical ideal limit is $\theta_B \approx 85°$. We return to a discussion of choosing analyser crystals in Section 14.6.2.

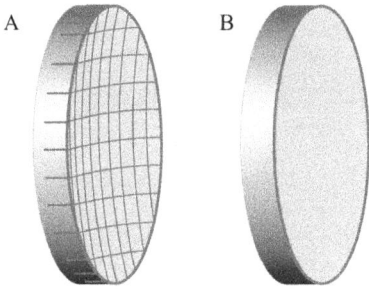

Figure 14.4: Schematic of diced and un-diced analyser crystals. A shows a diced analyser crystal, with the size of the dice deliberately exaggerated to allow visualization, and B shows an un-diced analyser crystal. Both are spherically bent to an internal radius of R_R.

Here we note that there are two basic types of analyser crystal, diced and un-diced (Figure 14.4), both spherically bent with a radius equal to the diameter of the Rowland circle. Un-diced analyser crystals are simply bent into a spherical figure. This bending inevitably causes undesirable elastic deformations of the crystal lattice, which in turn will increase the bandwidth of the analyser crystal, with a consequent decrease in energy resolution. This is more of a problem for spectrometers employing smaller Rowland circles (e.g. 0.5 m). Diced crystals have many cuts in their diffracting surface, and are divided into approximately cubic pieces, the exterior forming thousands of small flat diffracting crystal surfaces, each inclined so that their surfaces are tangential to the spherical bend. These crystal surfaces lack the elastic deformations of un-diced analysers, but the finite size of the individual crystal facets causes an additional geometrical contribution to the analyser resolution ΔE_g according to the following equation:

$$\left|\frac{\Delta E_g}{E}\right| = \frac{c}{R_R}\cot\theta_B \tag{14.2}$$

Here, c is the size of the crystal facet, which typically is 0.7 mm for a 100 mm analyser crystal, and R_R is the diameter of the Rowland circle. Both types of crystal thus have advantages and limitations. Moving now to discuss some applications of this technology, the first topic that we will address is high-resolution hard X-ray emission spectroscopy.

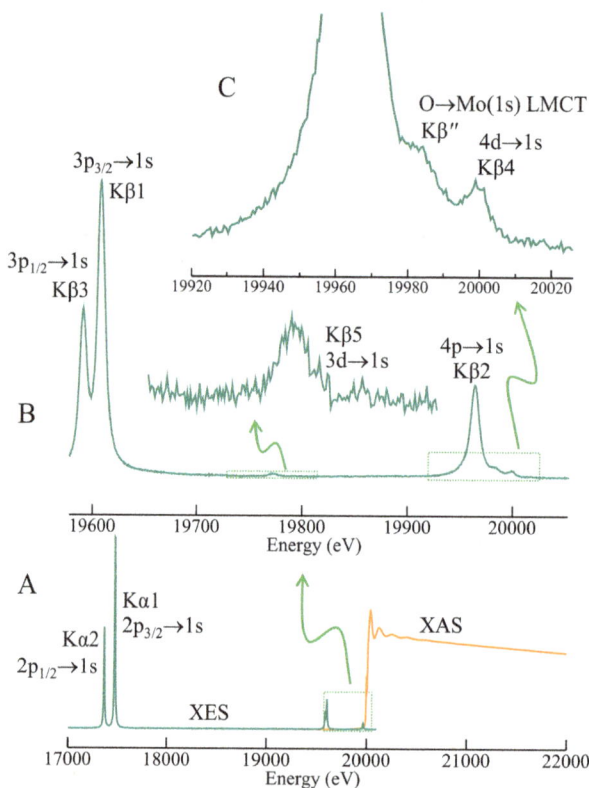

Figure 14.5: X-ray emission spectroscopy of potassium molybdate, K_2MoO_4 [3]. (A, bottom) A comparison of the XAS (orange) and the XES (green blue) with the major fluorescence lines indicated. (B) A zoomed-in plot of the Kβ region. (C) The minor line XES.

14.3 Hard X-ray emission spectroscopy (XES) and minor line XES

While the focus of this book is XAS, XES has intimate connections to XAS, and allows a range of variations. Minor line XES can potentially provide similar information to the EXAFS, but to date, has been very little used. An example, measured with a Johann-geometry array, is shown in Figure 14.5.

For large atoms, such as molybdenum, the Kα1 and Kα2, respectively originating from $2p_{3/2}{\rightarrow}1s$ and $2p_{1/2}{\rightarrow}1s$ transitions, are deep-atomic in nature and well-shielded

from chemical effects by the intervening shells of electrons. Little XES chemical sensitivity therefore is expected with these most intense fluorescence lines. The Kβ, on the
other hand, arises from less deeply buried levels and thus is expected to show chemical
sensitivity. Figure 14.5B shows this region expanded; the major lines are the formally
dipole-allowed $3p_{3/2} \rightarrow 1s$ and $3p_{1/2} \rightarrow 1s$ transitions giving rise to the Kβ1 and Kβ3, respectively. The next strongest line, Kβ2, arises from $4p \rightarrow 1s$ transitions, also formally dipole-
allowed. Mid-range in the expanded region of Figure 14.4B is the Kβ5, a minor line arising from the dipole-forbidden, quadrupole-allowed $3d \rightarrow 1s$ transition. Some asymmetry
has been observed in the Kβ2 line [4], which has been attributed to final-state multiple
interactions between the 4p and 4d manifolds. Slightly higher in energy than the Kβ2
are the Kβ″ and the Kβ4 minor lines (Figure 14.4C). These originate from a ligand-to-
metal-charge transfer transition (Kβ″), and from a formally dipole-forbidden, quadrupole-allowed $4d \rightarrow 1s$ transition (Kβ4). The Kβ″ is of particular interest and in the case of
Figure 14.5, derives from an $O(2s) \rightarrow Mo(1s)$ transition; it has also been called a crossover
or interatomic transition. The Kβ″ is absent in XES of the pure metal, and in compounds, is sensitive to both the bond length and ligand type [5]. The energy of the Kβ″
transition decreases with increasing atomic number of the ligand, while the intensity of
the Kβ″ line shows an approximately exponential relationship with bond length (longer
bonds give smaller Kβ″) [3, 5]. It has thus been suggested that the Kβ″ might serve as a
complement to EXAFS in allowing the identification of light ligands with similar atomic
numbers [5]. To date, however, minor line XES has been relatively little used, probably
because the X-ray fluorescence lines involved are very small (e.g. see Figure 14.5).

14.4 Tender X-ray XES

With tender X-rays, the number of crystals having d-spacings that allow a near-90° Bragg
geometry is limiting. Hence, a different approach can be adopted, and this is to use a Johansson analyser, with what is known as inside Rowland sample positioning [6, 7]. As we
discussed in Section 5.6.6, a cylindrically bent Johansson analyser is used with the sample
source inside the Rowland circle (Figure 14.6). The displaced source causes θ_B to change
along the surface of the analyser, with a corresponding energy dependence, enabling an
energy-dispersive mode of operation with the output recorded using a pixel detector. For
this mode of operation, the effects of source size on the energy resolution are minimal,
even for quite low values of θ_B [6, 7]. A schematic of the analyser geometry is shown in
Figure 14.6, with a photograph of the tender X-ray spectrometer at SSRL, in position inside
the beamline 6-2 hutch, shown in Figure 14.7.

In tender X-ray K-edge XES studies of second-period p-block elements, the Kβ lines
derive from the valence levels and are expected to be sensitive to electronic structure.
This recently has been exploited for sulfur compounds [8, 9] and for phosphorus [10].
Figure 14.8 shows some data measured with this experimental setup, together with a
comparison with conventional XAS for solutions of two sulfur compounds (elemental

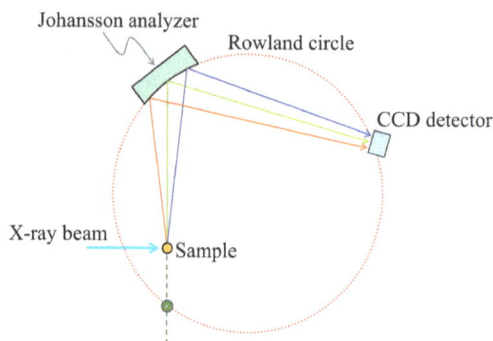

Figure 14.6: Schematic plan view of an inside Rowland sample geometry with a cylindrically bent Johannson analyser. The CCD detector is a position-sensitive device, giving a direct readout of the emission spectrum as a function of energy, represented by different coloured rays.

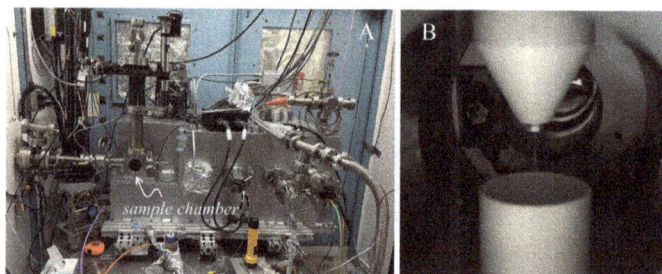

Figure 14.7: Photograph of the SSRL tender X-ray spectrometer. Panel A shows the exterior of the spectrometer installed inside the beamline 6-2 hutch. The sample chamber is indicated, with the detector and analyser encased in a large aluminium vacuum chamber. Panel B shows a liquid jet delivery system for use with solutions inside the sample chamber of the spectrometer in A.

sulfur and sulfate). The XES is of interest as a complement to XAS for chemical speciation and for insights into electronic structure. Qureshi et al. [8], used Monte Carlo methods to compare the speciation ability of sulfur K-edge XAS and sulfur Kβ XES for reduced sulfur compounds, finding that of the two methods, XAS is the more sensitive speciation tool for non-aromatic sulfur compounds, which included various thiols, sulfides and disulfides. This is because the highest occupied molecular orbital (HOMO) is a nonbonding orbital containing a pair of electrons, centred primarily on sulfur, with some (S–C)π* character. Transitions from this orbital give an intense high-energy Kβ XES feature that is relatively invariant between the different compounds. For the aromatic compound thiophene, however, the HOMO is a C=C π-bonding orbital with very little sulfur character, and it is the HOMO–1, a lone-pair type orbital with close to 50% sulfur 3p character that provides the highest energy XES transition. For conjugated more extended polycyclic aromatic compounds such as benzothiophene and dibenzothiophene, the polycyclic aromatic system rearranges the levels so that the orbital corresponding to the HOMO-1 of thiophene is now the HOMO, and this transition shifts in energy with the extent of conjugation, providing a potential probe of these compounds, so that for aromatic sulfur compounds, the Kβ XES may be more sensitive to speciation than the XAS [8].

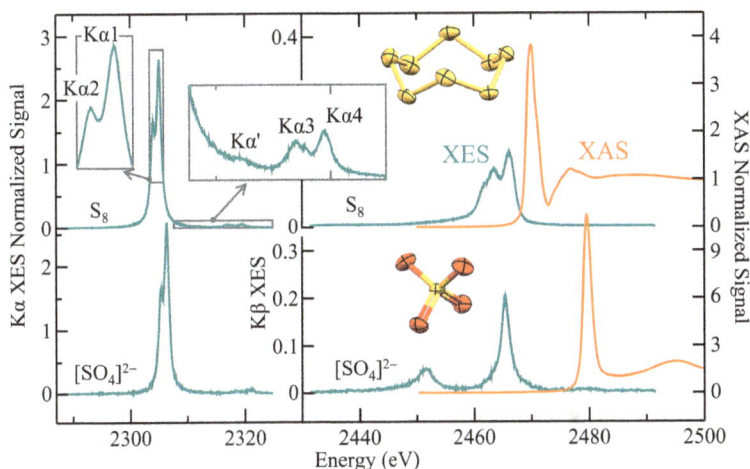

Figure 14.8: Comparison of sulfur K-edge XAS (orange) and XES (blue green) for elemental sulfur and ammonium sulfate. Samples were 100 mM sulfur solutions in water at pH 7.4 for $(NH_4)_2SO_4$ or in xylene for S_8. The left panel compares the Kα XES of the two compounds showing clear chemical shifts of the Kα1 Kα2 doublet, but no structure sensitivity to chemical form. The right panel shows the Kβ XES of the same two solutions, showing significant differences in the spectroscopic structure, paralleling that of the XAS. The Kα', Kα3 and Kα4 satellite peaks are 2p→1s transitions occurring in the presence of spectator 2p holes.

Ground-state density functional theory (DFT) calculations can be used with XES to obtain insights into electronic structure [8]. Figure 14.9 shows DFT simulations of the sulfur Kβ XES of dibenzothiophene, together with the molecular orbital isosurfaces corresponding to the three most intense emission lines (labelled A, B and C), which were computed using the Orca software package. Mainly because computations for XAS are less challenging, simulation of XES is substantially older than that of XAS, for example see Perera and LaVilla [11], who used both semi-empirical MNDO and *ab initio* STO-3G molecular orbital calculations to successfully simulate the sulfur Kβ XES of thiophene.

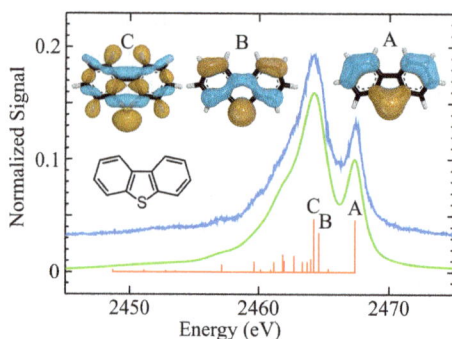

Figure 14.9: Sulfur Kβ XES of a toluene solution of dibenzothiophene. The blue curve shows the experimental data, with the red stick spectrum showing the transition positions and relative intensities, and the green curve following line shape convolution. The insets labelled A, B and C show the molecular orbital isosurfaces (0.03 e$^-$/au^3) corresponding to the three intense emission lines labelled A, B and C.

14.5 Soft X-ray XES

X-ray emission spectroscopy in the soft X-ray regime is typically detected either by a high-resolution grating spectrometer or by high-resolution solid-state detectors, such as transition edge sensor arrays (see Section 5.6.1) although these devices have lower resolution compared to grating spectrometers. Diffraction gratings were introduced in Section 3.3; many soft X-ray XES beamlines use variable line spacing (VLS) diffraction gratings with focusing optics to collect X-ray emissions from the sample, such as a spherical mirror. The VLS diffraction grating produces an energy dispersive output on a two-dimensional detector (Figure 14.10). Other spectrometer designs include what is called a Rowland circle grating spectrometer with the source (the sample illuminated by the incident X-ray beam), a spherically bent diffraction grating of bend radius equal to the diameter of the Rowland circle and an area detector, all on a Rowland circle. A major use of soft X-ray XES facilities is to record the RIXS, discussed in Section 14.7.

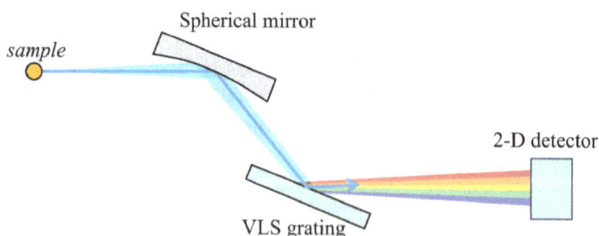

Figure 14.10: Schematic of a soft X-ray XES experiment using a VLS diffraction grating.

14.6 High-energy resolution fluorescence detected XAS (HERFD-XAS)

HERFD-XAS is an emerging technique. Not many years ago, it might have been considered esoteric and limited to a small range of highly concentrated samples, but recent applications have demonstrated a wider applicability, together with significant advantages over conventional XAS. The first of these advantages is that HERFD-XAS provides a way to overcome the lifetime broadening of XAS by measurement of the X-ray fluorescence, with a resolution that is better than the natural linewidth of a fluorescence line. As discussed in Section 9.6, two major factors contribute to the observed spectroscopic linewidth; the first is the resolution of the beamline optics (Chapter 3) and the second is what is called the lifetime broadening (Section 9.6). The optical resolution of the beamline provides a Gaussian broadening of the spectra, which we describe using $\Gamma_G(E)$, while the lifetime broadening provides a Lorentzian contribution Γ_{lifetime}, with the overall broadening expressed by a convolution of the two functions, which is known as a Voigt function:

$$\Gamma_{\text{expt}}(E) = \Gamma_{\text{lifetime}} \otimes \Gamma_G(E) \tag{14.3}$$

Here, the symbol $\otimes$ is used to signify the convolution of two functions. While with soft X-ray measurements, the beamline optics provide the dominant source of spectroscopic broadening, with hard X-ray spectroscopy, the lifetime broadening is the largest term, thus it is in this spectroscopic regime that HERFD-XAS can provide substantial gains. The optical broadening from the beamline optics can be measured, and for the lifetime broadening, we can refer to Heisenberg's uncertainty principle, which as already commented, can be cast in different ways, for example momentum and position, or time and energy, as follows:

$$\Gamma_{\text{lifetime}}\Delta t \geq \frac{\hbar}{2} \tag{14.4}$$

Here Δt is the lifetime of the core hole created by the primary photoexcitation event, which in the case of a K-edge would be a 1s core hole. Since the 1s electrons are the most strongly bound, a 1s core hole will be very short-lived. For a K-edge in the hard X-ray regime, Δt is typically less than 1 fs (10^{-15} s), for example for the selenium K-edge, Δt is about ¼ fs, which corresponds to a lifetime broadening of ~3 eV, which would compare with a typical beamline optical broadening of around 1 eV (see Table 3.1). Here, to be consistent with most of the literature, we will use a slightly different nomenclature than in previous chapters, employing Ω for the energy of the incident X-ray photon and ω for the energy of an emitted X-ray photon.

In Figure 14.11, we consider a simplified picture using atomic configurations using an incident X-ray, with Ω close to the K-absorption edge of a 4p block element such as selenium. The configuration of the initial state $|\psi_i\rangle$ can be written as $1s^2 2p^6 4p^n$. The inci-

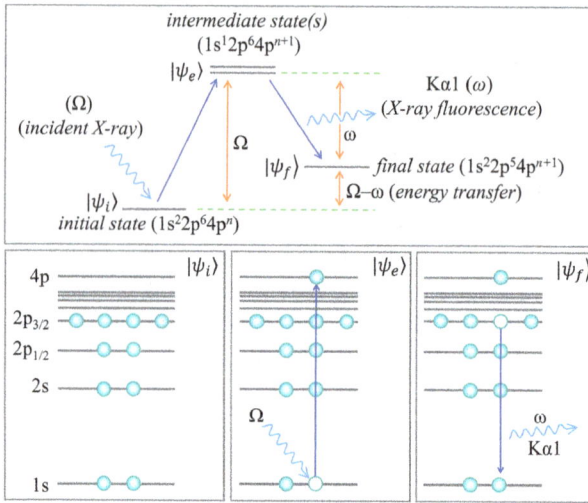

Figure 14.11: Relevant transitions for photon-in/photon-out spectroscopy. The upper panel shows a simplified description of X-ray absorption and emission using simple atomic configurations, consisting of initial state, intermediate state and final state, with the three bottom diagrams showing simplified schematics of these different states. These are not shown to scale and for simplicity, omit the n = 3, 4s or the 4p electrons, other than that being excited from the 1s.

dent X-ray excites a 1s electron, which for an intense transition might correspond to a dipole-allowed 1s→4p transition. This excitation creates a 1s core hole with an electron promoted to the 4p level, and results in an intermediate or excited state $|\psi_e\rangle$ with a configuration of $1s^1 2p^6 4p^{n+1}$. This intermediate state would correspond to the final state in conventional XAS, and as we have previously discussed, it might decay via fluorescent emission of an X-ray photon (of energy ω), which for Kα1 fluorescence, would correspond to a $2p_{3/2} \rightarrow 4p$ transition, filling the 1s core hole and creating a 2p core hole, to generate a final state $|\psi_f\rangle$, distinct from the initial state with a $1s^2 2p^5 4p^{n+1}$ configuration. The difference in energy between the initial state and final state is given by $\Omega - \omega$ and is usually called the energy transfer. We will return to the energy transfer when we discuss what is called the RIXS plane.

For simplicity, we focus our discussion here on K-edge XAS of a p-block element, measured using Kα1 fluorescence, although the same considerations will be important for other absorption edges and fluorescence lines. With conventional XAS, there are two core-hole lifetimes to consider: the energy broadening due to the short lifetime of the 1s core hole, which we will now refer to as Γ_K; and the broadening arising from the lifetime of the $2p_{3/2}$ hole created by the decay of a 2p electron to fill the 1s core hole, which we will call Γ_L. A conventional Kα1 X-ray fluorescence-detected XAS measurement corresponds to all possible transitions encompassing the energy uncertainties arising from Γ_K and Γ_L, with Γ_K being substantially larger than Γ_L. This is shown schematically in Figure 14.12A. Now, with the crystal spectrometers discussed above,

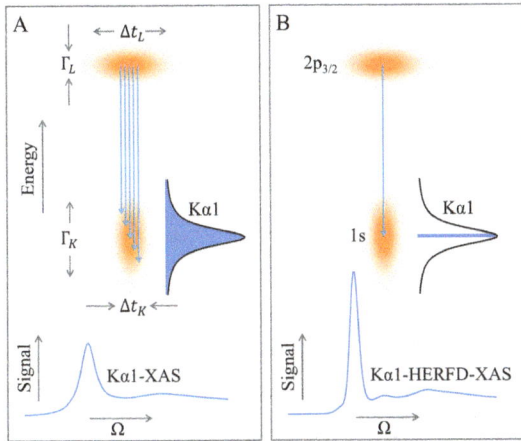

Figure 14.12: Schematic comparison of broadening processes in XAS (A) and HERFD-XAS (B). In both (A) and (B), the upper diagrams use the horizontal dimension to illustrate the different core-hole lifetimes Δt_K and Δt_L for the 1s and 2p levels, respectively, with the corresponding energetic uncertainties Γ_K and Γ_L shown in the vertical dimension. The areas of the resulting elliptical profiles are the same, equivalent to $\geq \hbar/2$ (eq. (14.4)). In (A), the fluorescence-detected XAS accepts the entire profile of the fluorescence line, including all possible transitions, while in (B), the HERFD-XAS accepts fluorescence corresponding to only a narrow band of transitions, eliminating much of the Γ_K broadening. The lower panels in (A) and (B), respectively show the resulting XAS and HERFD-XAS.

the X-ray fluorescence is measured with high-energy resolution, which effectively selects a narrow band of transitions from all of those that give rise to the fluorescence XAS. This is shown schematically in Figure 14.12B. In this case, the resulting spectrum is no longer dominated by the broadening from the 1s core-hole lifetime, Γ_K, although will still have the considerably smaller broadening due to the lifetime of the $2p_{3/2}$ hole, Γ_L. This gives a dramatic increase in spectroscopic resolution giving the high-energy resolution fluorescence-detected XAS or HERFD-XAS, helping to overcome one of the major limitations of near-edge spectroscopy, which is a lack of spectroscopic resolution. In agreement with this, analysis of the spectral peak shapes shows that the HERFD-XAS is comprised of features that appear mostly Gaussian, suggesting that now the beamline optics provide the limiting resolution, whereas the conventional XAS appear mostly Lorentzian, consistent with lifetime broadening being limiting [12]. This is shown for selenate in Figure 14.13.

Continuing with our example of a p-block element such as selenium and jumping ahead from the qualitative discussion to a more quantitative approach, the lifetime broadening of the HERFD-XAS can be shown to be given as follows:

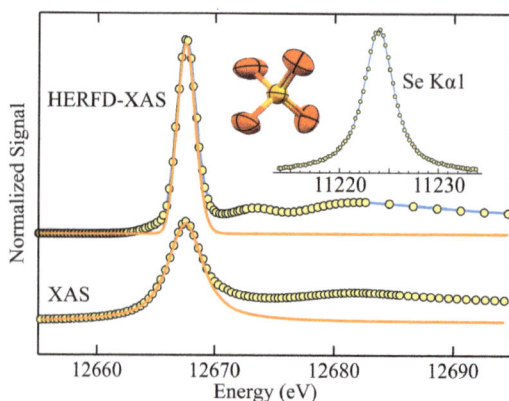

Figure 14.13: Comparison of peak deconvolution for XAS and HERFD-XAS. The sample is an aqueous solution (1 mM) of sodium selenate at pH 5.5. The result of peak deconvolution is shown by blue lines, with experimental data shown as points. The orange lines show the peak deconvolution result for the major feature in both spectra. For XAS, this fits best to an 82% Lorentzian with 3.55 eV FWHM, and for HERFD-XAS, a 100% Gaussian with 1.81 eV FWHM. The insets show the structure, and the Se Kα1 emission scan used to collect the HERFD-XAS.

$$\Gamma_{\text{HERFD}} = \frac{1}{\left(\frac{1}{\Gamma_L^2} + \frac{1}{\Gamma_K^2}\right)^{\frac{1}{2}}} \tag{14.5}$$

A comparison of the XAS and the HERFD-XAS for a range of biologically relevant reduced selenium compounds is shown in Figure 14.14. If one considers speciation analysis such as that discussed in Chapter 11, in many cases, the XAS spectra of Figure 14.14 might be hard to distinguish but, in each case, the additional spectroscopic resolution of the HERFD-XAS means that these are quite distinct.

14.6.1 HERFD-XAS nomenclature

A variety of nomenclature for HERFD-XAS has been suggested, including recommendations such as high-resolution XAS (HR-XAS or HR-XANES). Admittedly, while the point of HERFD-XAS experiment is to gain spectroscopic resolution, this has potential for confusion with high-resolution EXAFS measurements [14], and also infers that HERFD-XAS is more closely related to conventional XAS than it actually is. We discuss some of the differences and complexities in more detail below, but from our qualitative discussion above, it should be clear that HERFD-XAS can be considered as a subset of the total fluorescence yield XAS. One issue is that while HERFD-XAS can be associated with different fluorescence lines that will result in non-equivalent HERFD-XAS spectra, the HR-XAS terminology provides no information about which fluores-

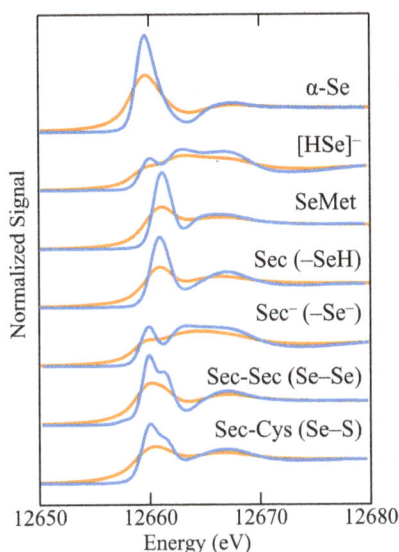

Figure 14.14: Improved resolution of HERFD-XAS. The figure compares the HERFD-XAS (blue lines) and the XAS (orange lines) for a series of biologically relevant selenium compounds. Aside from α-Se, samples were 1 mM Se aqueous solutions prepared, as previously described [13]. The abbreviations for selenocysteine (Sec) and selenomethionine (SeMet) are used in the figure.

cence line was used. For example, a study of the HERFD-XAS of Pb(II) compounds used the Lβ5 emission line, which is a minor line comprised of a pair of closely separated $5d_{5/2} \rightarrow 3p_{3/2}$ and $5d_{3/2} \rightarrow 3p_{3/2}$ transitions, respectively at 13,016.0 and 13,013.4 eV. This was selected over the most intense fluorescence line, the Lα1 ($3d_{5/2} \rightarrow 3p_{3/2}$) because the available analyser crystals [Si(880)] satisfied the near-90° condition for the Lβ5 emission, but not for others. We have suggested a nomenclature in which the line selected is part of the name, such as Se Kα1 HERFD-XAS or Pb Lβ5 HERFD-XAS. As before, we adopt the Siegbahn nomenclature, as it is standard in the field of X-ray spectroscopy, but we note here the IUPAC alternative (see Chapter 4) might be Se K-L3 HERFD-XAS or Pb L3-O5,4 HERFD-XAS.

14.6.2 Selection of HERFD-XAS analyser crystals

As we have already remarked, selection of the correct analyser with θ_B close to 90° is important for achieving improved energy resolution with HERFD-XAS using a Johann spectrometer. Table 14.1 lists some possible crystal cuts that satisfy this condition.

We note that some cuts in Table 14.1 are selected because they belong to a parent with good diffraction. For example, Pu Lα1 HERFD-XAS might be done using Si(11,5,3) analysers with $\theta_B = 84.5°$, but this is a very obscure cut with a relatively large ω_D;

Table 14.1: Selected HERFD-XAS analyser crystals.

Z	Element	Line	Analyser	Parent	θ_B (°)	ω_D (µrad)	F_{hkl}
22	Ti	Kα1	Ge(400)		76.3	266.4	163.5
25	Mn	Kα1	Ge(333)	Ge(111)	74.9	85.1	98.1
26	Fe	Kα1	Ge(440)		75.5	99.0	129.5
29	Cu	Kα1	Si(444)	Si(111)	79.3	29.2	39.9
30	Zn	Kα1	Si(642)		81.4	28.6	36.5
33	As	Kα1	Si(911)		80.5	9.5	19.9
34	Se	Kα1	Si(842)		85.3	21.1	25.2
42	Mo	Kα1	Ge(999)	Ge(111)	77.8	4.2	34.5
58	Ce	Lα1	Ge(331)		80.7	224.0	110.1
60	Nd	Lα1	Si(331)		72.1	43.6	40.7
62	Sm	Lα1	Si(422)		82.9	118.0	53.7
64	Gd	Lα1	Si(333)	Si(111)	78.3	43.1	36.5
66	Dy	Lα1	Si(440)	Si(220)	83.8	92.3	48.4
74	W	Lα1	Ge(642)		77.6	51.1	99.7
79	Au	Lα1	Ge(555)	Ge(111)	77.7	23.3	60.1
80	Hg	Lα1	Si(555)	Si(111)	81.8	13.0	21.3
82	Pb	Lα1	Si(911)		80.3	9.3	19.9
92	U	Lα1	Si(880)	Si(220)	71.6	3.2	20.4
94	Pu	Lα1	Si(777)	Si(111)	75.8	2.3	13.0

Si(777) is considerably more convenient because all HERFD-XAS beamlines will almost certainly possess Si(111) analysers as an entry level choice. Convenience is also part of the reason why Hg Lα1 HERFD-XAS using Si(555) analysers is often the first measurement conducted on new HERFD-XAS infrastructure, together with a very convenient X-ray energy and remarkable improvements in spectroscopic resolution. The importance of a correct choice of analyser crystal satisfying the near-90° condition cannot be overemphasized. Indeed, as we have previously commented [12, 13], the literature contains examples of HERFD-XAS experiments, where the spectroscopic resolution was no better [15], and in one case worse [16], than conventional XAS, although in fairness to the authors of these studies, we note that they were clearly aware of the issues.

The requirement for specific analyser crystals represents a limitation with respect to conventional XAS. While for XAS, changing elements can be a trivial matter, with HERFD-XAS, the process of changing and aligning crystals in the array detector means that different users examining the same element should probably be scheduled back-to-back. We note that some flexibility is afforded by the different orders of diffraction for some cuts, for example the Si(111) can be used for all of Gd Lα1, Cu Kα1, Hg Lα1 and Pu Lα1 HERFD-XAS.

14.6.3 Anatomy of a HERFD-XAS experiment

In this section, we review how a typical HERFD-XAS experiment might proceed. We assume that the HERFD-XAS measurements are of a dilute system, and use the most intense fluorescence line, such as the Kα1 for a K-edge or the Lα1 for a L_{III} edge (Table 14.1). Because, HERFD-XAS beamlines typically have high X-ray flux-densities, radiation damage is often an issue, with the sample typically placed in a liquid helium cryostat. We assume that appropriate crystal analysers (Table 14.1) have already been selected and aligned. An estimate of the energy resolution of the analyser is usually made by moving the incident X-ray energy to the energy of the fluorescence line of interest and measuring the width of the elastic scatter from a blank sample containing none of the element of interest. Once the sample has been aligned within the cryostat, the incident energy is moved well above the absorption edge; using this energy, the initial measurement is an emission scan such as shown in Figure 14.13 (inset), from which the analyser angle can be carefully calibrated to the experimentally measured peak of the X-ray emission line. A conventional XAS-type scan is then measured, usually acquired quite quickly in a total scan time of perhaps 1–2 min, with the structured portion of the spectrum being recorded in the first 30–60 s. With some beamlines, continuous scanning allows collection of a scan or sweep in even shorter time periods. The sample, which is typically much larger than the incident X-ray beam, is then translated so that the incident beam interrogates a fresh spot on the sample, and the next sweep is collected. Sometimes, two sweeps on one spot might be collected to monitor any changes that might be occurring, and if there are changes apparent, then faster scans might be used, and the average of all the first sweeps might be used.

14.6.4 Importance of accurate emission line determination

As we discussed in the last section, accurate determination of the emission line centroid is an important factor in a successful HERFD-XAS experiment. This is because if a slightly off-peak emission energy is used, then the HERFD-XAS will be shifted relative to the on-peak spectra, and relative to the conventional XAS. If the analyser energy is set to an emission energy that is +1 eV off-peak, then the resulting HERFD-XAS will be shifted by +1 eV, with some distortions of the spectra, although if the offset is small, then these will be subtle [13]. In many cases, automated software is available to fit the emission line, giving a peak position and FWHM with estimated standard deviations. Figure 14.15 shows some examples of such emission line fits.

As we have already mentioned (e.g. see Figure 14.8), the position of the emission line may also show chemical shifts that tend to be larger for smaller atoms. For elements that show appreciable chemical shifts, an emission scan must be recorded for each sample, and choosing the emission energy for mixtures of species becomes complicated, as we will discuss. Some early HERFD-XAS studies used a fixed emission energy

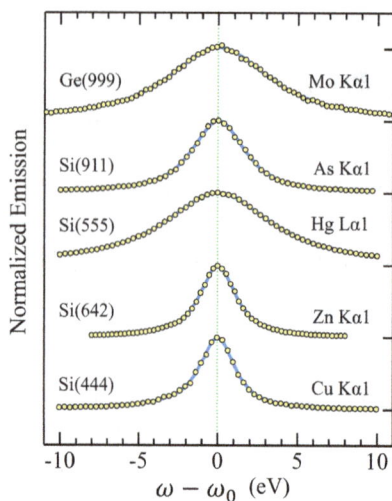

Figure 14.15: Examples of determination of emission line peak energies. A range of fluorescence lines for different elements are shown. Traces have been aligned on the abscissae by subtraction of the fitted emission peak position ω_0, and are arranged vertically in order of increasing emission energy. The experimental data are shown as the points with the least-squares fit to the peaks shown by the blue lines. The analyser crystal cut employed for each fluorescence line is shown adjacent to each curve. We note that the Lα1 emission shows an inherently larger natural linewidth, comparable to that of the Mo Kα1 line.

of a concentrated standard for all samples studied, and consequently reported spectra that were shifted. For lighter elements, larger shifts are expected. For sulfur, we observed a +1.36 eV Kα1 shift between α-sulfur (S_8) and sulfate $[SO_4]^{2-}$ [8], while between α-selenium (Se_8) and selenate $[SeO_4]^{2-}$, the Kα1 shifts by only +0.23 eV [13]. As might be expected (see Section 9.3.3), fluorescence line chemical shifts also occur between different chemical species that have the same formal oxidation states. Thus, α-sulfur (S_8) and organic sulfoxides both have formally zero-valent sulfur, but the sulfoxide Kα1 is shifted by +0.36 eV relative to that of α-sulfur [8]. For hard X-ray measurements of heavy elements, such as mercury, the chemical shifts in the major fluorescence line (e.g. the Lα1) can be so small as to be unobservable [12], which relieves the requirement for determination of emission lines on a per sample basis [12]. This invariance is related to the smaller chemical shifts observed with larger atoms, in that the intervening shells of s, p, d and f electrons (in the case of mercury) serve to shield the core level from any influence of changes in the valence levels.

14.6.5 HERFD-XAS of dilute systems

The solid angle afforded by Johann array detector systems is small relative to a conventional XAS setup. With a 1 m Rowland circle and analyser crystals of 100 mm diam-

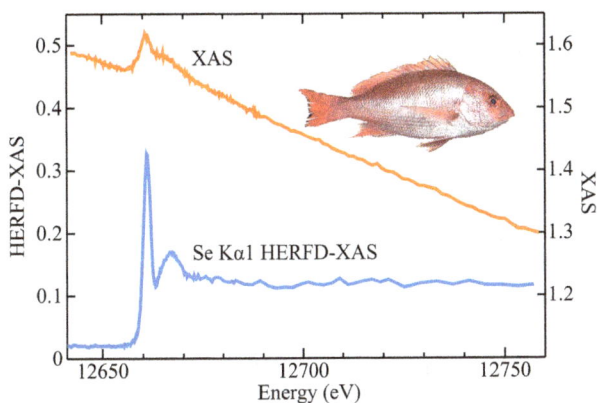

Figure 14.16: Spectroscopy of a very dilute sample. The plot compares raw averaged conventional XAS (orange) and Se Kα1 HERFD-XAS (blue) of the same sample of skeletal muscle from pacific red snapper *Lutjanus peru* (inset) containing ~5 µM Se. Similar conditions were used for both measurements. Data have been replotted from Nehzati et al. [13].

eter, a seven-analyser array will intersect only about 1/42 of the solid angle accepted in a conventional fluorescence XAS experiment using a 100-pixel germanium monolith. Challenges with signal to noise and dilute systems with HERFD-XAS therefore might be anticipated. In fact, in terms of signal to noise, the HERFD-XAS experiment compares quite favourably with standard XAS, for two reasons: HERFD-XAS provides vastly superior background rejection; and the HERFD-XAS experiment is very well behaved in a statistical sense.

Figure 14.16 compares the Se Kα1 HERFD-XAS with the conventional XAS of a very dilute sample. The XAS data in Figure 14.16 were collected using a 100-pixel germanium monolithic detector, with a resolution of ~250 eV (FWHM) under typical operating conditions. Since the electronic window was set to 400 eV to include the entire Se Kα12 fluorescence, considerable inelastic (Compton) X-ray scatter is accepted, which appears as a high sloping background in the XAS data of Figure 14.16, decreasing as the incident X-ray beam sweeps to higher energies. Conversely with HERFD-XAS, the fluorescence is measured at a vastly superior energy resolution, which might be less than 1 eV, depending on the element being studied. This effectively eliminates almost all the background signal from inelastic scatter. As discussed by Nehzati et al. [12, 13], various factors contribute to the background-related advantages, in particular the inelastic scattering extends to low X-ray energies, to an extent, which depends on the incident X-ray energy, as discussed in Chapter 4. For higher energy absorption edges, such as the Mo K-edge, this provides more significant background signal in the vicinity of the fluorescence line. Conversely, for lower energy absorption edges, such as the Fe K-edge, the profile of the inelastic X-ray scatter contributes much lower backgrounds in the vicinity of the fluorescence line. Moreover, if L-edges are being studied, the greater energy separation of the fluorescence line from the absorption edge, compared with a K-edge, also decreases the background from scattered X-rays in an XAS measurement. Thus, the low-concentration advantages of HERFD-XAS over XAS are most pronounced for the late first transition metal K-edges, for the second transition metal K-edges and for the K-edges of third row p-block elements.

In addition to improved background, the statistical behaviour of HERFD-XAS also more perfectly approaches the ideal with signal-to-noise increasing with the square root of the number of scans averaged [12]. With conventional fluorescence XAS, there are other issues that effectively limit the success of signal averaging for dilute and ultra-dilute samples. These include detector non-linearities, inadequacies of readout electronics and low-energy tails in the signal peak shapes from solid-state detectors. The overall result of these issues is that with conventional XAS, the noise levels of averaged spectra often show somewhat more noise than would be expected from photon-statistics alone.

14.6.6 HERFD-XAS of concentrated systems

As discussed in previous chapters, with conventional XAS, different methods of detection can be used, depending on the sample concentration. For concentrated samples, transmittance is often best as this avoids distortions from fluorescence self-absorption, while for dilute samples, fluorescence measurements can be used. Because HERFD-XAS can only be measured using the fluorescence (but see Section 14.6.8), self-absorption effects will be present in spectra from all concentrated samples. As discussed by Nehzati et al. [13], with HERFD-XAS, attenuation of both incident and fluorescent X-rays is governed by the conventional X-ray absorption, rather than any high-resolution effect. These workers derived a simple relation for a sample positioned at 45° to the incident X-ray beam, which can be used for self-absorption correction:

$$H'(E) \propto H(E) \frac{(A(E) + A_F)}{[1 - \exp(-[A(E) + A_F])]} \tag{14.6}$$

Here, $H'(E)$ is the self-absorption-corrected HERFD-XAS, $H(E)$ is the measured HERFD-XAS, $A(E)$ is the sample absorbance and A_F is the sample absorbance at the fluorescence energy. $A(E)$, which would be determined from incident and transmitted beam intensities measured simultaneously with the HERFD-XAS, and A_F would be determined by using a single point measurement by simply moving the incident X-ray energy to the fluorescent energy. The assumption here is that the angular offsets of the analysers in both the horizontal plane (due to positioning) and the vertical plane (due to the Bragg angle θ_B) are negligible, so the measurement of sample absorbance using the transmitted beam is approximately equal to the absorbance of the X-ray fluorescence along the beam trajectory from the sample to the analyser. For Se Kα1 HERFD-XAS, using SSRL's seven-crystal array with Si(844) analysers, Nehzati et al. calculated that $A_F = A(E_F) \times 0.9721$, so that $A_F \approx A(E_F)$ is a valid approximation [13].

14.6.7 HERFD-XAS speciation analysis of mixtures with fluorescence chemical shifts

In Chapter 11, we discussed speciation analysis of complex mixtures using XAS. For elements that show significant speciation-dependent X-ray fluorescence chemical shifts, the situation for HERFD-XAS is rather more complicated than for conventional XAS. Thus, with Se Kα1 HERFD-XAS, the Se Kα1 shifts by some 0.4 eV between formally Se(–II) selenomethionine and formally Se(VI) selenate. If a sample contains a mixture of species such as these, possibly in combination with others, then the use of single emission energy will mean that the resulting spectra for one or more components are shifted in energy and distorted, and that the methods discussed in Chapter 11 cannot be applied with any validity. Several strategies are open to the experimenter. The most obvious of these would be to collect data for all standard compounds and for the unknown, at the same emission energy corresponding to the emission peak energy of the unknown. This will mean that some standard spectra will inevitably be off-peak and thus shifted and distorted, and moreover, they cannot be transferred from one data analysis to the next, so a library of correctly recorded standard spectra will be less useful. Alternatively, the emission energy of a specific standard might be chosen for all samples; while this enables the library of standards to be transferred, it suffers the same issue that some standards, and even the unknown, will be measured off peak; this is particularly problematic if the unknown is dilute. An alternative strategy would be to collect all standard spectra using a range of different emission energies, incrementally spanning the full possible chemical shift range, which we have called the RIXS ribbon [13]. We have yet to discuss RIXS, but we will do so in Section 14.7. The unknown can then be measured at its optimized emission energy, with the appropriate standards corresponding to this emission energy employed in the analysis. A final strategy would be to collect the unknown using the RIXS ribbon method. The disadvantages of this method are that the unknown may be dilute and hence may require more precious beamtime to average; moreover, there will be no exact match with any linear combination of standard spectra, unless these are also collected using a RIXS ribbon. Finally, in investigation of biological samples, we have found that radiation damage of samples can sometimes be limiting with these RIXS ribbon methods, as they inevitably involve substantial beam exposure.

14.6.8 High-energy resolution electron yield detected X-ray absorption spectroscopy

In principle, the same advantages gained by measuring the fluorescence with high-energy resolution can be obtained by measuring the electron yield with high resolution. This has been demonstrated by Drube and co-workers, both for the $L_{III}M_{IV,V}M_{IV,V}$ Auger electron yield of palladium and silver metal [17], and for a series of silver-gold alloys [18]. These edges fall into the tender X-ray regime, and here, optical resolution of the beamline, in part, mitigates the advantages of high-resolution methods. Nonetheless, a clear increase in spectroscopic resolution is obtained with high-resolution Auger measurement, as shown in Figure 14.17 for palladium metal.

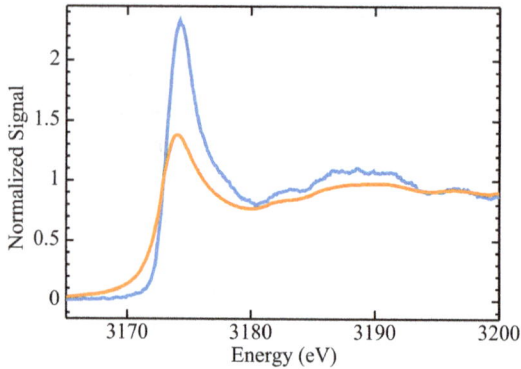

Figure 14.17: High-energy resolution electron yield detected XAS. The conventional total electron yield L_{III} edge XAS (orange) is compared with the $L_{III}M_{IV,V},M_{IV,V}$ Auger electron yield XAS, measured using high-energy resolution (blue). Data have been re-plotted from Drube et al. [17].

Despite the promise of this technique, and some advantages [18], at the time of writing, there are only a very few applications [18].

14.7 The RIXS plane

Returning now to photon-in/photon-out methods, we will now meet our promise to elaborate upon resonant inelastic X-ray scattering (RIXS). With the ability to detect X-ray emission with high-energy resolution and to scan the incident X-ray energy with high resolution, we have, in effect, the means to do two-dimensional spectroscopy. Such a measurement is called the resonant inelastic X-ray scattering plane or RIXS plane. The description of this photon-in/photon-out process as inelastic X-ray scattering (the IXS in RIXS) at first might seem confusing, but is not an incorrect description (see Chapter 4). The RIXS plane is usually defined as a plot of the intensity near a selected fluorescence line as a function of incident energy Ω and energy transfer $\Omega - \omega$, although sometimes it is plotted as incident energy Ω and emission energy ω. In general, the plot of Ω vs. $\Omega - \omega$ is preferred, as we elaborate below. Remaining with selenium spectroscopy, the Se Kα1 RIXS plane of zinc selenide (ZnSe) is shown in Figure 14.18A. The structure of the RIXS plane appears as a broad diagonally disposed range of peaks. If one takes a diagonal slice across the highest point in the RIXS plane, then one obtains the HERFD-XAS, which is shown in Figure 14.18B. RIXS is typically recorded by scanning the incident energy Ω for a series of emission energies ω, each scan corresponding to a diagonal cut on the RIXS plane (Figure 14.18A). The HERFD-XAS corresponds to the middle cut on the RIXS plane across the maximum of the structure and extending above the absorption edge into the so-called non-resonant region (i.e. see Section 14.6.3). Offsetting the HERFD-XAS emission energy, discussed in Section 14.6.4, corresponds to moving vertically off the central range of peaks in the RIXS plane, which both shifts and distorts the resulting spectra relative to the HERFD-XAS. Summing individual scans recorded to measure the RIXS plane – in Figure 14.18, summing vertically the values of $\Omega - \omega$ for a given Ω – effectively gives the conventional fluorescence XAS spectrum.

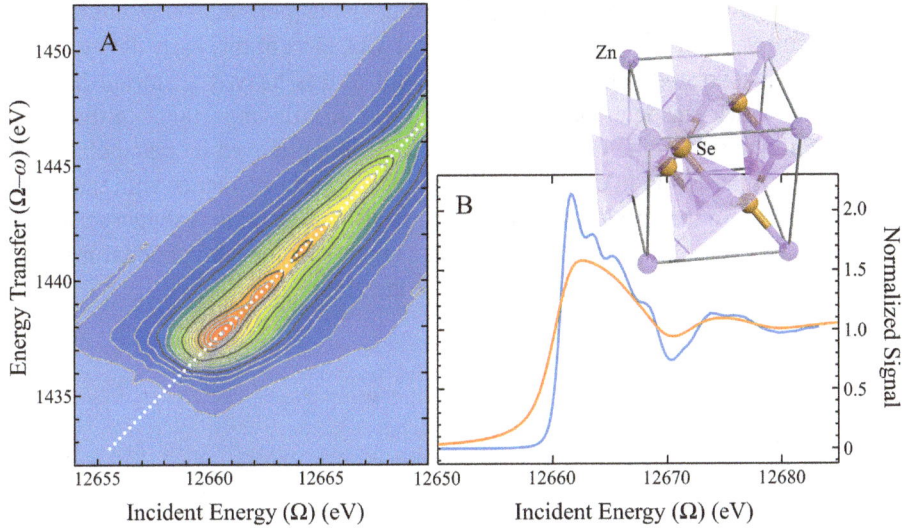

Figure 14.18: (A) The Se Kα1 RIXS plane for zinc selenide. The broken line shows the diagonal cut along the RIXS plane corresponding to the HERFD-XAS. (B) Se Kα1 HERFD-XAS (blue) and Se K-edge XAS (orange) for zinc selenide, with the inset showing the crystal structure. Summing all of the Ω scans comprising the RIXS plane in (A) gives the conventional XAS spectrum (orange).

The RIXS (and HERFD-XAS) described using the Kramers-Heisenberg dispersion relationship, again using the example of 1s K-edge photoexcitation, can be written as in eq. (14.7). Figure 14.11 (top) shows a diagram of the various states. The initial state is given by $|\psi_i\rangle$, the intermediate state by $|\psi_e\rangle$ and the final state by $|\psi_f\rangle$. The intermediate (excited) state $|\psi_e\rangle$ is reached from $|\psi_i\rangle$ via a transition operator $\hat{H}_1$ with terms $\langle\psi_e|\hat{H}_1|\psi_i\rangle$, and the final state $|\psi_f\rangle$ from $|\psi_e\rangle$ via a different transition operator $\hat{H}_2$ with terms $\langle\psi_f|\hat{H}_2|\psi_e\rangle$:

$$F(\Omega,\omega) = \sum_f \left| \sum_e \frac{\langle\psi_f|\hat{H}_2|\psi_e\rangle\langle\psi_e|\hat{H}_1|\psi_i\rangle}{E_i - E_e + \Omega - i\frac{\Gamma_K}{2}} \right|^2 \frac{\frac{\Gamma_K}{2\pi}}{\left(E_i - E_f + \Omega - \omega\right)^2 + \frac{\Gamma_L^2}{4}} \tag{14.7}$$

In eq. (14.7), $F(\Omega, \omega)$ describes the RIXS. E_i, E_e and E_f are the energies of the initial, intermediate and final states, respectively, and Γ_K and Γ_L are respectively the lifetime broadenings of the intermediate and final states.

With our first example, shown in Figure 14.18, all spectroscopic structure is aligned on the diagonal of the RIXS plane. In some cases, however, off-diagonal structure can be observed when there are a number of different intermediate states that can be excited with similar Ω; and different diagonal cuts across the RIXS plane can be used to examine different electronic states. Much early RIXS work focused on the dipole-forbidden, quadrupole-allowed 1s→3d transitions for first transition metals, for which off-diagonal RIXS are frequently observed, giving more advantageous separation of features when compared to conventional XAS. Most of this early data was collected on concentrated solids, with the major features heavily distorted by self-absorption, which was ignored. However, this was of little importance to the small 1s→3d, and in any case, the major absorption was often removed by manipulation of the data. With the 1s→3d RIXS, there is an interesting correspondence with the metal L-edge XAS, where the 2p→3d transitions are dipole-allowed. This is shown schematically in Figure 14.19, where the incident X-ray energy for the conventional metal L-edge XAS corresponds to the energy transfer in the RIXS.

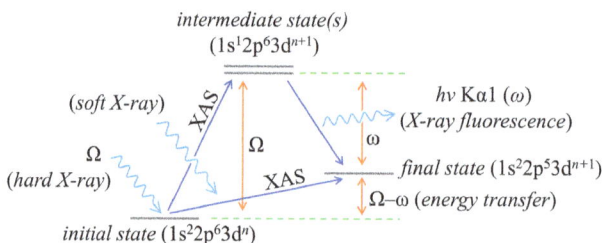

Figure 14.19: Schematic of first transition metal K-edge RIXS, with 1s→3d transitions. The same final state can be reached either via the photon-in/photon-out RIXS process involving quadrupole-allowed 1s→3d transitions or by the conventional $L_{III,II}$-edge XAS involving dipole-allowed 2p→3d transitions.

Quite complicated RIXS planes containing different diagonal contributions can be observed when the emission derives from the decay of valence electrons, such as with sulfur Kβ (3p→1s) or the oxygen Kα (2p→1s). The sulfur Kβ RIXS of diphenyldisulfide and the oxygen Kα RIXS of benzaldehyde and phenol are shown in Figures 14.20 and 14.21.

The RIXS provides detail that cannot be observed in the conventional XAS, for example in Figure 14.20, at $\Omega \approx 2{,}470$ eV, (S–S)σ* and (S–C)π* features overlap on the XAS but are separated along the energy transfer axis in the RIXS. The proximity of the Kβ to the absorption edge allows us to observe the resonant elastic X-ray scattering, which runs horizontally at zero energy transfer with $(\Omega - \omega) = 0$ eV. The REXS also shows chemical sensitivity [19], but to date this has been little explored.

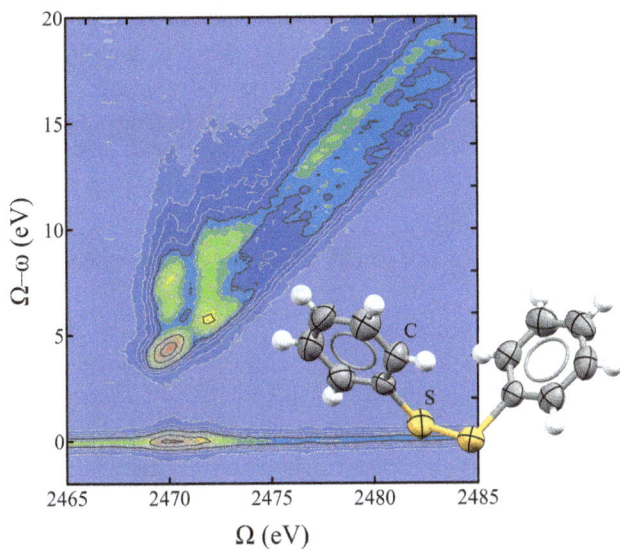

Figure 14.20: Sulfur Kβ RIXS of diphenyldisulfide [19]. The Kβ RIXS begins at energy transfer $(\Omega - \omega)$ values above ~2 eV, extending diagonally across the plot, separating features due to different intermediate states along the ordinate. The structure at $(\Omega - \omega) = 0$ eV, with intensity peaking at ~2,470 eV, is the resonant elastic X-ray scattering (REXS). The inset shows the structure of the compound.

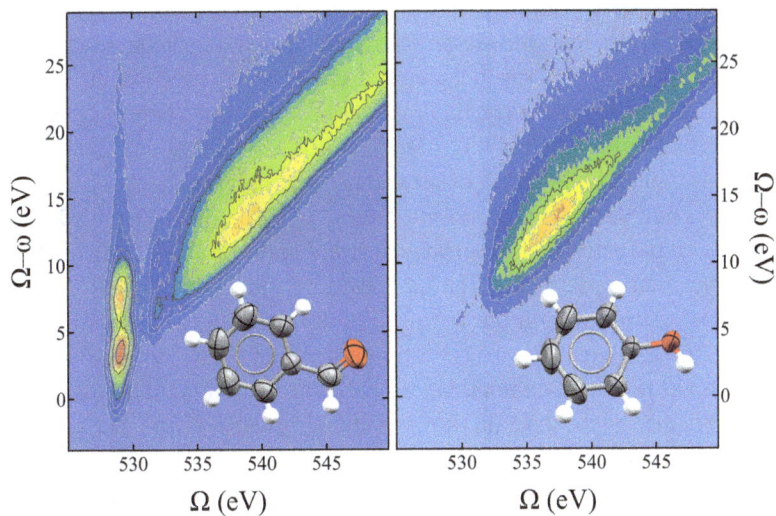

Figure 14.21: Oxygen Kα RIXS of benzaldehyde and phenol. The REXS has been removed from both plots to more clearly show the RIXS. The insets show the structures of the compounds.

Figure 14.21 shows the oxygen Kα RIXS of liquid benzaldehyde [20] recorded using a transition edge sensor (TES) array detector (see Section 5.6.1), showing two different (O=C)π* features, both at ~528 eV, separated along the energy transfer axis. In this case, the relatively poor resolution of the TES detector is apparent from the peak widths along the ordinate – a grating-based analyser would provide better spectroscopic resolution, while the TES accepts a considerably better solid angle, allowing studies of dilute systems.

One disadvantage of RIXS, without which it would probably be more extensively used, is the fact that it typically takes a long time to collect.

14.8 X-ray Raman scattering

The Raman effect, or Raman scattering, is an inelastic light scattering method that is part of the chemist's standard spectroscopy toolkit and is used as a probe of vibrational levels. In general, the Raman effect relates to the absorption and subsequent emission of a photon via an intermediate virtual state. Typically, a commercial Raman spectrometer uses a laser as an exciting source and provides information that is complementary to infrared spectroscopy because it is governed by different selection rules.[1] The Raman effect can be observed as small satellites to high and low energy of the elastically scattered light, called Stokes Raman and anti-Stokes Raman scattering, respectively. In the X-ray regime, Raman scattering can be observed as small structures within the Compton region, at energies lower than the elastic scattering, also known as non-resonant inelastic X-ray scattering or NIXS. X-ray Raman can decay either by radiative means, as shown in Figure 14.22, or non-radiatively via Auger electron emission. Nearly all applications of X-ray Raman exploit the radiative decay, which is therefore our focus. The spectrometer used is typically a Johann geometry array, and because the X-ray Raman signal can be very small, large arrays are often used [21]. Figure 14.2 shows X-ray Raman instruments on the extreme left (partially; forward scattering direction) and top right of centre (backscattering direction). The spherically bent analyser crystals all have intersecting Rowland circles at both the sample and detector, just as shown in Figure 14.1 for the HERFD-XAS array, but of course with more analyser crystals. Because these spectrometers are not designed to probe resonant effects, different analyser crystals are not required, and hence the SSRL instruments pictured in Figure 14.1 employ Si(440) analysers that use incident photon energies, typically close to 6,500 eV.

There are two ways to perform an X-ray Raman experiment. One could fix the incident energy Ω and scan the analyser energy ω in the same manner as one might

---

1 Infra-red spectroscopy needs a change in dipole moment for a vibrational transition to be excited, whereas Raman scattering occurs when the molecular polarizability differs between the initial and final state.

prepare for HERFD-XAS, resulting in the elastic scatter falling at the highest energy, with the Compton with Raman superimposed observed below this. Alternatively, the analyser energy ω can be fixed with the incident energy Ω scanned; the appearance would be reversed, with the elastic scatter at the low energy end and the inelastic features appearing above this. In practice, this latter approach is generally used since the X-ray monochromator, typically a double crystal device, is usually much easier and faster to scan than the analyser energy.

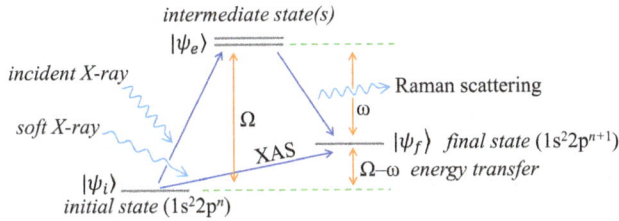

Figure 14.22: Schematic of X-ray Raman scattering and XAS for a first row p-block element. With Raman scattering, the incident X-ray of energy Ω excites a core electron from the initial state $|\psi_i\rangle$ to a high-level unoccupied state $|\psi_e\rangle$ and the excess energy is released as a Raman scattered photon of energy ω, to obtain the final state $|\psi_f\rangle$. The same final state is obtained with soft X-ray XAS, with a lower photon energy corresponding to the energy transfer $(\Omega - \omega)$.

The X-ray Raman can be described by the Kramers-Heisenberg formula given in eq. (14.6) and summarized in Figure 14.22.

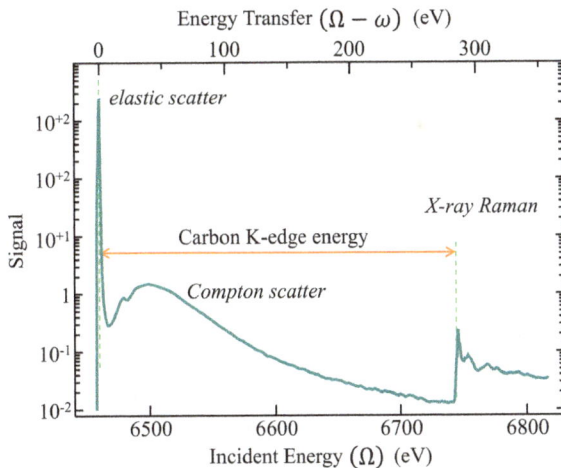

Figure 14.23: X-ray scattering of graphite. With this scan, the analyser energy ω was held at a fixed energy with the incident energy Ω scanned. The elastic scatter, the relatively broad Compton scattering and the X-ray Raman scattering, are all visible in the figure. Data courtesy of Dr Dimosthenis Sokaras, SSRL.

X-ray Raman spectroscopy, sometimes called XRS, effectively provides access to soft X-ray XAS using hard X-rays. The limitations of soft X-ray XAS, as previously discussed, for the most part relate to the lack of penetration of the soft X-ray beam. With X-ray Raman, the superior penetration of the hard X-ray beam means that many studies can be conducted that would otherwise be impossible or very difficult. For example, X-ray Raman might be used for operando studies of variations of lithium-ion electrical cells using the lithium K-edge Raman, or a pressure cell might be used to study a range of materials under extreme pressures or high temperatures. X-ray Raman is not limited to K-edges, for example first transition L-edges and early rare-earth M-edges are both possible subjects for study. Even soft X-ray XAS measurements of liquids under ambient conditions pose some challenges – for example to examine oxygen K-edge XAS of dilute materials. Vogt et al.[20] used sample cells equipped with 100 nm thick Si_3N_4 windows, cleaned of adsorbed oxygen by etching with HF and dried under argon gas, with a transition edge sensor 240 element array detector to separate the C, N and O Kα signals. A further complication relates to radiation damage of the sample. The fact that a soft X-ray beam will be essentially all absorbed within a very short path length means that soft X-ray XAS is somewhat prone to radiation damage. Oxygen K-edge XAS of aliphatic alcohols, for example, turns out to be challenging to collect, quantitatively converting to aldehydes, whereas no such difficulties exist for X-ray Raman scattering, as shown in Figure 14.24.

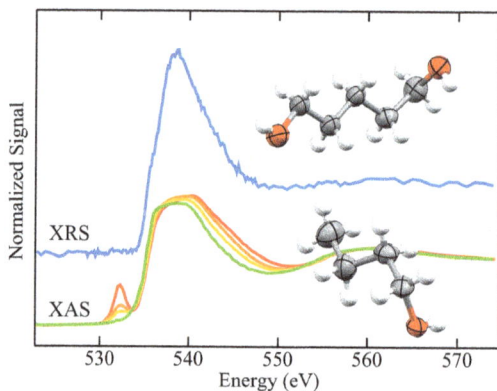

Figure 14.24: Comparison of X-ray absorption spectroscopy (XAS) and X-ray Raman spectroscopy. The figure compares oxygen K-edge XAS for an aliphatic thiol (*n*-decan-1-ol) (bottom, green to red) and the X-ray Raman for a closely related compound (*n*-pentane-1,5-diol) [22] (blue, top). The XAS changes rapidly in the X-ray beam due to radiation damage (green: first scan, red: last scan), with development of a low-energy feature at 532.2 eV, which is characteristic of an aldehyde. The two insets show the structure of *n*-pentane-1,5-diol (top) and *n*-butan-1-ol (bottom), which is similar to that of decanol. We thank Dr Dimosthenis Sokaras of SSRL for providing the X-ray Raman spectrum of *n*-penatane-1,5-diol.

A recent review provides a comprehensive review of applications of X-ray Raman scattering in the study of organic molecules [23]. The major disadvantage of the X-ray Raman method is the small signal size, requiring concentrated samples. Moreover, the infrastructure to perform the experiments is highly specialized and consequently access may be limited.

References

[1] Wang, X.; Grush, M. M.; Froeschner, A. G.; Cramer, S. P. High-resolution X-ray fluorescence and excitation spectroscopy of metalloproteins. *J. Synchrotron Rad.* **1997**, *4*, 236–242.

[2] Sokaras, D.; Weng, T.-C.; Nordlund, D.; Alonso-Mori, R.; Velikov, P.; Wegner, D.; Garachtenko, A.; George, M.; Borzenets, V.; Johnson, B.; Rabedeau, T.; Bergmann, U. A seven-crystal Johann-type hard X-ray spectrometer at the Stanford Synchrotron Radiation Lightsource. *Rev. Sci. Instrum.* **2013**, *84*, 053102/1–8.

[3] Doonan, C. J.; Zhang, L.; Young, C. G.; George, S. J.; Deb, A.; Bergmann, U.; George, G. N.; Cramer, S. P. High-resolution X-ray emission spectroscopy of molybdenum compounds. *Inorg. Chem.* **2005**, *44*, 2579–2581.

[4] Hoszowska, J.; Dousse, J.-Cl Enhanced X-ray emission from the valence states to the 1s and 2s levels in metallic Mo and several Mo compounds. *J. Phys. B: At. Mol. Opt. Phys.* **1996**, *29*, 1641–1653.

[5] Bergmann, U.; Horne, C. R.; Collins, T. J.; Workman, J. M.; Cramer, S. P. Chemical dependence of interatomic X-ray transition energies and intensities – a study of Mn Kβ″ and Kβ2,5 spectra. *Chem. Phys. Lett.* **1999**, *302*, 119–124.

[6] Kavčiča, M.; Budnar, M.; Mühleisen, A.; Gasser, F.; Žitnik, M.; Bučar, K.; Bohinc, R. Design and performance of a versatile curved-crystal spectrometer for high-resolution spectroscopy in the tender X-ray range. *Rev. Sci. Instrum.* **2012**, *83*, 033113/1–8.

[7] Nowak, S. H.; Armenta, R.; Schwartz, C. P.; Gallo, A.; Abraham, B.; Garcia-Esparza, A. T.; Biasin, E.; Prado, A.; Maciel, A.; Zhang, D.; Day, D.; Christensen, S.; Kroll, T.; Alonso-Mori, R.; Nordlund, D.; Weng, T.-C.; Sokaras, D. A versatile Johansson-type tender X-ray emission spectrometer. *Rev. Sci. Instrum.* **2020**, *91*, 033101/1–12.

[8] Qureshi, M.; Nowak, S.; Vogt, L. I.; Cotelesage, J. J. H.; Dolgova, N. V.; Sharifi, S.; Kroll, T.; Nordlund, D.; Alonso-Mori, R.; Weng, T.-C.; Pickering, I. J.; George, G. N.; Sokaras, D. Sulfur Kβ X-ray emission spectroscopy: Comparison with sulfur K-edge X-ray absorption spectroscopy of organosulfur compounds. *Phys. Chem. Chem. Phys.* **2021**, *23*, 4500–4508.

[9] Vogt, L. I.; Cotelesage, J. J. H.; Dolgova, N. V.; Boyes, C.; Qureshi, M.; Sokaras, D.; Sharifi, S.; George, S. J.; Pickering, I. J.; George, G. N. Sulfur X-ray absorption and emission Spectroscopy of organic sulfones. *J. Phys. Chem. A.* **2023**, *127*, 3692–3704.

[10] Mathe, Z.; McCubbin Stepanic, O.; Peredkov, S.; DeBeer, S. Phosphorus Kβ X-ray emission spectroscopy detects non-covalent interactions of phosphate biomolecules *in situ*. *Chem. Sci.* **2021**, *12*, 7888–7901.

[11] Perera, R. C. C.; LaVilla, R. E. Molecular X-ray spectra: S-Kβ emission and K absorption spectra of thiophene. *J. Phys. Chem.* **1986**, *84*, 4228–4234.

[12] Nehzati, S.; Dolgova, N. V.; Young, C. G.; James, A. K.; Cotelesage, J. J. H.; Sokaras, D.; Kroll, T.; Qureshi, M.; Pickering, I. J.; George, G. N. Mercury Lα1 high energy resolution fluorescence detected X-ray absorption spectroscopy: A versatile speciation probe for mercury. *Inorg. Chem.* **2022**, *61*, 5201–5214.

[13] Nehzati, S.; Dolgova, N. V.; James, A. K.; Cotelesage, J. J. H.; Sokaras, D.; Kroll, T.; George, G. N.; Pickering, I. J. High energy resolution fluorescence detected X-ray absorption spectroscopy: An analytical method for selenium speciation. *Anal. Chem.* **2021**, *93*, 9235–9243.

[14] Harris, H. H.; George, G. N.; Rajagopalan, K. V. High resolution EXAFS of the active site of sulfite oxidase: Comparison with density functional and X-ray crystallographic results. *Inorg. Chem.* **2006**, *45*, 493–495.

[15] Bissardon, C.; Proux, O.; Bureau, S.; Suess, E.; Winkel, L. H. E.; Conlan, R. S.; Francis, L. W.; Khan, I. M.; Charlet, L.; Hazemann, J. L.; Bohic, S. Sub-ppm level high energy resolution fluorescence detected X-ray absorption spectroscopy of selenium in articular cartilage. *Analyst.* **2019**, *144*, 3488–3493.

[16] Le Pape, P.; Blanchard, M.; Juhin, A.; Rueff, J.-P.; Ducher, M.; Morin, G.; Cabaret, D. Local environment of arsenic in sulfide minerals: Insights from high-resolution X-ray spectroscopies, and first principles calculations at the As K-edge. *J. Anal. At. Spectrom.* **2018**, *33*, 2070–2082.

[17] Drube, W.; Lessmann, A.; Materlick, G. Reduced core hole lifetime broadening in Auger final state yield spectra. *Jpn. J. Appl. Phys.* **1993**, *32*, 173–175.

[18] Drube, W.; Treusch, R.; Sham, T. K.; Bzowski, A.; Soldatov, A. V. Sublifetime-resolution Ag L$_3$-edge XANES studies of Ag-Au alloys. *Phys. Rev. B.* **1998**, *58*, 6871–6876.

[19] Qureshi, M.; Nowak, S.; Vogt, L. I.; Cotelesage, J. J. H.; Dolgova, N. V.; Sharifi, S.; Kroll, T.; Nordlund, D.; Alonso-Mori, R.; Weng, T.-C.; Pickering, I. J.; George, G. N.; Sokaras, D. *Unpublished observations.*

[20] Vogt, L. I.; Cotelesage, J. J. H.; Titus, C. J.; Sharifi, S.; Butterfield, A. E.; Hillman, P.; Pickering, I. J.; George, G. N.; George, S. J. *Unpublished observations.*

[21] Sokaras, D.; Nordlund, D.; Weng, T.-C.; Alonso Mori, R.; Velikov, P.; Wenger, D.; Garactchenko, A.; George, M.; Borzenets, V.; Johnson, B.; Rabedeau, T.; Bergmann, U. A high resolution and large solid angle X-ray Raman spectroscopy end-station at the Stanford Synchrotron Radiation Lightsource. *Rev. Sci. Instrum.* **2012**, *83*, 043112/1–9.

[22] Sokaras, D. *Unpublished observations.*

[23] Georgiou, R.; Sahle, C. J.; Sokaras, D.; Bernard, S.; Bergmann, U.; Rueff, J.-P.; Bertrand, L. X-ray Raman scattering: A hard X-ray probe of complex organic systems. *Chem. Rev.* **2022**, *15*, 12977–13005.

15 XAS using diffraction and reflection

15.1 Introduction

In this chapter, we explore some methods that exploit X-ray diffraction or reflection in combination with X-ray absorption spectroscopy (XAS). Unlike earlier chapters focused on phenomena involving absorption of X-rays by matter, in this chapter we will initially discuss diffraction anomalous fine structure (DAFS), which involves elastic X-ray scattering and not absorption, although as we will see it is intimately related to XAS.

15.2 Diffraction anomalous fine structure

In Chapter 14, we discussed X-ray Raman spectroscopy, which exploits inelastically scattered X-rays to obtain spectroscopic information. The technique of DAFS exploits coherent elastic scattering – diffraction – to obtain spectroscopic information. Here we will initially discuss the DAFS of powdered samples [1] because this may be more generally useful. We have discussed the basics of Bragg's diffraction in Section 3.2 including Bragg's law $n\lambda = 2d_{hkl}\sin\theta_B$ (eq. (3.1)) in which n is the order of diffraction, d_{hkl} is the crystal lattice spacing for the plane described by the Miller indices h, k and l, λ is the X-ray wavelength, and θ_B is the Bragg angle. X-ray diffraction of a powdered sample might be measured using a set-up similar to that shown in Figure 15.1.

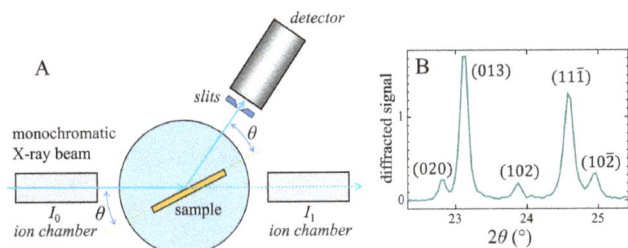

Figure 15.1: Powder diffraction basics. (A) Schematic of a simple experimental set-up for powder X-ray diffraction. (B) Example powder diffraction pattern of $K_2Ni(CN)_4$ showing discrete diffraction peaks with Miller indices (*hkl*) indicated. The angle θ is shown in (A), with diffraction in (B) conventionally plotted versus 2θ, the angle between diffracted and transmitted beams. Many modern diffraction beamlines employ area detectors to collect powder rings, which allow more rapid data acquisition than the scanning system shown.

In X-ray diffraction, abscissae are typically plotted either as the diffractometer angle 2θ, as in Figure 15.1B or as the magnitude of the X-ray scattering vector, Q, typically

https://doi.org/10.1515/9783110570441-015

measured in Å^{-1}. This accounts for X-ray wavelength λ and is given by eq. (15.1), in which c is the velocity of light and h is Planck's constant:

$$Q = 4\pi \frac{1}{\lambda} \sin \theta = 4\pi \frac{E}{hc} \sin \theta \tag{15.1}$$

If one measures the X-ray diffraction of a sample while scanning the incident X-ray energy through an X-ray absorption edge of one of its constituent elements, changes in the intensity of diffraction peaks are observed. This is shown in Figure 15.2 for a sample of the spinel Co_3O_4 in a pressed disk of lithium carbonate. Three energies are shown: 7,420 eV, which is well below the Co K X-ray absorption edge; 7,720 eV, which is mid-way up the rise of the Co K-edge step of Co_3O_4 and 8,020 eV, which is well above the Co K-edge. Figure 15.2 shows two diffraction peaks from Co_3O_4 and four peaks from the Li_2CO_3 matrix. While the latter are essentially invariant in intensity with incident X-ray energy, the Co_3O_4 diffraction peaks show remarkable changes in intensity, with a striking minimum at 7,720 eV. This change in diffracted intensity is called the anomalous effect. While it is most common to observe anomalous effects that show a minimum in diffracted intensity close to the energy of the absorption edge, it is also possible to observe maxima.

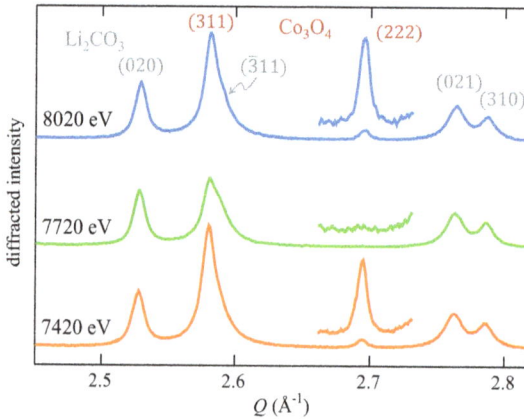

Figure 15.2: Energy-dependent X-ray diffraction for powdered Co_3O_4 in Li_2CO_3 disk. The indices of the Li_2CO_3 diffraction peaks are labelled in grey and the Co_3O_4 peaks in red.

For quantitative analysis accurate diffraction peak areas are needed and diffraction data such as those shown in Figure 15.2 must be peak-deconvoluted using a non-linear optimization procedure. In practice corrections must also be applied for bulk X-ray absorption variations with θ, $A(\theta)$, which would be measured using the readings from I_0 and I_1 in the experiment of Figure 15.1, using $A(\theta) = \mu t / \sin \theta = \log(I_0/I_1)$, where μ is the bulk X-ray absorption coefficient and t is the perpendicular thickness of the sample. With a thin plate, as in Figure 15.2, using a diffractometer scanning in the $\theta - 2\theta$

geometry, the attenuation of diffraction will be given by eq. (15.2), in which $A(\theta)$ is the measured X-ray absorption at a particular X-ray energy:

$$D = \frac{1 - \exp[-2A(\theta)]}{A(\theta)\sin\theta} \qquad (15.2)$$

The intensity of the diffracted beam will also be affected by what is known as the Lorentz factor, L, which corrects for the change in observable crystallites with θ and is given as follows:

$$L = 2\sin^2\theta\cos\theta \qquad (15.3)$$

When using synchrotron radiation from a plane wiggler, undulator or a bending magnet source, the beam will be close to 100% plane polarized in the horizontal plane. And with an upward reflecting geometry no polarization corrections are needed, and measured diffracted intensities can be corrected by multiplying by L/D (eqs. (15.2) and (15.3)). If the peak fitting and correction is applied to the Co_3O_4 (222) diffraction peak, then one can plot, what is known as the DAFS, is shown in Figure 15.3.

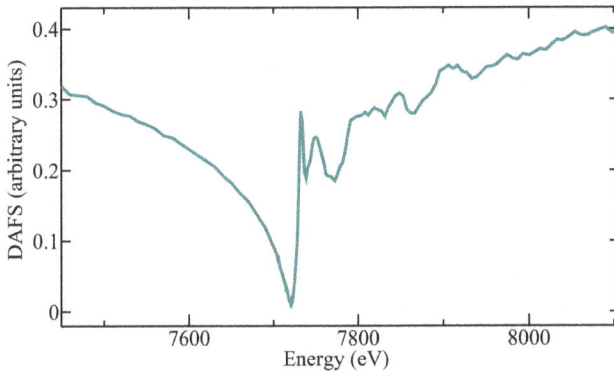

Figure 15.3: The Co_3O_4 (222) diffraction anomalous fine structure (DAFS). Data were extracted by peak-deconvolution and corrected as described in the text.

Before discussing both how one might analyse such data and why we might choose to do so, a brief discussion of nomenclature is appropriate. Among the first observations of DAFS was by Arčon et al. [2], who examined single crystals of copper sulfate and suggested the method be called Bragg reflectivity extended fine structure or BREFS. Subsequently, Pickering et al. [3] suggested diffraction-XAFS or DIFFRAXAFS, but later adopted DAFS, the nomenclature suggested by Stragier et al. [4], in the interests of not creating confusion in the literature [1]. To understand the analysis of DAFS we must first discuss X-ray diffraction in more quantitative detail.

15.2.1 X-ray diffraction and anomalous scattering

The intensity of a diffraction peak hkl is related to $|F_{hkl}|^2$, where F_{hkl} is the structure factor. This in turn is related to the phase factor ϕ_r and the atomic scattering factor f_r of individual atoms r by eqs. (15.4) and (15.5), in which the summation is over all atoms in the crystallographic unit cell:

$$F_{hkl} = \sum_r f_r \exp(2\pi i \phi_r) \tag{15.4}$$

$$F_{hkl} = \sum_r f_r \cos(2\pi\phi_r) + i \sum_r f_r \sin(2\pi\phi_r) \tag{15.5}$$

Here, we use the simplifying convention of $\phi_r = hx_r + ky_r + lz_r$, where x_r, y_r and z_r are the fractional unit cell coordinates of atom r.

When the X-ray energy E is near an X-ray absorption edge, the scattering factor changes with energy, exhibiting what is called **anomalous scattering**. Here, $f(E)$ now includes an energy-independent term f_0 as well as real and imaginary parts, respectively, called $f'(E)$ and $f''(E)$ (eq. (15.6)), which are also called $f_1(E)$ and $f_2(E)$:

$$f(E) = f_0 + f'(E) + if''(E) \tag{15.6}$$

The imaginary part of the anomalous scattering factor is proportional to the X-ray absorption cross-section $\sigma(E)$, multiplied by the X-ray energy, as follows:

$$f''(E) = \left(\frac{m_e c}{2e^2 h}\right) E\sigma(E) \tag{15.7}$$

where m_e is the electron rest mass and e is the charge. The real and imaginary parts of the anomalous scattering are related by mutual Kramers-Kronig relationships:

$$f'(E) = +\frac{2}{\pi} \int_{E'=0}^{\infty} \frac{E'}{E^2 - E'^2} f''(E') dE' \tag{15.8}$$

$$f''(E) = -\frac{2}{\pi} \int_{E'=0}^{\infty} \frac{E}{E^2 - E'^2} f'(E') dE' \tag{15.9}$$

These relationships are reversible – if one computes $f'(E)$ using eq. (15.8) from $f''(E)$ then feeding the result into eq. (15.9) will regenerate the original $f''(E)$ function. We note that there are infinities in these integrals when $E = E'$, which must be excluded from numerical evaluation of the integrals, and that the integrals also extend to infinite energy, which can only be approximated by numerically integrating over a very wide energy range. Early crystallographic codes to compute anomalous scattering factors approximated $f''(E)$ using a measured absorption spectrum and computed $f'(E)$ using eq. (15.8) with extrapolation of $f''(E)$ with computed atomic functions. This is an ungainly and time-consuming approach. As we will see, spectroscopic applications required repeated computation of Kramers-Kronig transforms for both eqs. (15.8) and (15.9), and consequently computer code was developed using fast Fourier transforms to compute the Kramers-Kronig transform [1], which hugely streamlined the calculation. Figure 15.4 compares the anomalous scattering factors $f'(E)$ and $f''(E)$ for cobalt metal with an atomic calculation (broken lines in Figure 15.4) that has been convoluted with a suitable broadening function to account for core-hole lifetime and spectroscopic resolution.

There is an almost universal convention that defines the Thomson scattering amplitude f_0 as a positive quantity. Feynman's theory of quantum electrodynamics has shown that the Thomson scattering for electrons is in fact negative, and that consequently $f'(E)$ is a positive cusp-shaped function, inverted relative to the $f'(E)$ plot in Figure 15.4. That f_0 is negative is not inconsistent with an observed overall increase in the scattering amplitude with energy because the term is squared when calculating intensities. However, the convention of $f'(E)$ as a negative cusp is sufficiently entrenched in the diffraction literature that this usage is essentially universal and because of this we use the conventional practice for the display of $f'(E)$ and $f''(E)$ in Figure 15.4.

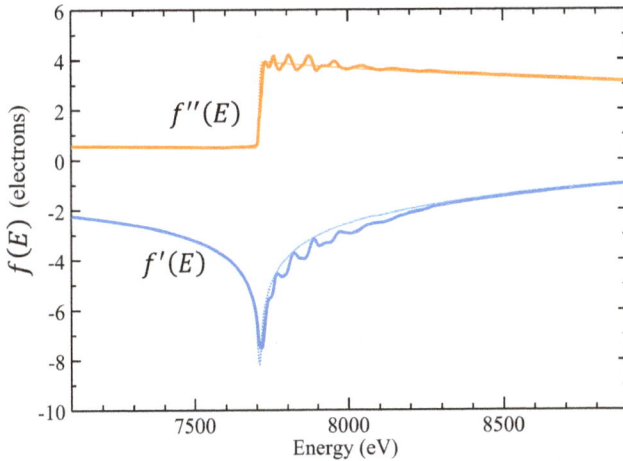

Figure 15.4: Anomalous scattering factors for metallic cobalt. The solid lines show $f'(E)$ (blue) and $f''(E)$ (orange), with the broken lines showing the corresponding smooth atomic functions $f'_a(E)$ and $f''_a(E)$.

We can call the smooth atomic functions of Figure 15.4 $f'_a(E)$ and $f''_a(E)$, which differ from the realistic functions by the fine structure function $\delta f'(E)$ and $\delta f''(E)$, which tend to zero at the ends of the spectra:

$$f'(E) = f'_a(E) + \delta f'(E) \tag{15.10}$$

$$f''(E) = f_a''(E) + \delta f''(E) \tag{15.11}$$

The computation of $\delta f'(E)$ from the Kramers-Kronig transform of $\delta f''(E)$ and vice versa are much more tractable because the integrals only need to be computed over a limited energy range, with $f'(E)$ or $f''(E)$ then simply obtained using eqs. (15.8) and (15.9). Moreover, the aforementioned fast Fourier transform method of evaluating Kramers-Kronig transforms has the further advantage that the infinities in eqs. (15.8) and (15.9) do not occur using this treatment.

As already mentioned, the intensity of a diffracted X-ray $I(hkl)$ is proportional to the square of the structure factor, $I(hkl) \propto |F_{hkl}|^2$. The consequence is that the intensities from diffraction peaks corresponding to hkl and $\overline{hkl}$ are equal, although with opposite phases, and this is known as Friedel's law, $I(hkl) = I(\overline{hkl})$ [5]. The resulting symmetrical pairs of diffraction peaks with identical intensities, which occur even if a centre of symmetry is not present in the contents of the unit cell, are called Friedel pairs or Bijvoet pairs. It was realized relatively early that using X-rays in the vicinity of an absorption edge of a constituent atom, the anomalous effects cause the breaking of Friedel's law, hence $I(hkl) \neq I(\overline{hkl})$ with an anomalous difference in intensity, and that this could be used to solve crystal structures [6]. In modern synchrotron X-ray crystallography of macromolecules, anomalous scattering is exploited to solve new crystal structures using a method called multiple-wavelength anomalous dispersion (MAD) [7]. We will return to consideration of proteins in Section 15.2.4.

15.2.2 Analysis of the DAFS

From eq. (15.5) it is shown that

$$|F|^2 = (A^2 + B^2) + 2(A\alpha + B\beta)f_m'(E) + 2(B\alpha - A\beta)f_m''(E) + (\alpha^2 + \beta^2)\left\{ [f_m'(E)]^2 + [f_m''(E)]^2 \right\} \tag{15.12}$$

where m refers to the anomalous atoms in the unit cell, and A, B, α and β are given by the following equations:

$$A = \sum_r f_{0,r} \cos(2\pi\phi_r) \tag{15.13}$$

$$B = \sum_r f_{0,r} \sin(2\pi\phi_r) \tag{15.14}$$

$$\alpha = \sum_m \cos(2\pi\phi_m) \tag{15.15}$$

$$\beta = \sum_m \sin(2\pi\phi_m) \tag{15.16}$$

Here, the summation over m includes all anomalous atoms in the unit cell. A and B are, respectively, the real and imaginary parts of the structure factor in the absence of any anomalous contributions and do not depend upon energy. The terms α and β are the sums of the phases of the anomalous atoms and are also independent of energy.

In the centrosymmetric case, the sine terms cancel and B and β are zero, with eq. (15.12) reducing to the simpler form:

$$|F|^2 = A^2 + 2A\alpha f'_m(E) + \alpha^2 \left\{ \left[f'_m(E) \right]^2 + \left[f''_m(E) \right]^2 \right\} \qquad (15.17)$$

For several distinct anomalous atoms in a centrosymmetric cell,

$$f'_m(E) = \frac{1}{\alpha} \sum_i \alpha_i f'_i(E) \text{ and } f''_m(E) = \frac{1}{\alpha} \sum_i \alpha_i f''_i(E) \qquad (15.18)$$

$$\text{where } \alpha = \sum_i \alpha_i.$$

Here the summations are over each distinct type of anomalous atom i. With more than one type of anomalous atom, the final $f''_m(E)$ will be a weighted mean. The weighting will vary for different diffraction peaks so that some diffraction peaks will provide sensitivity to different individual sites. Thus, with our previous example, Co_3O_4, a spinel having cobalt atoms in octahedral and tetrahedral coordination geometry in the ratio 2:1, the DAFS from different diffraction peaks will have different contributions from tetrahedral and octahedral sites.

As initially demonstrated by Pickering et al. [1], eq. (15.17) can be solved iteratively to obtain $f''_m(E)$, from which site-specific XAS can be obtained. A major advantage of this method is that it requires no foreknowledge of structure and hence can potentially be applied to a very wide range of research problems. The iterative solution of eq. (15.7) begins by using the smooth computed atomic functions $f'_a(E)$ and $f''_a(E)$ and fitting eq. (15.7) (plus a scaling factor) to the relatively unstructured low and high energy ends of the DAFS spectrum, excluding the highly structured portion near to the absorption edge. This gives an estimate of the ratio α/A, and eq. (15.7) is then solved to obtain $f'_m(E)$ and $f''_m(E)$ from its Kramers-Kronig transform. The new estimates of $f'_m(E)$ and $f''_m(E)$ are then used to obtain updated estimates of A and α, and the procedure is repeated until the residual between experimental and calculated DAFS approaches a minimum. At this point eq. (15.7) is used to calculate an XAS spectrum corresponding to a particular diffraction peak, which will have preferential contributions from a particular set of anomalous atoms. A major advantage of this method is that all the conventional and well-established tools for analysis of XAS can now be used on the DAFS-derived data, which are in essence equivalent to site-specific XAS spectra.

Figure 15.5 shows such an analysis for our previous example, that of Co_3O_4, in which the (222) diffraction peak has anomalous contributions from the octahedral sites and the

(422) diffraction peak is sensitive only to the tetrahedral sites. The near-edge portion of the spectra can be seen to be characteristic of their respective geometries, and the EXAFS shows strong resemblance to predicted site-specific data, using the ab initio code FEFF for simulation. This proof-of-concept work represents the first site-specific DAFS analysis [1].[1]

As discussed and demonstrated by Pickering et al. [1], an almost trivial use of DAFS could be to effectively separate the XAS of individual components in mixtures. Perhaps more interesting, at least from the point of view of analysis is that DAFS can provide polarized spectra from powdered samples.

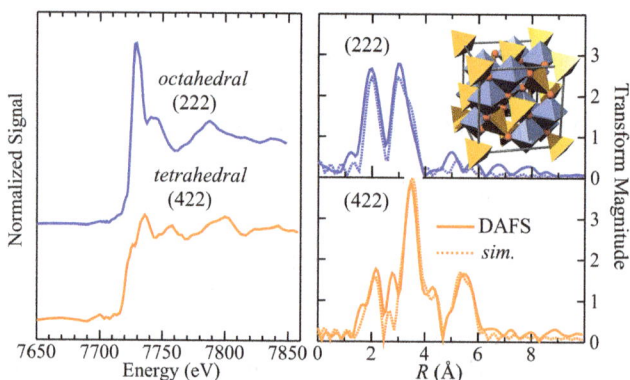

Figure 15.5: DAFS analysis of Co_3O_4 for octahedral and tetrahedral sites. The left plot shows the near-edge spectra, while the right shows the EXAFS Fourier transforms compared with the transforms of simulations of the octahedral and tetrahedral sites based on the crystal structure, which is shown in the inset.

15.2.3 Extraction of polarized spectra from powder DAFS

The basic idea behind the extraction of polarized spectra from DAFS is that the X-ray e-vector must always be within the [hkl] plane from which the X-ray diffraction is occurring, as shown in Figure 15.6. By choosing different specific diffraction peaks, which correspond to different molecular orientations for a specific molecule within the crystal structure, it is possible to derive a set of polarized spectra from a powder.

---

1 The development of DAFS as a practical method was a competition between two different research teams – the team led by Bouldin and co-workers (National Institute of Standards and Technology and the University of Washington) and Pickering et al. (Corporate Research Laboratories, Exxon Research and Engineering Company). Both teams used the NSLS-I facility using different beamlines (X23A2 and X10C, respectively) but were initially unaware of each other's work. The NIST/U.W. team studied epitaxial films, such as Cu(111), with their analysis methods requiring foreknowledge of structure. Pickering and George studied powder DAFS, initially considered intractable by their competitors and developed the iterative Kramers-Kronig method of data analysis, which required no *a priori* knowledge of structure.

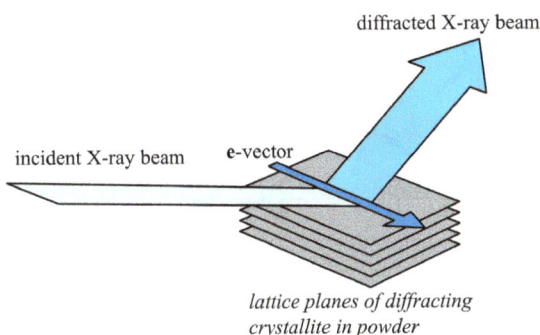

Figure 15.6: Principle of polarized DAFS. The incident X-ray beam is plane-polarized, and the X-ray e-vector must always be within the [*hkl*] plane from which the X-ray diffraction occurs.

Pickering et al. [1] demonstrated this with a powdered sample of $K_2Ni(CN)_4$ that contains the square planar $[Ni(CN)_4]^{2-}$ anion. This compound was selected because Kosugi et al. [8] had already shown substantial anisotropy in polarized Ni K-edge XAS from the hydrate $K_2Ni(CN)_4 \cdot H_2O$, although this study did not use crystallographic alignment, but instead assumed molecular orientation relative to a well-defined crystal cleavage plane. As discussed in Chapter 9, dipole-allowed transitions with conventional XAS show a $3\cos^2\vartheta$ angular dependence relative to the un-polarized powder intensity, where ϑ is the angle between the X-ray e-vector and the transition dipole vector. Polarized DAFS should show a polarization dependence relative to the powder XAS of $(3\sin^2\zeta)/2$, where ζ is the angle between the normal to the [*hkl*] plane (in which the e-vector must lie) and the transition dipole vector. For the square-planar $[Ni(CN)_4]^{2-}$ anion, Kosugi et al. [8] had shown intense and strongly polarized features in the XAS that they assigned to out-of-plane 1s → π^* transitions, in which case the angle ζ would represent the angle between the normal to the $[Ni(CN)_4]^{2-}$ plane and the normal to the [*hkl*] plane [1].

The X-ray crystal structure of $K_2Ni(CN)_4$ shows two different $[Ni(CN)_4]^{2-}$ orientations within the unit cell, inclined at 18.55° to each other and at 80.73° to the [010] plane [9]. Hence the analysis required two polarized DAFS angles ζ_1 and ζ_2. Figure 15.7 shows the polarized DAFS-derived spectra from six different diffraction peaks, selected because they provided a good range of ζ_1 and ζ_2, together with the conventional powder XAS of $K_2Ni(CN)_4$. Five of the six diffraction peaks could be collected in a single DAFS experiment because they have similar d-spacings (3.847–3.523 Å), and the X-ray diffraction from a single energy point is shown in Figure 15.1B.

As expected from the assignments of Kosugi et al. [8], peak A increases when $\zeta \to 90°$, when the X-ray e-vector is perpendicular to the $[Ni(CN)_4]^{2-}$ plane and is close to zero when $\zeta \to 0°$. The peaks B and C, clearly visible in the powder XAS spectrum, show less obvious anisotropy than A. For a detailed discussion of the assignments the reader is referred to Pickering et al. [1], with the addition that calculations with modern codes sug-

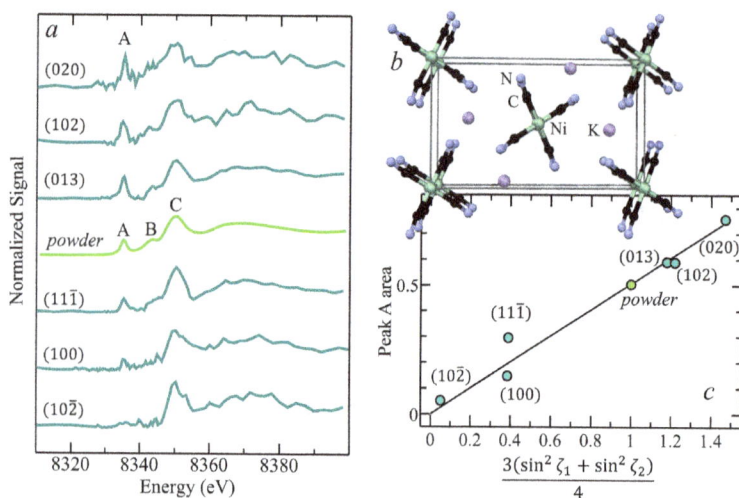

Figure 15.7: Polarized DAFS of $K_2Ni(CN)_4$. Panel *a* shows the spectra extracted from different diffraction peaks indicated, arranged in order of decreasing $(3\sin^2\zeta)/2$ averaged over the two $[Ni(CN)_4]^{2-}$ orientations within the unit cell. Panel *b* shows the crystal structure of $K_2Ni(CN)_4$ viewed approximately along the crystallographic *a*-axis. Panel *c* shows a plot of the area of peak A (panel *a*) obtained by peak deconvolution versus the average $(3\sin^2\zeta)/2$, showing that the intensity tends to a maximum when the e-vector is normal to the $[Ni(CN)_4]^{2-}$ plane.

gest that the assignments of Kosugi et al. [8] are essentially correct, with the low-energy peak A arising from a transition to a π^* orbital involving metal $4p_z$ with contributions from all four cyanide donors.

DAFS experiments tend to be photon-limited, and it is notable that all of the proof-of-principle data reported by Pickering et al. [1] were collected using a bend magnet beamline on a second-generation synchrotron radiation facility. A modern insertion device beamline operating on a third or fourth-generation facility with an area detector would give improved photon flux by several orders of magnitude and consequently should give considerably better data than the early experiments.

15.2.4 Metalloprotein and single-crystal DAFS

Two approaches to X-ray diffraction detected XAS of metalloproteins have been reported. Macromolecules have challenges that are absent in systems with small unit cells. With typically many thousands of atoms giving many thousands of diffraction peaks, the contribution from a small number of heavy element atoms can be small for many of the peaks, however some of the diffraction peaks will have a substantial contribution from the target atoms.

The first method is really an extension of conventional crystallography [10] and has been called spatially resolved anomalous dispersion or SpReAD [11]. The method basically uses crystallographic structure refinement at a series of carefully chosen X-ray energies that span the X-ray absorption edge of a heavy element and evaluation of its $f''(E)$. Thus, the first such application, examined the ferredoxin Fd4 from the hyperthermophilic bacterium *Aquifex aeolicus*,[2] which contains a two-iron, two-sulfide cluster externally bound by four-cysteine thiolates, which can exist in two different oxidation states, the formally all-ferric $[Fe_2S_2]^{2+}$ and the ferrous-ferric $[Fe_2S_2]^+$. Previous Mössbauer spectroscopy had already shown that the ferrous-ferric $[Fe_2S_2]^+$ core had the additional electron localized on one Fe atom, formally ferrous, rather than being delocalized over both, as is the case with other types of iron-sulfide cluster. The goal of the proof-of-principle work reported by Einsle et al. [10] was to determine which of the two irons possessed formally ferric and ferrous oxidation states. To do this these workers measured complete diffraction data sets at nine different incident X-ray energies including seven spaced at 1 eV increments spanning the structured part of the Fe K-near-edge XAS.

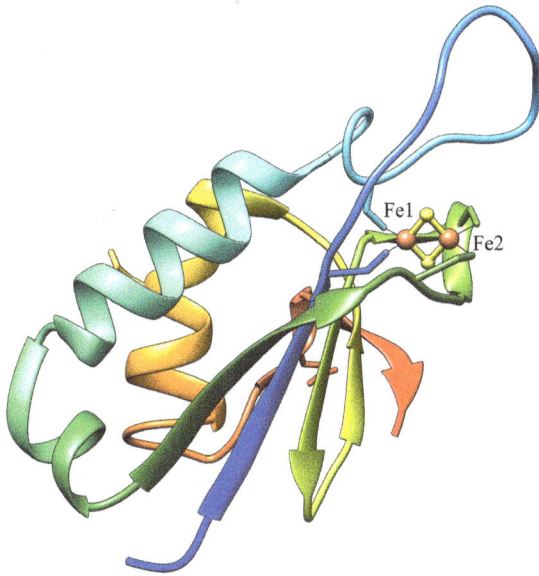

Figure 15.8: Structure of *Aquifex aeolicus* ferredoxin Fd4. The protein is shown using a cartoon rendition with rainbow colour progressing from blue to red corresponding to the C- to N-terminus. The two different iron atoms comprising the [2Fe-2S] cluster are shown in brown.

---

2 Hyperthermophiles are organisms with a very high preferred temperature for growth, which in some cases can exceed 100 °C. Almost all hyperthermophiles are Archaea, but *A. aeolicus* is one of a handful of known bacterial hyperthermophiles, with a preferred temperature for growth of between 85 °C and 95 °C, it is thought to be evolutionarily ancient in that it diverged early from the rest of the bacteria, possessing an unusually small genome.

The analysis method proceeds from eqs. (15.4) and (15.5) and requires that the structure is already known. The contributions from the anomalous atoms (in this case iron), which we denote by m, can then be separated from contributions of the rest of the contents of the crystallographic unit cell, denoted as n, as in eq. (15.19). The first term is the protein scattering, while the energy-dependent scattering from the iron atoms is contained in the second term:

$$F_{hkl} = \sum_n f_n \exp(2\pi i \phi_n) + \sum_m f_m(E) \exp(2\pi i \phi_m) \tag{15.19}$$

The approach was to use non-linear optimization to refine the anomalous parts of scattering factors $f_m(E)$. Einsle et al. [10] found the evaluation of $f'_m(E)$ to be challenging, but the refinements of $f''_m(E)$ are described as very robust. *Aquifex aeolicus* ferredoxin Fd4 contains two protein molecules in the asymmetric unit (only one is shown in Figure 15.8), with consistent results from both. The iron atom closer to the surface is the reduced ferrous iron, with an $f''(E)$ shifted to lower energy, while the less exposed iron atom shown is the ferric iron, with an $f''(E)$ shifted to higher energy. Einsle et al. [10] report the shift as approximately 2 eV, somewhat larger than the close to 1 eV shift expected (e.g. see Figure 9.4) although subsequent re-evaluation of the data with improved normalization suggested a smaller shift, closer to 0.5 eV. Possibly the shift differs from the true value because the $f'(E)$ contribution was not fully reflected in the analysis. Notwithstanding the size of the shift, the conclusion is clear – the external iron atom (Fe2 in Figure 15.8) is the site that becomes reduced while the internal iron atom (Fe1 in Figure 15.8) remains oxidized. While this specific result was already suspected based on nuclear magnetic resonance studies of related proteins [12], Einsle et al. [11] were primarily aiming at developing a technique suitable for addressing a much larger question, that of understanding the redox chemistry of substantially larger proteins.

Variants of this SpReAD method have recently been used by Bartholomew and co-workers to estimate site-specific $f'(E)$ and $f''(E)$ functions [13]. The authors of this work called their method multiwavelength anomalous X-ray diffraction or MAD for short, perhaps an unfortunate choice because it employs the same acronym as MAD phasing which is a mainstay of macromolecular crystallography [7]. The two methods, although related, differ in the details. Subsequent applications of the method have used the DAFS acronym [14, 15], but the analysis method used is closely related to that of Einsle et al. [10]. The methods used by Bartholomew et al. [13] and later workers first determined the crystal structures of their small molecule metal cluster complexes using conventional methods employing X-ray energies that were far from any absorption edges. They then collected crystallographic data sets using 20–30 different incident X-ray energies across the absorption edge and refined $f'_m(E)$ and $f''_m(E)$ on a per-energy and site-dependent basis. The refined $f'_m(E)$ and $f''_m(E)$ were treated as independent parameters, with no attempt to analytically relate them. In the obverse choice made by Einsle et al. [10], this work [13–15] examined only $f'_m(E)$ with $f''_m(E)$ relegated to supporting material.

Returning now to methods for X-ray diffraction detected XAS of metalloproteins, the second method [16] was explored using crystals of the same protein as the previous study by Einsle et al. [10] *A. aeolicus* ferredoxin Fd4 (Figure 15.8). The work by Sherrell [16] used crystals that were a gift from the laboratory of Prof. Einsle. The goal of the study was to help develop metalloprotein DAFS as a tool that could be used on a standard protein crystallography beamline. The approach involved more energy points than the first study, with 67 (compared to 9), obtained after processing 67 crystallographic data sets and using three energy sweeps, each data set being composed of ~7,000 diffraction peaks. Using the known structure, all the data were screened for preferential anomalous contributions from either Fe1 or Fe2, with 727 reflections obtained for Fe1 and 352 for Fe2. From these, principal component analysis (PCA) (see Chapter 11) was used to extract $f'(E)$ and $f''(E)$. PCA effectively isolates $f'(E)$ and $f''(E)$ because they are orthogonal functions. The first eigenvector shows a near-linear slope with essentially no structure, while the second and third eigenvectors are $f'(E)$ and $f''(E)$, respectively. This method confirmed the results reported by Einsle et al. [10] and previous studies [12], showing that the Fe2 or outer iron is the site of reduction while the inner Fe1 iron remains oxidized, but showed energy shifts between both $f'(E)$ and $f''(E)$ for the two sites, which were somewhat smaller at about 1 eV [16], but more in line with what might be expected for a four-coordinate iron with sulfur donors. However, signal to noise proved a substantial challenge [16], and metalloprotein DAFS using the methods developed by Sherrell [16] may not be useful until this has been overcome.

15.3 The Borrmann effect

In X-ray diffraction, as we have already discussed, it is common practice to correct for absorption by the crystal. However, some crystals, when turned in an X-ray beam, become nearly transparent to transmitted X-rays at very particular orientations. This phenomenon was first observed by W. L. Bragg as early as 1921 [17] and later studied by Borrmann [18, 19] for whom the effect is named,[3] in addition to other workers [20]. This strange-seeming high transmission of X-rays in crystals under diffraction conditions arises from interference between incident and diffracted X-ray beams, which can create a standing wave with nodes at the strongly absorbing atoms. At the nodes, the electric field is greatly reduced and consequently dipole-allowed absorption of X-rays by the atoms on the nodes becomes dramatically reduced. The crystal, which can be many ab-

---

3 Both W.H. Bragg and W. L. Bragg are certainly not under-represented in terms of physical nomenclature, and another Bragg effect might have been somewhat confusing.

sorption lengths in thickness, then becomes nearly transparent to X-rays. Of interest to us is what happens when the incident X-ray energy is scanned through the absorption edge. While the intense dipole-allowed transitions in the near-edge are substantially diminished due to the near-zero electric field, the field gradient which stimulates quadrupole-allowed transitions remains. Consequently, the Borrmann effect gives a striking relative enhancement of the quadrupole transitions in measured X-ray transmission spectra [21, 22]. Low temperatures enhance the Borrmann effect by further decreasing the intensity of dipole transitions, which is expected because of the freezing of vibrational modes, while not changing the intensity of the quadrupole transitions. The Borrmann effect has been used to resolve the 2p → 4f dipole-forbidden, quadrupole-allowed transitions at the L_{III} and L_{II}-edges of Gd [21, 22]. An example of the Borrmann effect is shown in Figure 15.9 showing the enhancement of the Ti K-edge quadrupole 1s → 3d transition for a single crystal of the cubic perovskite $SrTiO_3$, doped with iron. $SrTiO_3$ contains titanium in the formal Ti(IV) oxidation state, which is in a 3d [1] configuration. In an octahedral ligand field the 3d manifold splits into t_{2g} and e_g orbitals, with the single d-electron residing in the t_{2g}. The ligand field splitting for a TiO_6 octahedral coordination will be close to 2 eV and hence resolved in the Ti K-edge XAS. What is of particular interest with the data of Figure 15.9 is that the lower energy 1s → 3d(t_{2g}) transition, to orbitals having lobes directed between the six oxygen atoms, shows relative enhancement from the Borrmann effect, while the higher energy 1s → 3d(e_g) transition, to orbitals with lobes directed at the oxygen atoms, is relatively suppressed. Such final-state specific Borrmann effects would be expected to be more pronounced for lighter transition elements such as Ti. At present the Borrmann effect remains explored by only a handful of researchers, with few applications to problems in the chemical or physical sciences despite substantial potential in a significant number of systems.

15.4 Grazing incidence X-ray absorption spectroscopy

As the name suggests, the technique of grazing incidence X-ray absorption spectroscopy (GIXAS) uses a grazing incidence X-ray illumination of a sample. As we will discuss, it provides a tool for the study of surfaces and interfaces. To recap upon Snell's law, which we discussed in Chapter 3, when X-rays fall upon the interface between two media such as the surface of a sample, the light can be reflected and refracted, depending upon the refractive indices of the two media, which we call η_0 and η, and the angle of incidence of the X-ray beam:

$$\eta_0 \cos \theta = \eta \cos \theta' \tag{15.20}$$

In eq. (15.20) (also eq. (3.11)) θ is the angle between the direction of the incident X-ray and the sample surface and θ' is the angle of the refracted X-ray beam within the sample. We call angles of θ for which $\theta' = 0$ the critical angle θ_c. When $\theta < \theta_c$ the X-ray beam is reflected, and when $\theta > \theta_c$ the refracted X-rays propagate into the sample.

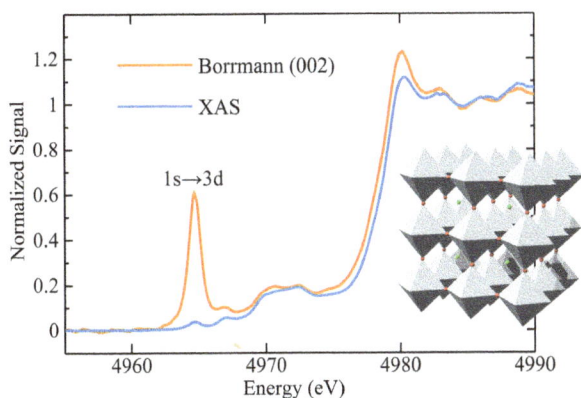

Figure 15.9: The Borrmann effect. Relative enhancement of the Ti K-edge quadrupole 1s → 3d transition by diminution of the dipole transitions for a single crystal of iron-doped $SrTiO_3$ measured in Laue mode for the (002) diffraction at a sample temperature of 6 K. The inset shows the crystal structure of $SrTiO_3$ (polyhedral are TiO_6 octahedra, O atoms shown red, Sr atoms shown in green). Data are replotted from Collins and coworkers [23], with permission from Dr Steve Collins.

When $\theta = \theta_c$ the refracted X-rays propagate along the sample surface in what is known as an **evanescent wave**, which travels immediately below and parallel to the surface, with no energy propagation normal to the surface. The evanescent wave also propagates along the surface when $\theta < \theta_c$, and this affords sensitivity to approximately the first 20–30 Å of a sample. If the intensity of the reflected beam is monitored as a function of incident X-ray energy (a measurement of reflectivity) then this shows a dip corresponding to the absorption edge, and an inverted structure related to conventional XAS [24], which can be analysed to give structural information about the surface or interfaces within the sample. Such measurements are sometimes referred to as ReflEXAFS. In a variation of the GIXAS method, Keil et al. [25] have investigated the use of the so-called Yoneda non-specular peak. The Yoneda peak occurs at $\theta > \theta_c$ for exit angles close to θ_c and is related to lateral surface roughness or interfaces within the sample. It may have even greater surface sensitivity than conventional GIXAS with $\theta < \theta_c$ [25]. Methods that measure the reflected beam can be used to study surface loadings that correspond to near-monolayers, but many samples of interest will have much less than a monolayer of the element of interest. For example, in environmental geochemistry, the nature of sorption of contaminants on mineral surfaces is of critical interest for understanding the mechanisms of transport. In such cases GIXAS using the X-ray fluorescence of the element of interest has been used to obtain surface-sensitive information [26]. In these experiments an optically flat single crystal sample of a geochemically relevant mineral, such as α-Al_2O_3, cleaved along some specific crystallographic plane and polished to be suitably flat, might be studied under an aqueous film containing some environmentally important contaminant in solution. When using synchrotron radiation from a plane wiggler or undulator the X-ray light will be plane polarized with the e-vector in the horizontal plane so that additional information can often be obtained from polarized measurements. The experimental set-up is typically arranged so that the sample can be rotated either about the direction of the incoming X-ray beam (typically called χ) or about the normal to the plane of the sample (typically called φ) to change the polarization. Figure 15.10 shows a schematic of a GIXAS set-up, and Figure 15.11 shows a photograph of an experiment.

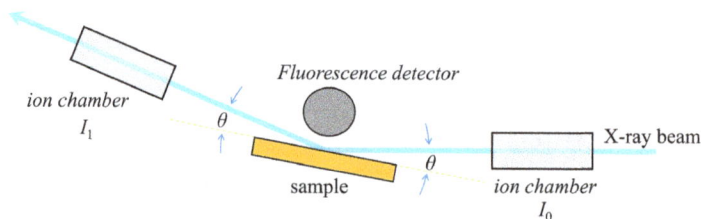

Figure 15.10: Schematic of experimental set-up for GIXAS. A side view of the experiment is shown. Reflected intensity can be measured with the I_1 ion chamber if close to monolayer surface coverage is available, or the fluorescence can be monitored for lower surface coverages. The fluorescence detector is in the horizontal plane at 90° to the incident beam.

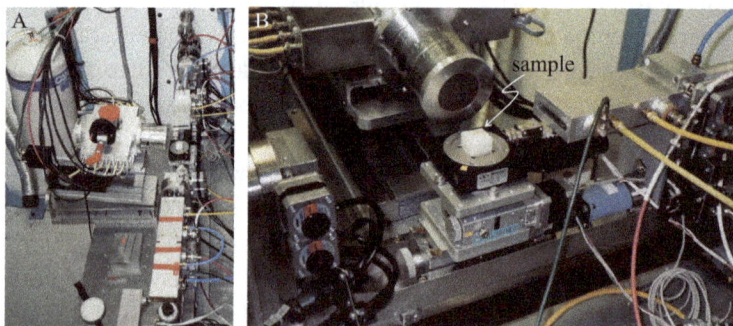

Figure 15.11: Photographs of an experimental set-up for GIXAS. Panel A shows an overall view of the experiment within the hutch, with the beam entering from the top right and a germanium solid-state detector array on the left. Panel B shows a close-up of the sample and GIXAS positioning stages, with the beam entering from the right side and the fluorescence detector behind the sample.

The experiments pictured in Figure 15.11 were conducted at the niobium K-edge at 18,986 eV, which required quite shallow angles of incidence to satisfy the condition $\theta < \theta_c$. With soft X-rays and tender X-rays, however, it becomes increasingly simpler to measure GIXAS at least from an X-ray optics point of view because θ_c increases as the X-ray energy decreases.

References

[1] Pickering, I. J.; Sansone, M.; Marsch, J.; George, G. N. Diffraction anomalous fine structure: A new technique for probing local atomic environment. *J. Am. Chem. Soc.* **1993**, *115*, 6302–6311.

[2] Arčon, I.; Kodre, A.; Glavič, D.; Hribar, M. Extended fine structure of Bragg reflectivity of copper sulfate in the vicinity of copper K-edge. *J. de Physique Colloq.* **1987**, *48(C9)*, 1105–1108.

[3] Pickering, I. J.; Sansone, M.; Marsch, J.; George, G. N. Site-specific X-ray absorption spectroscopy using DIFFRAXAFS. *Jpn. J. Appl. Phys.* **1993**, *32*, Suppl. *32–2*, 206–208.

[4] Stragier, H.; Cross, J. O.; Rehr, J. J.; Sorenson, L. B.; Bouldin, C. E.; Woicik, J. C. Diffraction anomalous fine structure: A new X-ray structural technique. *Phys. Rev. Lett.* **1992**, *69*, 3064–3067.

[5] Friedel, G. Sur les symétries cristallines que peut révéler la diffraction des rayons Röntgen. *C. R. Acad. Sci.* **1913**, *157*, 1533–1536.

[6] Phillips, J. C.; Wlodawer, A.; Yevetz, M. M.; Hodgson, K. O. Applications of synchrotron radiation to protein crystallography: Preliminary results. *Proc. Natl. Acad. Sci. USA.* **1976**, *73*, 128–132.

[7] Guss, J. M.; Merritt, E. A.; Phizackerley, R. P.; Hedman, B.; Murata, M.; Hodgson, K. O.; Freeman, H. C. Phase determination by multiple-wavelength X-ray diffraction: crystal structure of a basic "blue" copper protein from cucumbers. *Science.* **1988**, *241*, 806–811.

[8] Kosugi, N.; Yokoyama, T.; Kuroda, H. Polarization dependence of XANES of square-planar $Ni(CN)_4^{2-}$ ion. A comparison with octahedral $Fe(CN)_6^{4-}$ and $Fe(CN)_6^{3-}$ ions. *Chem. Phys.* **1986**, *104*, 449–453.

[9] Vannerberg, N.-G. The crystal structure of $K_2Ni(CN)_4$. *Acta Chem. Scand.* **1964**, *18*, 2385–2391.

[10] Einsle, O.; Andrade, L. A.; Dobbek, H.; Meyer, J.; Rees, D. C. Assignment of individual metal redox states in a metalloprotein by crystallographic refinement at multiple X-ray wavelengths. *J. Am. Chem. Soc.* **2007**, *129*, 2210–2211.

[11] Einsle, O.; Rees, D. C. Structural enzymology of nitrogenase enzymes. *Chem. Rev.* **2020**, *120*, 4969–5004.

[12] Dugad, L. B.; La Mar, G. N.; Banci, L.; Bertini, I. Identification of localized redox states in plant-type two-iron ferredoxins using the nuclear Overhauser effect. *Biochemistry.* **1990**, *29*, 2263–2271.

[13] Bartholomew, A. K.; Teesdale, J. J.; Sánchez, R. H.; Malbrecht, B. J.; Juda, C. E.; Ménard, G.; Bu, W.; Iovan, D. A.; Mikhailine, A. A.; Zheng, S.-L.; Sarangi, R.; Wang, S. G.; Chen, Y.; Betley, T. A. Exposing the inadequacy of redox formalisms by resolving redox inequivalence within isovalent clusters. *Proc. Natl. Acad. Sci. USA.* **2019**, *116*, 15836–15841.

[14] Alayoglu, P.; Chang, T.; Victoria Lorenzo Ocampo, M.; Murray, L. J.; Chen, Y.-S.; Mankad, N. P. Metal site-specific electrostatic field effects on a tricopper(i) cluster probed by resonant diffraction anomalous fine structure (DAFS). *Inorg. Chem.* **2023**, *62*, 15267–15276.

[15] Alayoglu, P.; Chang, T.; Yan, C.; Chen, Y.-S.; Mankad, N. P. Uncovering a CF_3 effect on X-ray absorption energies of $[Cu(CF_3)_4]^-$ and related copper compounds by using resonant diffraction anomalous fine structure (DAFS) measurements. *Angew. Chem. Int. Ed.* **2023**, *62*, e202313744/1–7.

[16] Sherrell, D. A. Diffraction spectroscopy of metalloproteins. PhD Thesis, University of Saskatchewan, **March 2014**. (Supervisors, George, G. N.; Pickering, I. J.).

[17] Bragg, W. L.; James, R. W.; Bosanquet, C. H. The intensity of reflexion of X-rays by rock-salt – Part II. *Phil. Mag. S.* 6, **1921**, *42*, 1–17.

[18] Borrmann, G. Über extinktionsdiagramme von quarz. *Physikalische Zeitschrift.* **1941**, *42*, 157–162.

[19] Borrmann, G. Die absorption von Röntgenstrahlen im fall der interferenz. *Zeitschrift für Physik.* **1950**, *127*, 297–323.

[20] Campbell, H. N.; X-ray absorption in a crystal set at the Bragg angle. *J. Appl. Phys.* **1951**, *22*, 1139–1142.

[21] Pettifer, R. F.; Collins, S. P.; Laundy, D. Quadrupole transitions revealed by Borrmann spectroscopy. *Nature.* **2008**, *454*, 196–199.

[22] Collins, S. P.; Tolkiehn, M.; Pettifer, R. F.; Laundy, D. Borrmann spectroscopy. *J. Phys. Conf. Ser.* **2009**, *190*, 012045/1–9.

[23] Collins, S. P.; Taurus, T. Tolkiehn. Borrmann spectroscopy of Fe-doped SrTiO₃. *DESY Photon Science Annual Report*, **2010**, 20101342/1–2.

[24] Heald, S. M.; Chen, H.; Tranquanda, J. M. Glancing-angle extended X-ray-absorption fine structure and reflectivity studies of interfacial regions. *Phys. Rev. B.* **1988**, *38*, 1016–1026.

[25] Keil, P.; Lützenkirchen-Hecht, D.; Frahm, R. Grazing incidence XAFS under non-specular conditions. *Physica B.* **2005**, *357*, 1–5.

[26] Waychunas, G. A.; Grazing-incidence X-ray absorption and emission spectroscopy. *Rev. Mineral. Geochem.* **2002**, *49*, 267–315.

16 New and future sources for XAS

16.1 Introduction

In this book we have attempted to provide a comprehensive but accessible text with an emphasis on X-ray absorption spectroscopy (XAS). In doing so we have focused almost entirely on synchrotron radiation derived from storage ring sources. While we have no doubt that storage ring sources will remain the XAS workhorse for the foreseeable future, some special capabilities are available with new sources, which we will discuss briefly in this last chapter of our book.

16.2 Inverse Compton X-ray sources

We remember from Chapter 4 (Section 4.3.2) that in Compton scattering the electron is at rest when part of the kinetic energy of the photon is imparted to the electron, which then recoils, giving a photon of lower energy and longer wavelength. In the inverse Compton effect, the electron is moving and imparts part of its energy to the photon, thereby increasing its energy and shortening its wavelength, hence, the moniker **inverse Compton effect**. For our discussion here, we examine an equivalent but more convenient viewpoint, which is to consider an accelerated electron interacting with laser light that provides a periodic electromagnetic wave that interacts with the electron very like an undulator. From Chapter 2 (eq. (2.9)), we recall that the wavelength of first harmonic undulator radiation, λ_u, is given by eq. (16.1), when the angle of observation is θ_{obs}, where λ_p is the undulator period, K is the insertion device deflection parameter (eq. (2.8)) and $\gamma = 1/\sqrt{1 - (v/c)^2}$, the Lorentz factor (eq. (2.2)) corresponding to the velocity v of the electrons with respect to the speed of light c:

$$\lambda_u = \frac{1}{2\gamma^2}\lambda_p\left(1 + \frac{K^2}{2} + \gamma^2\theta_{obs}^2\right) \tag{16.1}$$

Now, an electron interacting with a laser beam will experience alternating electric and magnetic fields of period $\lambda_L/2$, where λ_L is the wavelength of the laser. The electron will be driven so that its trajectory deviates in the transverse plane in the same manner as it is caused by an undulator magnet, with the differences that K is very small and can be neglected, and that $\lambda_L/2 \ll \lambda_p$ so that the photons resulting from the inverse Compton effect will have a wavelength of λ_C as follows:

$$\lambda_C = \frac{1}{4\gamma^2}\lambda_L\left(1 + \gamma^2\theta_{obs}^2\right) \tag{16.2}$$

The source of accelerated electrons can be either a compact storage ring or a linear accelerator, but because $\lambda_L/2 \ll \lambda_p$ the value of γ for the electrons can be much lower

https://doi.org/10.1515/9783110570441-016

than the combination of a conventional storage ring plus a conventional undulator used as a source of synchrotron radiation. This gives the possibility of what has been called a compact light source; the one commercial version of this has been given the abbreviation CLS.[1] Inverse Compton sources have several potential advantages over synchrotron light: the resulting X-ray beam is nearly monochromatic, resulting in minimal thermal load on the X-ray optics and much less stringent measures required for radiation shielding. Until recently, inverse Compton-based instruments were commercially available with one having been used for XAS measurements at the silver K-edge [1]. For now, however, in practical terms, the possibility of XAS with inverse Compton sources must remain academic.

16.3 Laser-wakefield X-ray sources

Laser-wakefield X-ray sources use a high-power femtosecond pulsed laser focused in a gaseous medium. The laser light ionizes the medium and excites a plasma wave that follows in the wake of the laser pulse. Electrons are trapped by oscillations in this plasma wave and experience extremely large electric fields, accelerating the electrons to relativistic velocities and causing the emission of an X-ray pulse. The laser-wakefield source shares many of the properties of synchrotron radiation, producing a broadband tunable source of X-rays that can extend to high photon energies. Additionally, this is an ultrafast source with each shot having a duration in the femtosecond regime, with perhaps 10^9 photons per shot and up to 10^6 photons per electronvolt. Proof-of-principle XAS experiments using laser-wakefield sources have been reported [2]. However, the signal-to-noise ratio is very challenging, and while specialized experiments might in future benefit from these sources, they are currently not competitive with synchrotron radiation from storage ring-based sources. Moreover, the complexity of the nonlinear laser-plasma interaction results in a lack of shot-to-shot reproducibility that will be challenging to control. Thus, at the time of writing, laser-wakefield technology only shows some potential as a future source for XAS.

16.4 XAS with free electron laser sources

The use of free electron lasers that can operate in the X-ray regime provides some new opportunities for X-ray spectroscopy. To recap, a bend magnet source produces the number of radiated photons approximately proportional to the number of electrons in the stored beam, n_e, a wiggler with N periods produces photons approxi-

---

1 Somewhat unfortunately the same abbreviation, CLS, is used for the Canadian Light Source, the "home" synchrotron source of the authors, located at the University of Saskatchewan.

mately proportional to the product n_eN and an undulator to n_eN^2. Free electron lasers use a linear accelerator to produce electron bunches passing through a very long undulator. The first free electron laser to operate in the X-ray regime was Stanford's Linac Coherent Light Source (LCLS). As we write, the second-generation facility, LCLS-II, has just commenced operations. This new facility employs two alternative undulators between which the beam can be switched. The undulators are both made up of 32 segments each 3.4 m in length, with one undulator designed for soft X-ray work and the other for hard X-ray experiments.

The LCLS facility, and those like it, exploit what is known as self-amplified spontaneous emission (SASE). In the first sections of a long undulator, the electron bunch, which is of small cross section and high peak current, shows a uniform density distribution; at the start of its journey through the undulator, the electron bunch emits photons in the normal manner of an undulator. Here, the electrons are not in-phase and hence produce radiation that lacks coherence. As the electrons pass through the undulator, their speed is just slightly less than the speed of the emitted photons, and travelling almost together the electrons and the photons interact. This interaction causes the electrons to gain or lose energy, grouping at the radiation wavelength: a microbunching of the electrons within the electron bunch, developing coherence along the undulator. Because of this microbunching, the electrons emit with a high degree of coherence and provide a greatly amplified production of X-rays. Thus, with a free electron laser, the photons are approximately proportional to $(n_eN)^2$, providing enormously greater peak photon fluxes than conventional synchrotron radiation sources. The X-ray free electron laser (XFEL) is also a femtosecond pulsed source, and it is the time structure that gives rise to some of the unique capabilities available from XFEL sources. Important characteristics are short pulse lengths of 1–10 fs, tunable photon energy and remarkably high intensity. The astounding X-ray intensity of XFEL sources, with power densities close to 10^{20} W/cm^2, will effectively destroy a sample, drilling a hole through whatever lies in its path and turning it into a tiny puff of plasma. Thus, any experimental data must be acquired before this process can happen. Fortunately, as we have discussed in Chapter 4, the time scales of both X-ray photoabsorption and fluorescence emission are sufficiently fast. Therefore, this is not a problem as long as the data is collected and the sample refreshed before the next XFEL shot arrives.

Typically, XFEL XAS experiments proceed by collecting one incident energy point per XFEL SASE shot with the X-ray monochromator tuned to the energy required, and with the sample being delivered in small droplets or in a continuous stream of solution. A conventional laser pulse might be used to excite the sample, with the timing between the laser pulse and the XFEL shot varied to give a time course. Typically, many repeatable shots are required for the collection of adequate data, which can number in the tens of thousands. It has been shown recently that if dispersive X-ray optics are used, the near-edge portion of the XAS can be collected with just a single shot [3], albeit with a somewhat limited energy range, although possibilities do exist for expanding this [3].

The quantities of sample required for XFEL XAS experiments are relatively high compared to conventional XAS. For example, a typical XFEL-based experiment at the iron K-edge might require 10–20 mL of a 5–50 mM solution, whereas conventional XAS with similar signal-to-noise ratio would require perhaps 0.1 mL of 0.5 mM (see Section 7.5). Consequently, the types of systems that can be studied with XFEL-based XAS are at present limited to models or other samples that are conducive to high concentrations or are available in large quantities. Moreover, to our knowledge, there are no reports of any extended X-ray absorption fine structure experiments using an XFEL, although these are certainly possible [1, 3]. At the time of writing, there are only five XFEL sources operating worldwide, each of which represents a considerable cost investment, which is justified by the unique science that they can enable. This small number of operating XFEL sources contrasts with more than 10 times the number of storage ring sources, each providing X-ray light to many beamlines. Notwithstanding these limitations, XFEL XAS can certainly provide unique information [4, 5], available via no other method, which makes their capabilities very exciting for those interested in the fundamentals of chemical processes, catalysis and electronic structure.

References

[1]   Huang, J.; Günther, B.; Achterhold, K.; Cui, Y.; Gleich, B.; Dierolf, M.; Pfeiffer, F. Energy-dispersive X-ray absorption spectroscopy with an inverse Compton source. *Sci. Rep.* **2020**, *10*, 8772/1–10.

[2]   Kettle, B.; Gertsmayr, E.; Streeter, M. J. V.; Albert, F.; Baggot, R. A.; Bourgeois, N.; Cole, J. M.; Dann, S.; Falk, K.; Gallerdo González, I.; Hussein, A. E.; Lemos, N.; Lopes, N. C.; Lundh, O.; Ma, Y.; Rose, S. J.; Spindloe, C.; Symes, D. R.; Šmíd, M.; Thomas, A. G. R.; Watt, R.; Mangles, S. P. D. Single-shot multi-keV X-ray absorption spectroscopy using an ultrashort laser-wakefield accelerator source. *Phys. Rev. Lett.* **2019**, *123*, 254801/1–6.

[3]   Harmand, M.; Cammarata, M.; Chollet, M.; Krygier, A. G.; Lemke, H. T.; Zhu, D. Single-shot X-ray absorption spectroscopy at X-ray free electron lasers. *Sci. Rep.* **2023**, *13*, 18203/1–11.

[4]   Miller, N. A.; Deb, A.; Alonso-Mori, R.; Glownia, J. M.; Kiefer, L. M.; Konar, A.; Michocki, L. B.; Sikorski, M.; Sofferman, D. L.; Song, S.; Toda, M. J.; Wiley, T. E.; Zhu, D.; Kozlowski, P. M.; Kubrarych, K. J.; Penner-Hahn, J. E.; Sension, R. J. Ultrafast X-ray absorption near edge structure reveals ballistic excited state structural dynamics. *J. Phys. Chem. A.* **2018**, *122*, 4963–4971.

[5]   Sension, R. J.; McClain, T. P.; Lamb, R. M.; Alonso-Mouri, R.; Lima, F. A.; Ardana-Lamas, F.; Biednov, M.; Chollet, M.; Chung, T.; Deb, A.; Dewan, P. A., Jr; Gee, L. B.; Huang Ze En, J.; Jiang, Y.; Khakhulin, D.; Li, J.; Michocki, L. B.; Miller, N. A.; Florian, O.; Yohei, U.; Van driel, T. B.; Penner-Hahn, J. E. Watching excited state dynamics with optical and X-ray probes: The excited state dynamics of aquocobalamin and hydroxocobalamin. *J. Am. Chem. Soc.* **2023**, *145*, 14070–14086.

Closing remarks

During our research careers, X-ray absorption spectroscopy (XAS) has grown from a specialized technique practiced by a small number of experts, to a mainstream spectroscopy with thousands of applications worldwide. The number of synchrotron radiation facilities capable of supporting XAS experiments also has increased from just a few to approximately seventy. Just as the advent of theoretical EXAFS standards transformed the method three decades ago, the continuous refinement of modern codes for computing near-edge spectra promises to further transform this spectroscopy into a highly flexible and powerful tool. Modern photon-in/photon-out methods hold promise for speciation of complex mixtures and for probing in-depth electronic structure information on a range of systems. Related technological advances, coupled with increased brightness of fourth generation sources, are facilitating access to dilute and ultra-dilute concentration regimes and revealing speciation with unprecedented spatial detail. X-ray free electron lasers have the potential to unlock the time domain to probe the unfolding of chemical reactions. We hope this book conveys our enthusiasm for XAS and its community of users, along with our optimism for a bright future for this technique.

https://doi.org/10.1515/9783110570441-017

Appendix A Complex numbers

A.1 Introduction

This text is intended to serve either as an introduction to or a reminder of complex numbers. We will not derive expressions, but instead provide a collection of useful relations.

Complex numbers involve both **real** and **imaginary** parts; an **imaginary number** is one that when squared has a negative result. We typically write imaginary numbers using i, which is defined by $i^2 = -1$, so that $i = \sqrt{-1}$. Essentially, complex numbers are simply a way of specifying both the value and direction of some quantity. Irrespective of the background, the ideas behind complex numbers are in fact simple, and perhaps most importantly once it is observed how useful they can be, the notion of an imaginary number often becomes much less problematic.

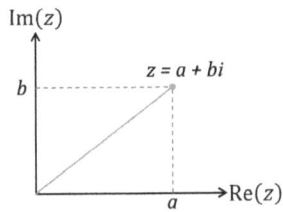

Figure A.1: The complex plane.

A.2 The complex plane

Complex numbers are useful **abstract quantities** that can be used mathematically and, importantly, result in **physically meaningful solutions**. A complex number z can be written as a sum of its real part a and its imaginary part b:

$$z = a + bi \tag{A.1}$$

The real part a and the imaginary part b can be written as $\mathrm{Re}(z)$ and $\mathrm{Im}(z)$, respectively. Complex numbers are considered as points in a Cartesian system known as the **complex plane**, shown in Figure A.1.

If we have a complex number $z = a + bi$, then if we mirror this point along the imaginary direction, we obtain what is known as the **complex conjugate** of z, or $\bar{z}$, as shown in Figure A.2:

$$\bar{z} = a - bi \tag{A.2}$$

This can be written in different ways, for example:

https://doi.org/10.1515/9783110570441-018

Im(z)

b ⋯⋯⋯⋯⋯⋯⋯ z = a + bi

a →Re(z)

-b ⋯⋯⋯⋯⋯⋯⋯ z̄ = a - bi

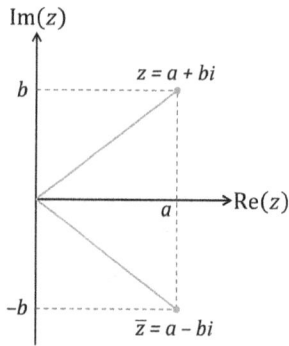

Figure A.2: The complex conjugate.

$$\bar{z} = z^* = (a+bi)^* = \overline{(a+bi)} \tag{A.3}$$

The complex conjugate of a complicated expression may often be written simply as "c.c." (e.g. see eq. (8.12)) to avoid writing out the expression in full. One general rule is that the product of conjugates is equal to the conjugate of the product.

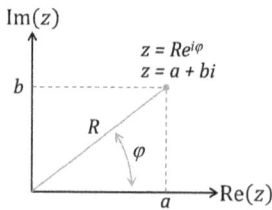

Im(z)

z = Re^{iφ}
z = a + bi

b ⋯⋯⋯⋯⋯

R

φ

a →Re(z)

Figure A.3: Exponential formulation.

A.3 Exponential formulation

Euler's formula provides the exponential form for complex expressions and their complex conjugates:

$$\exp(ix) = \cos x + i \sin x \tag{A.4}$$

$$\exp(-ix) = \cos x - i \sin x \tag{A.5}$$

When $x = \pi$ we obtain what is known as Euler's identity:

$$\exp(i\pi) + 1 = 0 \tag{A.6}$$

For general complex numbers, we can write

$$\exp(a+bi) = (\exp a)(\cos b + i \sin b) \tag{A.7}$$

Using Euler's formula, we can take a complex number $z = a + bi$ and cast it in the "phasor" or polar form, as shown in Figure A.3:

$$z = a + bi = R\exp(i\varphi) = R(\cos\varphi + i\sin\varphi) \tag{A.8}$$

Here, R is called the **complex modulus**, **complex norm** or **amplitude** and φ is called the complex angle, phase or argument of z.

Relationships (A.9)–(A.12) are worth noting:

$$R = \sqrt{z^*z} = |z| = \sqrt{a^2 + b^2} \tag{A.9}$$

$$z = |z|\exp(i\varphi) = \exp(\ln|z|)\exp(i\varphi) = \exp(\ln|z| + i\varphi) \tag{A.10}$$

$$\ln z = \ln|z| + i\varphi \tag{A.11}$$

$$\varphi = \tan^{-1}\left(\frac{b}{a}\right) \tag{A.12}$$

The complex conjugate can be conveniently defined in polar form as follows:

$$\overline{R\exp(i\varphi)} = R\exp(-i\varphi) \tag{A.13}$$

The argument and modulus of a complex number are sometimes abbreviated as the functions $\arg(z)$ and $\mathrm{mod}(z)$, respectively, as in the following equations:

$$\arg(z) = \varphi \tag{A.14}$$

$$\mathrm{mod}\,(z) = R \tag{A.15}$$

Cosine and sine quantities can also be written in the complex exponential form:

$$\cos\varphi = \frac{e^{i\varphi} + e^{-i\varphi}}{2} \tag{A.16}$$

$$\sin\varphi = \frac{e^{i\varphi} - e^{-i\varphi}}{2i} \tag{A.17}$$

The exponential form is also useful for multiplication, division and exponentiation (but not for addition and subtraction), for example:

$$\text{Multiplication:} \quad A\exp(i\vartheta)B\exp(i\varphi) = AB\exp(\vartheta + \varphi) \tag{A.18}$$

$$\text{Division:} \quad \frac{A\exp(i\vartheta)}{B\exp(i\varphi)} = \frac{A}{B}\exp(\vartheta - \varphi) \tag{A.19}$$

$$\text{Exponentiation:} \quad [A\exp(i\vartheta)]^x = A^x\exp(ix\vartheta) \tag{A.20}$$

A.4 Physical functions

In many cases, a complex function is a function of another variable, which here we will call x so that z depends on x and $z(x) = a(x) + ib(x)$. This can be plotted as a three-dimensional rendition with $\text{Re}(z)$ and $\text{Im}(z)$ versus x. We can plot three different two-dimensional projections, $\text{Re}(z)$ versus x, $\text{Im}(z)$ versus x, and $\text{Re}(z)$ versus $\text{Im}(z)$. The projection of this plot onto the complex plane (where we plot $\text{Re}(z)$ versus $\text{Im}(z)$) with x as a running variable is known as an Argand diagram (Figure A.4).

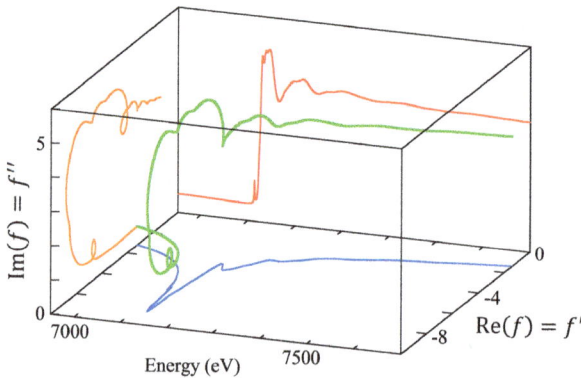

Figure A.4: An example of a physical function with complex components: the X-ray anomalous scattering factor, with real and imaginary parts f' (blue) and f'' (red), respectively, with orange showing the Argand diagram and green the total function.

Appendix B Fourier transforms

B.1 Introduction

This text is intended to be a practical guide to Fourier transforms (FTs) and some of their properties. It is intended as a general but not a comprehensive text. Therefore, we will treat many parts of the topic in only superficial detail and not provide detailed derivations of expressions to keep this appendix relatively concise. We have also designed this appendix to be a generally useful text, and not directed specifically to X-ray absorption spectroscopy.

Many of the everyday phenomena that surround us are wave-like in nature; a very familiar example is found in sound waves. All such waveforms can be described as the sum of simple sine waves having different frequencies. The **FT** allows us to decompose a waveform into component sinusoids. To set the stage for understanding the FT, we first examine the **Fourier series**, which is the FT for periodic functions.

B.2 Fourier series for periodic functions

Let us examine a square wave of amplitude ±1. If we sum the sine waves shown in Figure B.1, we can construct a square wave.

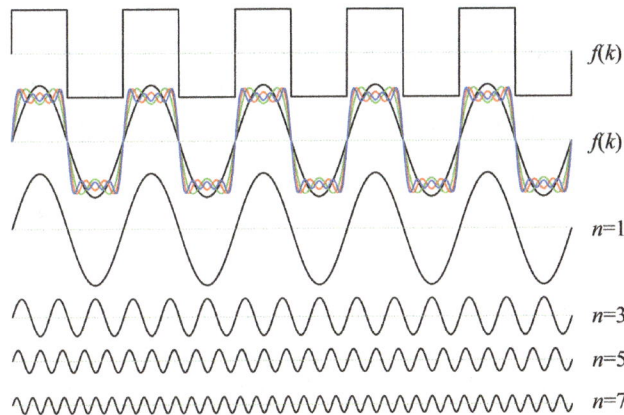

Figure B.1: Fourier series reconstruction of a square wave. The top curve shows the series summed to high order, with the coloured lines summed to n_{max} = 1, 3, 5 and 7 with the component sine waves corresponding to eq. (B.1) shown in the order below. As we include more sine waves in the sum, the representation of the square wave improves.

https://doi.org/10.1515/9783110570441-019

Expressed analytically, this series is given as follows:

$$f(k) = \frac{4}{\pi} \sum_{n=1,3,5,\ldots}^{\infty} \frac{1}{n} \sin\left(\frac{n\pi k}{L}\right) \tag{B.1}$$

Here L is the period of the square wave and k is the abscissa. We note that if eq. (B.1) is computed with all n rather than just the odd values of n (i.e. $n = 1, 2, 3, 4, \ldots$), then a saw tooth is generated rather than a square wave. For any periodic function, we can write an expression for the Fourier series in terms of sine and cosine functions:

$$f(k) = \frac{a_0}{2} + \sum_{n=1}^{\infty} (a_n \cos nk + b_n \sin nk) \tag{B.2}$$

where the coefficients a_n and b_n, respectively, give the amplitudes of each of the sines and cosines in the series.

If we use Euler's formulae employing complex numbers, $\cos\varphi = [\exp(i\varphi) + \exp(-i\varphi)]/2$ and $\sin\varphi = [\exp(i\varphi) - \exp(-i\varphi)]/2i$, we can write this expression using complex numbers:

$$f(k) = \sum_{n=-\infty}^{\infty} A_n \exp\left(i\frac{2\pi nk}{L}\right) \tag{B.3}$$

Here, $A_0 = a_0/2$, $A_n = (a_n - ib_n)/2$ and $A_{-n} = (a_n + ib_n)/2$. The coefficients A_n are called the complex Fourier coefficients. The complex form of the Fourier series is more symmetric and algebraically more concise than the series using eq. (B.2). Therefore, it is the complex form of eq. (B.3) that is more often employed in science.

The process of deriving the values of the complex Fourier coefficients A_n to represent different periodic functions is part of what is known as the Fourier analysis. We can determine the values of A_n using the integral given in the following equation:

$$A_n = \frac{1}{L} \int_{-\frac{L}{2}}^{\frac{L}{2}} f(k) \exp\left(-i\frac{2\pi nk}{L}\right) dk \tag{B.4}$$

Let us consider a real periodic function $f(k)$ consisting of square pulses at regular k intervals, as shown in Figure B.2. In this case, $f(k)$ is real and the Fourier series coefficients A_n are also real, and we can plot these as a function of n. This is shown in Figure B.3 for differing periodicities, specifically 2, 4, 8 and 16.

As the periodicity increases, the spacing of the A_n values in n decreases, and the coefficients start to appear as a continuous curve.

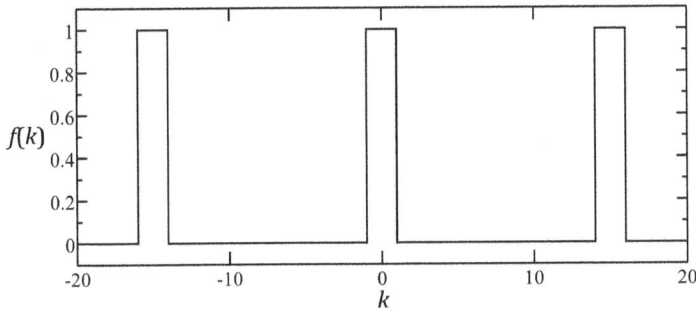

Figure B.2: Periodic square pulse function $f(x)$ versus k.

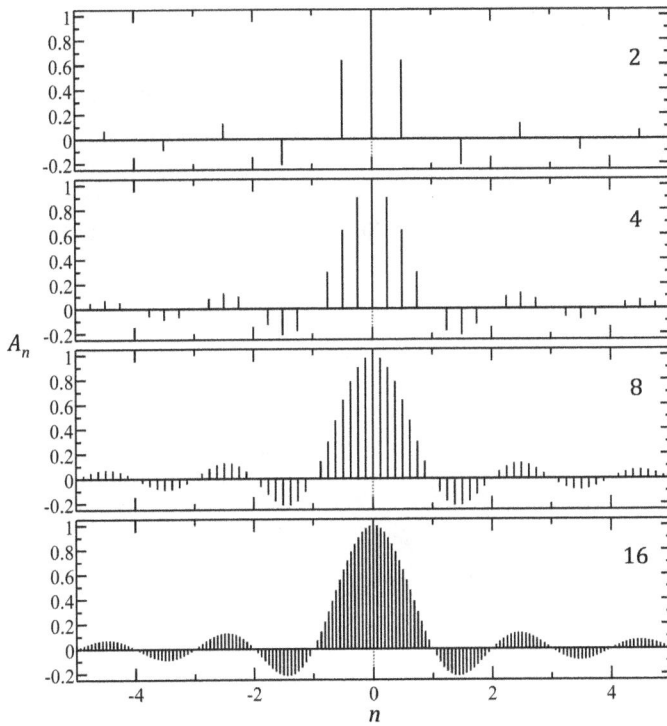

Figure B.3: Fourier coefficients for square pulse functions $f(k)$ with periodicity 2, 4, 8 and 16.

B.3 The Fourier transform

If we consider the transformation of A_n from a discrete variable to a continuous function, which we specify as $F(R)$, with the length $L \to \infty$, then we have what amounts to a generalization of the complex Fourier series – the FT:

$$F(R) = \int_{-\infty}^{\infty} f(k)\exp(-2\pi iRk)dk \tag{B.5}$$

We call this the forward FT, often written as $F(R) = \mathcal{F}[f(k)](R)$ with the inverse FT written as $f(k) = \mathcal{F}^{-1}[F(R)](k)$ or in full as follows:

$$f(k) = \int_{-\infty}^{\infty} F(R)\exp(2\pi iRk)dR \tag{B.6}$$

Thus, if one computes $F(R)$ and then the forward FT of this, but with the exchange of real and imaginary parts, then this will effectively be an inverse FT, re-computing $f(k)$. We note in passing that the variables k and R are used here for convenience and signify real and Fourier space, respectively, although the resemblance to k and R in EXAFS is not a coincidence. Analogous everyday-world variables might be time and frequency so that the periodic function of Figure B.2 could be a switch being engaged at regular time intervals.

The FT is thus a **reversible transform**, which is a very important property.

Returning now to our example, with a small amount of effort, it can be shown analytically that the FT of our example of periodic square pulse is given as follows (see Figure B.4):

$$F(R) = \frac{\sin(\pi R)}{\pi R} = \text{sinc}\,(R) \tag{B.7}$$

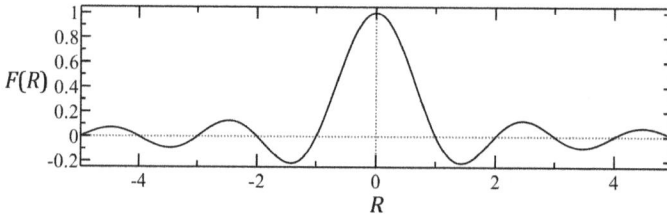

Figure B.4: Fourier transform of the square pulse function $f(k)$.

This is the well-known sinc function , for which we note that $\text{sinc}\,(0) = 1$ because of the limit $\lim_{x\to 0}(\sin x/x) = 1$, where x now refers to a generic variable.

The FT of a pulse function, often called rect, or $\Pi\,(k)$, is a sinc function:

$$\Pi\,(k) = \begin{cases} 1 & |k| < \dfrac{1}{2} \\[2mm] 0 & |k| \geq \dfrac{1}{2} \end{cases} \tag{B.8}$$

$$\mathcal{F}[\Pi(k)] = \text{sinc}(R) \tag{B.9}$$

As a second example, we consider the simple cosine function and its FT:

$$f(k) = \cos(2\pi ak) \tag{B.10}$$

$$F(R) = \mathcal{F}[f(k)] = \frac{\delta(R-a)}{2} + \frac{\delta(R+a)}{2} \tag{B.11}$$

where $\delta(x)$ is a delta function, which has the value of zero except at $x = 0$ (where again x is now a generic variable), as shown in Figure B.5.

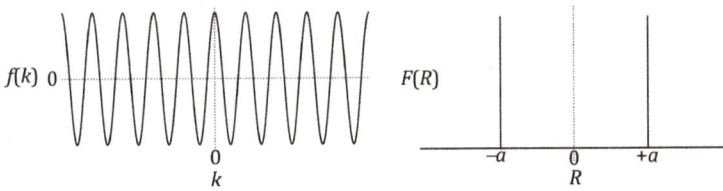

Figure B.5: $f(k) = \cos(2\pi ak)$ and its corresponding Fourier transform $F(R)$.

When numerical rather than analytical methods are used, then FTs must be computed over finite data ranges, rather than $-\infty$ to $+\infty$, and instead of a delta function, a complex sinc-type function is obtained, again centred at $\pm a$. Figure B.6 shows the FT of the exact x range plotted in Figure B.5, showing both real and imaginary parts of $F(R)$, together with the transform magnitude, often called the **power spectrum**, given by the following equation:

$$|F(R)| = \left(\text{Re}[F(R)]^2 + \text{Im}[F(R)]^2\right)^{\frac{1}{2}} \tag{B.12}$$

The plot of $F(R)$ in Figure B.6 is essentially an approximation of the plot of $F(R)$ in Figure B.5 with any differences caused by the finite extent of the data in k. The ripples in $F(R)$ are variously known as truncation artefacts, or series termination effects.

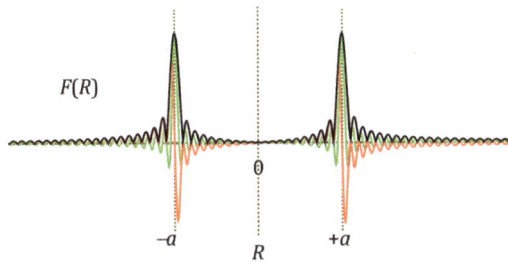

Figure B.6: Fourier transform of $f(k) = \cos(2\pi aR)$ computed over a finite date range, showing transform magnitude (black), real (green) and imaginary (red) components of the complex function.

With more extended data ranges, the transform peak will become sharper and the termination ripples less noticeable, but with actual computations, data ranges must always be finite; therefore, these are always present.

B.4 The discrete Fourier transform

The discrete FT (DFT) applies to the case of a discrete function, and includes all frequencies up to the limit imposed by the point spacing in k, such that the extent of $F(R)$ will be given by $R_{max} = \pi/\Delta k$, and includes both positive and negative frequencies, so effectively ranging from $-R_{max}$ to $+R_{max}$. For any real-valued input data, the resulting DFT will be **Hermitian**, which means that the real part of the transform is an **even function** and the imaginary part of the transform is an **odd function**. This is evident in Figure B.6, and means that the negative frequencies of the transform provide no additional information beyond that from the positive frequencies as they can easily be inferred from the positive frequencies.

B.5 Some properties of the Fourier transform

The FT has some properties that can be very useful in numerical computing, and a selection of these is summarized in Table B.1.

Table B.1: Properties of Fourier transforms.

	Property	$f(k)$	$F(R)$
1	Linearity	$af_1(k) + bf_2(k)$	$aF_1(R) + bF_2(R)$
2	Convolution theorem	$f_1(k) \otimes f_2(k)$	$F_1(R)F_2(R)$
3	Product theorem	$f_1(k)f_2(k)$	$F_1(R) \otimes F_2(R)$
4	k shift	$f(k - \Delta k)$	$F(R)\exp(-2\pi i R\Delta k)$
5	R shift	$f(k)\exp(-2\pi i k\Delta R)$	$F(R - \Delta R)$
6	Scaling	$f(ak)$	$|a|^{-1}F(R/a)$
7	Parseval's theorem	$\int_{-\infty}^{\infty} |f(k)|^2 dk = \int_{-\infty}^{\infty} |F(R)|^2 dR$	

The symbol $\otimes$ is used to signify the convolution of two functions. Thus, if one wanted to numerically compute a convolution of two functions, then a simple way would be to compute the product of the FTs of the two functions and then the inverse FT. Thus, we can write

$$f_1(k) \otimes f_2(k) = \mathcal{F}^{-1}\{\mathcal{F}[f_1(k)]\mathcal{F}[f_2(k)]\} \tag{B.13}$$

Similarly, if a deconvolution is desired, then this can be affected by computing the ratio of the two functions in Fourier space, followed by an inverse FT. For example, if an experimental signal $a(k)$ contains a broadening $b(k)$, this will typically manifest as a convolution of the unbroadened signal $c(k)$ with the broadening function:

$$a(k) = c(k) \otimes b(k) \tag{B.14}$$

If $b(k)$ is known, perhaps through some independent measurement, then in principle we can obtain the unbroadened signal $c(k)$ by calculating the ratio in the Fourier space:

$$c(k) = \mathcal{F}^{-1}\left\{\frac{\mathcal{F}[a(k)]}{\mathcal{F}[b(k)]}\right\} \tag{B.15}$$

In practice, the value of $\mathcal{F}[b(k)]$ at high R will often be small, so that computing the ratio in eq. (B.15) tends to magnify any high-frequency noise in an experimental signal, which can complicate the deconvolution procedure, requiring the application of an additional filter in Fourier space to counteract the magnification. This approach finds use in many fields, for example in image processing. Other entries in Table B.1 also find a variety of uses in various signal and data processing techniques, and the reader might get tempted to imagine new applications.

B.6 Calculations of numerical derivatives and integrals

The FT also provides a simple means by which derivatives can be evaluated; for a general function $f(k)$, we can evaluate the first derivative by simply multiplying by $2\pi i R$ in Fourier space, followed by an inverse transform to obtain the derivative:

$$\mathcal{F}\left[\frac{df(k)}{dk}\right] = 2\pi i R \mathcal{F}[f(k)] \tag{B.16}$$

$$\frac{df(k)}{dk} = \mathcal{F}^{-1}\{2\pi i R \mathcal{F}[f(k)]\} \tag{B.17}$$

This method also holds for higher derivatives, with the general formula for the nth derivative applying

$$\mathcal{F}\left[\frac{d^n f(k)}{dk^n}\right] = (2\pi i R)^n \mathcal{F}[f(k)] \tag{B.18}$$

$$\frac{d^n f(k)}{dk^n} = \mathcal{F}^{-1}\{(2\pi iR)^n \mathcal{F}[f(k)]\} \tag{B.19}$$

Likewise, integrals can also be computed:

$$\mathcal{F}\left[\int_{k=-\infty}^{k} f(k')dk'\right] = \frac{\mathcal{F}[f(k)]}{2\pi iR} \tag{B.20}$$

$$\int_{k=-\infty}^{k} f(k)dk = \mathcal{F}^{-1}\left\{\frac{\mathcal{F}[f(k)]}{2\pi iR}\right\} \tag{B.21}$$

As we have mentioned above, in addition to its analytical uses, the FT finds a great deal of utility in numerical applications. The integrals of eqs. (B.5) and (B.6) can be evaluated numerically, or the fast FT (FFT) method may be used, as we will discuss below.

B.7 Fourier transforms of some simple functions

It is convenient to know the FTs of some simple functions. The FT of a simple one-sided exponential function (B.22) gives a Lorentzian function for the square of the transform magnitude (B.23):

$$f(k) = \begin{cases} 0 & k \leq 0 \\ e^{-ak} & k > 0 \end{cases} \tag{B.22}$$

$$F(R) = \frac{1}{2\pi iR + a} \qquad |F(R)|^2 = \frac{1}{a^2 + 4\pi^2 R^2} \tag{B.23}$$

The curves for $|F(R)|^2$ become broader and decrease in amplitude as a increases (Figure B.7).

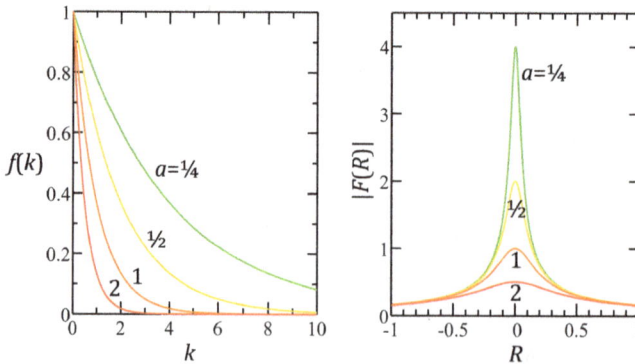

Figure B.7: A series of simple exponentials (B.18) and their corresponding Fourier transform magnitudes $|F(R)|$ (eq. (B.19)). The values of a used are indicated adjacent to the curves for $f(k)$ and $|F(R)|$.

We now consider a simple Gaussian function given in eq. (B.24), which we have chosen so that its total area is unity[1]:

$$f(k) = \exp\left(-\pi k^2\right) \tag{B.24}$$

$$\int_{-\infty}^{\infty} \exp\left(-\pi k^2\right) dk = 1 \tag{B.25}$$

It is fairly straightforward to show analytically that the FT is given by the following equation:

$$F(R) = \exp\left(-\pi R^2\right) \tag{B.26}$$

Therefore, the FT of a Gaussian is a Gaussian.

B.8 Fast Fourier transforms

To compute an FT, the integral equation (B.5) needs to be evaluated for every k point. For example, with 256 points in each R and k mean that 256×256 = 65,536 operations need to be computed, which, in the days before high-speed computers, took significant time. To get around this, the FFT was developed, a clever algorithm that can use $N\log_2 N$ operations, where N is the number of points. This would mean that our hypothetical transform calculation with $N = 256$ would require 256 × 8 = 2,048 operations. Hence, computing an FFT is many times faster than the FT. FFT algorithms are also very accurate, but nothing is perfect; hence, FFT has some computational compromises. There are various FFT algorithms, but all suffer from the same two restrictions:

i)   **Number of data points and k point spacing**: All FFT algorithms need equally spaced points in k, and are usually restricted to specific number of total points N. For example, for the algorithm to function, N must be equal to the power of 2, or be divisible by three prime numbers, based on the specific FFT algorithm used. The requirement for equally spaced k-points often means that data must be interpolated, which is undesirable as it represents non-essential manipulation of the data. If we are to consider EXAFS data, even if the data acquisition uses equal k- spacing, the process of recalibration will change this, as in a stepper motor to control the monochromator crystal position.
ii) **Range and R point spacing**: All FFT algorithms compute DFT (Section B.4). As discussed, the DFT extends to $R_{max} = \pi/\Delta k$, where Δk is the point separation of the k-space data. The DFT will also include both positive and negative values for R, and with most algorithms, the negative values are appended to the high R end of the transform; for example, the left-hand half of Figure B.6 would appear on the right-

---

1 This holds from the well-known relationship $\int_{-\infty}^{\infty} \exp\left(-x^2\right) dx = \sqrt{\pi}$.

hand high R side. A consequence of this is that when an FFT is computed, the point spacing in R can seem sparse. Thus, for a transform of EXAFS data extending to a maximum k of 12.8 Å^{-1}, with 256 data points having a Δk of 0.05 Å^{-1}, the FFT would compute DFT to $R_{\max} = 2\pi/\Delta k$, or 62.8 Å. As this includes negative frequencies, the point spacing in R would be 0.245 Å. The transform would completely describe the k-space data in R-space, but such sparse points would appear unpleasant when plotted. Even though this is a purely cosmetic consideration, almost all computer codes use a practice called zero filling to remedy the point spacing issue. Zero filling extends the data at the high-k end with zeros in order to get a finer R point spacing in FFT. Thus, for example, zero filling to 4,096 points would extend the data range to a k of 205.6 Å^{-1}, with the same Δk, and the FFT would yield a point spacing of 0.015 Å. When plotted the transform of the zero-filled data would be more visually appealing than the unpleasant-looking transform of the original data. This example is shown in Figure B.8. In this figure, we have deliberately chosen a function of $f(x)$ that has close to zero amplitude at the ends of the data, so that termination effects are minimal. With an FT, zero filling is not needed, as the integral can be evaluated at any chosen R points. As we have mentioned, the reasons for carrying out zero filling are purely aesthetic, and actually add nothing of any numerical value.

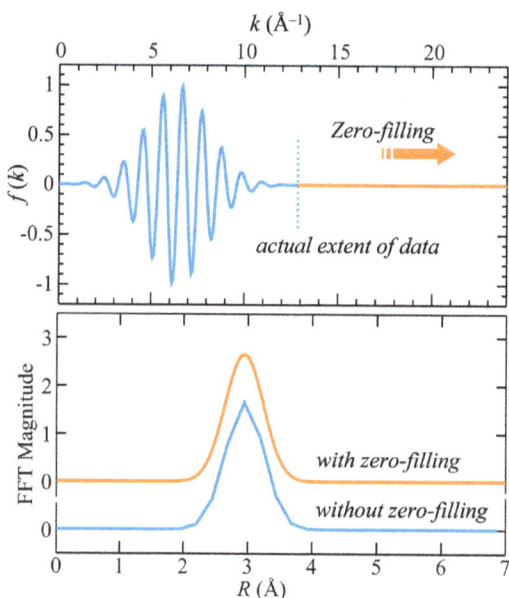

Figure B.8: Example of an FFT with and without zero filling. The upper plot shows the data to be Fourier transformed as a blue line (256 data points) with part of the zero filling shown by the orange line (4,096 points). The result is shown in the lower plot: the zero-filled data appears as a smooth curve while without zero filling the coarse point spacing is less attractive.

Appendix C Elements of quantum mechanical nomenclature

C.1 Introduction

Our goal for this appendix is to provide an overview of some of the mathematical symbolism and nomenclature used in quantum mechanics. Our purpose is to provide those lacking a background in physics and/or chemistry a brief familiarity with these topics. We do not aim to provide a detailed review of quantum mechanics and spectroscopy as there are many excellent texts that cover this topic at various levels of detail. A second goal is to provide a brief reference for use if one encounters unfamiliar mathematical notation – an answer to "what **is** that?" might be found in this appendix.

C.2 Wavefunctions and operators

Wavefunctions are central to quantum mechanics and contain a mathematical description of the system being considered. In essence, a wavefunction is simply an equation that can be used to describe the behaviour of quantum particles such as electrons. The wavefunction may contain expressions for charge, position and time, and can be of widely differing complexity. The wavefunction is typically represented by the Greek character ψ and will contain all the measurable information about the particle. Its square, at a particular position and time, will give the probability of finding the particle at that position and at that time. To give a simple example of a wavefunction, Schrödinger's hydrogen-like atomic wavefunction for the 1s orbital (excited in a K-edge XAS experiment) is given as follows:

$$\psi_{1s} = \frac{1}{\sqrt{\pi}} \left(\frac{Z}{a_0}\right)^{\frac{3}{2}} \exp\left(-\frac{Z}{a_0}r\right) \tag{C.1}$$

Here, Z is the atomic number, a_0 is a constant (called the Bohr radius) with a value of 0.5292 Å and r is the distance of the electron from the nucleus.

An operator, in the mathematical sense, is simply a symbol that describes a process that will transform a function to give another function. Even those with little background in mathematics will probably already be familiar with operators, but not necessarily by the name "operators". For example, the square root of a function $f(x)$ would be described by the square root operator (called a radical) $\sqrt{}$ so that $(f(x))^{1/2} = \sqrt{f(x)}$. Other operators that are probably familiar are the summation operator Σ, the product Π, the cosine cos and the derivative d/dx. If more than one operator is to operate on a function, then the one immediately adjacent and to the left of the function is used first, giving a new function. The next opera-

https://doi.org/10.1515/9783110570441-020

tor, going left again, operates on that new function, followed by the next one, and so on. If the order of the operators does not change the end result then the operators are said to **commute**. However, in general, operators cannot be ordered in the same manner as variables or functions; for example, while $\cos(\theta)x$ is the same as $x\cos(\theta)$, $\cos(\theta)(d/dx)$ is not the same as $(d/dx)\cos(\theta)$; hence, these operators do not commute.

Operators are frequently given a little hat to distinguish them from variables, for example, to distinguish between a variable A and an operator, we might use $\hat{A}$. Alternatively, operators might be written using a script font, for example, the Hamiltonian operator is often written as $\mathcal{H}$. In quantum mechanics, operators represent things that might be observed, such as position or energy, and each observable typically has its own associated operator. Many quantum mechanical operators are the same as their counterparts in classical mechanics. For example, the potential energy of two charges q_1 and q_2 separated by a distance r (i.e. an electron near a nucleus) is described by Coulomb's law, and is proportional to $q_1 q_2/r$. Significantly, however, the momentum operator is completely different in quantum mechanics; in the case of just one spatial dimension x, it is $\hat{p} = -i\hbar\partial/\partial x$, where $\hbar$ is the reduced Planck's constant, $\hbar = h/2\pi$. The form of the momentum operator is one of the foundational postulates of quantum mechanics. The partial derivative operator $\partial/\partial x$ is often specified by the symbol ∇, which is also called Del. Most frequently, this is used for the three-dimensional Euclidean case so that $\nabla f = (\partial f/\partial x + \partial f/\partial y + \partial f/\partial z)$, and the momentum operator can be written as $\hat{p} = -i\hbar\nabla$. The square of the ∇ operator, ∇^2, is a second-order differential operator and is called the **Laplacian**, $\nabla^2 = (\partial^2/\partial x^2 + \partial^2/\partial y^2 + \partial^2/\partial z^2)$.

C.3 Eigenfunctions and eigenvalues

As we have discussed, operators in general operate on a function to give another function. However, in some cases, the original function is regenerated by the application of the operator, and ends up being scaled by some value. Functions of this type are called **eigenfunctions** and the scaling values are called **eigenvalues**. For example, remembering that operators are frequently given a little hat to distinguish them from variables, the Hamiltonian operator is frequently encountered in quantum mechanics. It is often written as $\hat{H}$, which gives the energy levels E_n of a system as follows:

$$\hat{H}\psi_n(r) = E_n\psi_n(r) \tag{C.2}$$

Here, n is a constant (there will be several different energy levels), the wavefunction $\psi_n(r)$ is the **eigenfunction** and the energy E_n is the **eigenvalue**.

C.4 Vectors and bra-ket notation

A vector can be described conveniently using Cartesian components, using the matrix formulation. A matrix $\mathbf{M}$ is simply an array of numbers called matrix elements, which may be complex quantities (see Appendix A). There are various notations that can be used for vectors, including $\mathbf{M}$, $\bar{M}$ or $\vec{M}$. To describe a vector $\mathbf{a}$ in three dimensions of Euclidian space x, y and z, we could use its projections on to each of the x, y and z axes, a_x, a_y and a_z, and specify it as a 3×1 matrix, as follows:

$$\mathbf{a} = \begin{pmatrix} a_x \\ a_y \\ a_z \end{pmatrix} \tag{C.3}$$

Quantum pioneer Paul Dirac introduced his **bra-ket** notation in 1939, which has since become the standard notation in quantum mechanics. Dirac's bra-ket provides another way of writing vectors. The main reason for its widespread use is that the Dirac notation allows us to briefly state what would otherwise be quite complicated expressions. Using Dirac's notation, we could write the matrix of (C.3) as a **ket**, in the following equation:

$$|a\rangle = \begin{pmatrix} a_x \\ a_y \\ a_z \end{pmatrix} \tag{C.4}$$

In quantum mechanics, the ket is used to specify a **quantum state**.[1] For example, we might use it to specify an initial quantum state $|\psi_i\rangle$, where the elements of the ket will be complex (Appendix A). The **bra** is similar to the ket, except that it is a row matrix, and the elements are the complex conjugates of ket's elements (see Appendix A). Therefore, it could be called the **complex conjugate transpose** of the ket. To use a trivial example, we might have a ket shown as follows:

$$|g\rangle = \begin{pmatrix} a+ib \\ c-id \end{pmatrix} \tag{C.5}$$

which would have the bra given by eq. (C.6), remembering that the complex conjugate of $a+ib$ is $a-ib$:

$$\langle g| = \begin{pmatrix} a-ib & c+id \end{pmatrix} \tag{C.6}$$

We will return to the quantum mechanical uses of the Dirac notation after briefly considering some more aspects of nomenclature.

---

1 A quantum state represents the condition in which a quantum mechanical system exists, usually described by a wavefunction together with a set of quantum numbers.

C.5 The inner product

The inner product, also called the **dot product** or the **scalar product**, takes two equal sized matrices and returns a single number (a scalar). In vector nomenclature, the inner product of a vector **a** with another vector **b** would be written as $\mathbf{a} \cdot \mathbf{b}$. In Dirac's notation, this would be written as a bra adjoining a ket or $\langle a|b \rangle$. The inner product effectively projects one vector onto a second vector and then multiplies their lengths:

$$\langle a|b \rangle = (a_1 \quad a_2 \quad a_3) \begin{pmatrix} b_1 \\ b_2 \\ b_3 \end{pmatrix} = a_1 b_1 + a_2 b_2 + a_3 b_3 \tag{C.7}$$

When the inner product of two vectors is zero, we say that the two vectors are **orthogonal**, which is the case in Euclidean space when they are at right angles to each other. The inner product of a vector with itself gives the length of the vector squared.

C.6 Basis vectors

We can consider a vector $|a\rangle$ in three-dimensional Euclidian space:

$$|a\rangle = \begin{pmatrix} a_x \\ a_y \\ a_z \end{pmatrix} = a_x \begin{pmatrix} 1 \\ 0 \\ 0 \end{pmatrix} + a_y \begin{pmatrix} 0 \\ 1 \\ 0 \end{pmatrix} + a_z \begin{pmatrix} 0 \\ 0 \\ 1 \end{pmatrix} = a_x|e_x\rangle + a_x|e_y\rangle + a_z|e_z\rangle \tag{C.8}$$

The kets $|e_x\rangle$, $|e_y\rangle$ and $|e_z\rangle$ are called **basis vectors**. Here, these are simple unit vectors, but basis vectors can be any set of independent vectors. In physics, it is quite common to have more than three dimensions in which case a vector $|a\rangle$ in an N-dimensional vector space would be conveniently represented by a similar linear combination of basis vectors:

$$|a\rangle = \begin{pmatrix} a_1 \\ a_2 \\ \vdots \\ a_N \end{pmatrix} = \sum_{i=1}^{N} a_i|e_i\rangle \tag{C.9}$$

Frequently an **orthonormal** basis is needed, in which the basis vectors are orthogonal, so that $i \neq j \leftrightarrow \langle e_i|e_j \rangle = 0$, and normalized so that each basis vector has a length of 1; hence, $\langle e_i|e_i \rangle = 1^2 = 1$.

C.7 The outer product

Dirac also defined the outer product, the opposite of the inner product, which is specified by $|a\rangle\langle b|$ and is sometimes written as $\mathbf{a} \otimes \mathbf{b}$. Using a finite-dimensional N vector space, the outer product can be evaluated using matrix multiplication to give an $N \times N$ matrix:

$$|a\rangle\langle b| = \begin{pmatrix} a_1 \\ a_2 \\ \vdots \\ a_N \end{pmatrix} \begin{pmatrix} b_1^* & b_2^* & \cdots & b_N^* \end{pmatrix} = \begin{pmatrix} a_1 b_1^* & a_1 b_2^* & \cdots & a_1 b_N^* \\ a_2 b_1^* & a_2 b_2^* & \cdots & a_2 b_N^* \\ \vdots & \vdots & \ddots & \vdots \\ a_N b_1^* & a_N b_2^* & \cdots & a_N b_N^* \end{pmatrix} \tag{C.10}$$

C.8 Bra-ket in quantum mechanics

The mathematics of matrices and vectors is intimately linked to the mathematics of operators and functions. In quantum mechanics, almost every calculation involves vectors or linear operators; hence, bra-ket notation is very useful. We have already noted that quantum states (wavefunctions) are written as kets, for example $|\psi\rangle$. A wavefunction ψ_f can be written as $|\psi_f\rangle$ or just $|f\rangle$ for short, and the complex conjugate of the wavefunction is written as the bra $\langle \psi_f |$. In quantum mechanics, if a bra is shown on the left of an expression and a ket on the right, then integration over $d\tau$ is implied. In quantum mechanics, this $d\tau$ notation means that we integrate over **the full range of relevant variables**. Often, but not always, this would be three-dimensional space, so that it would mean integration over x, y and z from $-\infty$ to $+\infty$. Thus, to summarize, we can write the following equation:

$$\langle \psi || \phi \rangle = \int (\psi^*) \phi \, d\tau = \int\int\int_{-\infty}^{\infty} (\psi^*) \phi \, dx \, dy \, dz \tag{C.11}$$

Frequently, the middle vertical lines between the bra and the ket are merged, so that $\langle \psi || \phi \rangle \equiv \langle \psi | \phi \rangle$. As with the inner product, two functions ψ and ϕ are orthogonal if the following equation (C.12) is true:

$$\langle \psi | \phi \rangle = \int (\psi^*) \phi \, d\tau = 0 \tag{C.12}$$

The function ψ is said to be normalized if the following equation is true:

$$\langle \psi | \psi \rangle = \int (\psi^*) \psi \, d\tau = 1 \tag{C.13}$$

The eigenfunctions of an operator will be orthogonal if they have different eigenvalues.

In quantum mechanics, integrals of the type given in the following equation are often found:

$$\int \psi_i^* \hat{A} \psi_j d\tau = \left\langle \psi_i \middle| \hat{A} \middle| \psi_j \right\rangle = \left\langle i \middle| \hat{A} \middle| j \right\rangle \tag{C.14}$$

This integral is a **matrix element** (the ij th element) of the operator $\hat{A}$. We will return to this type of expression in Section C.10.

Our eigenvalue equation (C.2) using the Hamiltonian operator, when written in bra-ket nomenclature, is similarly compact:

$$\hat{H}|\psi\rangle = E|\psi\rangle \tag{C.15}$$

Here is a reasonable place to mention more about the Hamiltonian operator, which is shown in (C.15) and (C.2). The Hamiltonian operator is the sum of the kinetic energy operator $\hat{T}$ and the potential energy operator $\hat{V}$, both of which we denote with a little hat to distinguish them as operators, so that $\hat{H} = \hat{T} + \hat{V}$. The potential energy operator $\hat{V}$ corresponds to a classical mechanical description, while the kinetic energy operator $\hat{T}$ does not as it must include the momentum operator $\hat{p}$, which we have already noted is distinct from the classical case. Here we can write $\hat{T} = \hat{p}^2/2m$, where m is the mass of the particle, and using the expression for the momentum operator $\hat{p} = -i\hbar\nabla$, we can write $\hat{T} = -(\hbar^2/2m)\nabla^2$. Combining this with eq. (C.15) would give us the Schrödinger wave equation.

C.9 Expectation values

In quantum mechanics, one often wishes to relate the results of quantum mechanical treatment to some parameter that can be experimentally measured. The expectation value of this measurable parameter can be calculated and can be thought of as the average value that one would expect to obtain from a very large number of measurements, weighted by the probability of each result. Note that the expectation value will not be the same as the most probable value resulting from the measurement. For example, for the one-dimensional case with a position x, over time t, we would call the expectation value $\langle x \rangle$, and this would be defined as follows:

$$\langle x \rangle = \frac{\int\limits_{-\infty}^{\infty} \psi^*(x,t)x\psi(x,t)dx}{\int\limits_{-\infty}^{\infty} \psi^*(x,t)\psi(x,t)dx} = \frac{\langle \psi|x|\psi\rangle}{\langle \psi|\psi\rangle} \tag{C.16}$$

Here again the Dirac nomenclature is much more succinct. In many treatments, the normalization condition $\langle \psi|\psi \rangle$ is assumed, so that we might write $\langle x \rangle = \langle \psi|x|\psi \rangle$. In general, for a system with a wavefunction ψ, for some observable quantity belonging to an operator $\hat{Q}$, the quantum mechanical **expectation value** $\langle \hat{Q} \rangle$ is defined as follows:

$$\left\langle \hat{Q} \right\rangle = \frac{\left\langle \psi | \hat{Q} | \psi \right\rangle}{\left\langle \psi | \psi \right\rangle} \tag{C.17}$$

C.10 Fermi's golden rule and XAS

Here we expand upon some of what is discussed in Chapters 8 and 9. We consider a photon being absorbed by an electron in a system, promoting a transition from an initial quantum state $|\psi_i\rangle$, consisting of the photon and the electron, to a final state $|\psi_f\rangle$. The transition rate between initial and final states depends on the degree of coupling between the states and the number of ways the transition can take place. To quantify photoabsorption we can use a relationship that was first described by the quantum pioneer Paul Dirac, which describes the transition rate of one quantum state to another induced by a relatively weak time-dependent perturbation (in our case, due to the photon). A fellow quantum pioneer Enrico Fermi described Dirac's relation in his textbook as one of two "golden rules" (actually, the second golden rule), as a result of which it has become widely known as Fermi's golden rule, despite its origin with Dirac. Fermi's golden rule describes the transition probability w_{if} between the two quantum states, $|\psi_i\rangle$ and $|\psi_f\rangle$, with the transition probability being proportional to the square of the matrix element M_{if}:

$$w_{if} = \frac{2\pi}{\hbar} |M_{if}|^2 \rho_f \tag{C.18}$$

In eq. (C.18), $\hbar$ is the reduced Planck constant and ρ_f is the density of final states. If there are multiple $|\psi_f\rangle$ with identical energies, called **degenerate states**, then this will increase w_{if}. In cases where there is a continuum of final states, the density of final states will be expressed as a function of energy, which for a photoelectron's final state will be proportional to the photoelectron wave vector k. The matrix element M_{if} can be expressed by eq. (C.19) in which the interaction with the photon[2] that stimulates the transition is expressed through the operator H' that couples $|\psi_i\rangle$ and $|\psi_f\rangle$ which is called the light-matter interaction operator. Note that here we have chosen to leave off the little hat on the operator to be consistent with Chapters 8 and 9:

$$M_{if} = \left\langle \psi_f | H' | \psi_i \right\rangle \tag{C.19}$$

---

2 We note that some other texts erroneously suggest that ψ_i expresses the initial electron state and the photon. When using Fermi's golden rule, the photon is modelled in the operator H'.

This kind of integral approach is the same as what we used for expectation values, above. For XAS, we can sum over the available final states to approximate the X-ray absorption cross section, $\sigma(E)$:

$$\sigma(E) \propto \sum_f \left|\left\langle \psi_f | H' | \psi_i \right\rangle\right|^2 \delta(E - E_f + E_i) \tag{C.20}$$

The δ in eq. (C.20) is often called Dirac's delta, which we have already mentioned in Appendix B. If we expressed the Dirac delta relative to a real variable x, it has a value of zero everywhere except at zero, and its integral over the entire range is equal to one, so that we might write the following equations:

$$\int_a^b \delta(x)\,dx = \begin{cases} 1, & a \leq 0 \leq b \\ 0, & a \nleq 0 \nleq b \end{cases} \tag{C.21}$$

$$\int_{-\infty}^{\infty} \delta(x)\,dx = 1 \tag{C.22}$$

The Dirac delta is in some ways a curious concept; it is sometimes stated that $\delta(0) = \infty$, but its important properties are the unit integrals in (C.21) and (C.22). It is very useful for modelling sudden events, such as an impulse like a hammer strike, or a transition between two quantum states. Dirac's delta is included in (C.20) so that the transition will happen when the photon energy E is equal to the difference between the initial and final state energies, E_i and E_f, respectively.

For XAS, the initial state $|\psi_i\rangle$ is a deep-core state, for example $|1s\rangle$ (eq. (C.1)), and will be highly localized near the nucleus, while $|\psi_f\rangle$ is the final state containing the excited electron in a previously unoccupied level in the presence of a core hole. We approximate the inner-shell excitation by the photon within the operator H' as being induced by an electromagnetic wave with an electric field vector $\mathbf{e}$ and vector potential $\mathbf{A}(\mathbf{r})$. H' is given by the product of the electron momentum vector $\mathbf{p}$ and the vector potential of the wave $\mathbf{A}(\mathbf{r})$, $H' = \mathbf{p} \cdot \mathbf{A}(\mathbf{r})$, and using the expression for a classical wave $\mathbf{A}(\mathbf{r}) = \mathbf{e}\exp(i(\mathbf{k} \cdot \mathbf{r}))$, where $\mathbf{k}$ is the X-ray forward propagation vector of the photon, and H' can be written as follows:

$$H' = (\mathbf{e} \cdot \mathbf{p})\exp(i(\mathbf{k} \cdot \mathbf{r})) \tag{C.23}$$

The exponential term can be subjected to a series expansion, such that

$$\exp(x) = \sum_{k=0}^{\infty} \frac{x^k}{k!}$$

The first term in the series expansion of the exponential term is 1 and this gives $H' \approx H'_D = (\mathbf{e} \cdot \mathbf{p})$ which is the so-called dipole approximation, which accounts for what are called dipole-allowed transitions. This can be thought of as the transition being stimulated by an oscillating plane wave and would correspond to $\Delta l = \pm 1$. If we include the second term in the series expansion of the exponential, then we have $\exp(i(\mathbf{k} \cdot \mathbf{r})) \approx 1 + i(\mathbf{k} \cdot \mathbf{r})$ and hence $H' \approx H'_D + H'_Q$ where $H'_Q = i(\mathbf{e} \cdot \mathbf{p})(\mathbf{k} \cdot \mathbf{r})$. The operator H'_Q gives the $\Delta l = \pm 2$ quadrupole-allowed transitions, so that when including both, eq. (C.19) becomes

$$M_{if} = \left\langle \psi_f | (\mathbf{e} \cdot \mathbf{p}) + i(\mathbf{e} \cdot \mathbf{p})(\mathbf{k} \cdot \mathbf{r}) | \psi_i \right\rangle \qquad (C.24)$$

Quadrupole-allowed transitions can be thought of as being stimulated by changes in the electromagnetic field gradient, which becomes significant when the wavelength of the photon is short, as is the case with X-rays. However, in general, quadrupole transitions are always much weaker than the dipole transitions. We note that some treatments give eq. (C.19) in reverse, as $\langle \psi_i | H' | \psi_f \rangle$, rather than $\langle \psi_f | H' | \psi_i \rangle$, but the results are of course the same.

Appendix D The EXAFSPAK analysis code

In this book, whenever possible, we have deliberately avoided mentioning specific computer codes. We have attempted to provide general statements about strengths and weaknesses without criticism or endorsement of specific X-ray spectroscopy analysis packages. Here, however, we note that all figures and data analyses related to XAS in this book were computed using the exafspak computer code (https://exafspak.com/).

D.1 History of the exafspak analysis code

One of us (George) had his first taste of XAS data analysis following his first beamtime in April 1982 at DESY, Hamburg, Germany, with the OTOKO programme, developed in Hamburg by Peter Bendall, Michelle Koch and Joan Bordas. OTOKO was a one-size-fits-all solution that was used for X-ray diffraction and scattering in addition to XAS. On his return home after his beamtime and inspired by some of the ideas in OTOKO, George immediately began writing his own computer programmes for the analysis of XAS data on the University of Sussex School of Molecular Sciences' DEC VAX 780 computer. In 1985, George spent 6 months at SSRL, working with Steve Cramer. One of the tasks assigned to George was to move Steve Cramer's analysis code and all his experimental data from an IBM 360 mainframe to SSRL's VAX computer. This code was a loose collection of programmes and subroutines, many of which had been initially authored by Tom Eccles while both Steve and Tom were graduate students in the laboratory of Keith Hodgson in the Department of Chemistry at Stanford University. Steve had also substantially added to the code himself. George used much of this code, his own fledgling code from the University of Sussex, a great deal of new code, and an entirely new graphical treatment as the basis for what is now exafspak. The original exafspak code was written in a mixture of VAX Macro Assembler and VAX Fortran, with the assembler routines later being converted to the C programming language by Dr Simon George, who ported exafspak to Microsoft Windows, Linux and Apple Macintosh. The current exafspak code contains recent contributions from Graham George, Ingrid Pickering and Simon George.

https://doi.org/10.1515/9783110570441-021

Index

https://doi.org/10.1515/9783110570441-022

www.ingramcontent.com/pod-product-compliance
Lightning Source LLC
Chambersburg PA
CBHW080649220326
41598CB00033B/5151

9 783110 570373